Micromachining of Engineering Materials

MECHANICAL ENGINEERING
A Series of Textbooks and Reference Books

Founding Editor

L. L. Faulkner

Columbus Division, Battelle Memorial Institute
and Department of Mechanical Engineering
The Ohio State University
Columbus, Ohio

1. *Spring Designer's Handbook*, Harold Carlson
2. *Computer-Aided Graphics and Design*, Daniel L. Ryan
3. *Lubrication Fundamentals*, J. George Wills
4. *Solar Engineering for Domestic Buildings*, William A. Himmelman
5. *Applied Engineering Mechanics: Statics and Dynamics*, G. Boothroyd and C. Poli
6. *Centrifugal Pump Clinic*, Igor J. Karassik
7. *Computer-Aided Kinetics for Machine Design*, Daniel L. Ryan
8. *Plastics Products Design Handbook, Part A: Materials and Components; Part B: Processes and Design for Processes*, edited by Edward Miller
9. *Turbomachinery: Basic Theory and Applications*, Earl Logan, Jr.
10. *Vibrations of Shells and Plates*, Werner Soedel
11. *Flat and Corrugated Diaphragm Design Handbook*, Mario Di Giovanni
12. *Practical Stress Analysis in Engineering Design*, Alexander Blake
13. *An Introduction to the Design and Behavior of Bolted Joints*, John H. Bickford
14. *Optimal Engineering Design: Principles and Applications*, James N. Siddall
15. *Spring Manufacturing Handbook*, Harold Carlson
16. *Industrial Noise Control: Fundamentals and Applications*, edited by Lewis H. Bell
17. *Gears and Their Vibration: A Basic Approach to Understanding Gear Noise*, J. Derek Smith
18. *Chains for Power Transmission and Material Handling: Design and Applications Handbook*, American Chain Association
19. *Corrosion and Corrosion Protection Handbook*, edited by Philip A. Schweitzer
20. *Gear Drive Systems: Design and Application*, Peter Lynwander
21. *Controlling In-Plant Airborne Contaminants: Systems Design and Calculations*, John D. Constance
22. *CAD/CAM Systems Planning and Implementation*, Charles S. Knox
23. *Probabilistic Engineering Design: Principles and Applications*, James N. Siddall
24. *Traction Drives: Selection and Application*, Frederick W. Heilich III and Eugene E. Shube
25. *Finite Element Methods: An Introduction*, Ronald L. Huston and Chris E. Passerello

26. *Mechanical Fastening of Plastics: An Engineering Handbook*, Brayton Lincoln, Kenneth J. Gomes, and James F. Braden
27. *Lubrication in Practice: Second Edition*, edited by W. S. Robertson
28. *Principles of Automated Drafting*, Daniel L. Ryan
29. *Practical Seal Design*, edited by Leonard J. Martini
30. *Engineering Documentation for CAD/CAM Applications*, Charles S. Knox
31. *Design Dimensioning with Computer Graphics Applications*, Jerome C. Lange
32. *Mechanism Analysis: Simplified Graphical and Analytical Techniques*, Lyndon O. Barton
33. *CAD/CAM Systems: Justification, Implementation, Productivity Measurement*, Edward J. Preston, George W. Crawford, and Mark E. Coticchia
34. *Steam Plant Calculations Manual*, V. Ganapathy
35. *Design Assurance for Engineers and Managers*, John A. Burgess
36. *Heat Transfer Fluids and Systems for Process and Energy Applications*, Jasbir Singh
37. *Potential Flows: Computer Graphic Solutions*, Robert H. Kirchhoff
38. *Computer-Aided Graphics and Design: Second Edition*, Daniel L. Ryan
39. *Electronically Controlled Proportional Valves: Selection and Application*, Michael J. Tonyan, edited by Tobi Goldoftas
40. *Pressure Gauge Handbook*, AMETEK, U.S. Gauge Division, edited by Philip W. Harland
41. *Fabric Filtration for Combustion Sources: Fundamentals and Basic Technology*, R. P. Donovan
42. *Design of Mechanical Joints*, Alexander Blake
43. *CAD/CAM Dictionary*, Edward J. Preston, George W. Crawford, and Mark E. Coticchia
44. *Machinery Adhesives for Locking, Retaining, and Sealing*, Girard S. Haviland
45. *Couplings and Joints: Design, Selection, and Application*, Jon R. Mancuso
46. *Shaft Alignment Handbook*, John Piotrowski
47. *BASIC Programs for Steam Plant Engineers: Boilers, Combustion, Fluid Flow, and Heat Transfer*, V. Ganapathy
48. *Solving Mechanical Design Problems with Computer Graphics*, Jerome C. Lange
49. *Plastics Gearing: Selection and Application*, Clifford E. Adams
50. *Clutches and Brakes: Design and Selection*, William C. Orthwein
51. *Transducers in Mechanical and Electronic Design*, Harry L. Trietley
52. *Metallurgical Applications of Shock-Wave and High-Strain-Rate Phenomena*, edited by Lawrence E. Murr, Karl P. Staudhammer, and Marc A. Meyers
53. *Magnesium Products Design*, Robert S. Busk
54. *How to Integrate CAD/CAM Systems: Management and Technology*, William D. Engelke
55. *Cam Design and Manufacture: Second Edition*; with cam design software for the IBM PC and compatibles, disk included, Preben W. Jensen
56. *Solid-State AC Motor Controls: Selection and Application*, Sylvester Campbell
57. *Fundamentals of Robotics*, David D. Ardayfio
58. *Belt Selection and Application for Engineers*, edited by Wallace D. Erickson
59. *Developing Three-Dimensional CAD Software with the IBM PC*, C. Stan Wei
60. *Organizing Data for CIM Applications*, Charles S. Knox, with contributions by Thomas C. Boos, Ross S. Culverhouse, and Paul F. Muchnicki

61. *Computer-Aided Simulation in Railway Dynamics*, by Rao V. Dukkipati and Joseph R. Amyot
62. *Fiber-Reinforced Composites: Materials, Manufacturing, and Design*, P. K. Mallick
63. *Photoelectric Sensors and Controls: Selection and Application*, Scott M. Juds
64. *Finite Element Analysis with Personal Computers*, Edward R. Champion, Jr., and J. Michael Ensminger
65. *Ultrasonics: Fundamentals, Technology, Applications: Second Edition, Revised and Expanded*, Dale Ensminger
66. *Applied Finite Element Modeling: Practical Problem Solving for Engineers*, Jeffrey M. Steele
67. *Measurement and Instrumentation in Engineering: Principles and Basic Laboratory Experiments*, Francis S. Tse and Ivan E. Morse
68. *Centrifugal Pump Clinic: Second Edition, Revised and Expanded*, Igor J. Karassik
69. *Practical Stress Analysis in Engineering Design: Second Edition, Revised and Expanded*, Alexander Blake
70. *An Introduction to the Design and Behavior of Bolted Joints: Second Edition, Revised and Expanded*, John H. Bickford
71. *High Vacuum Technology: A Practical Guide*, Marsbed H. Hablanian
72. *Pressure Sensors: Selection and Application*, Duane Tandeske
73. *Zinc Handbook: Properties, Processing, and Use in Design*, Frank Porter
74. *Thermal Fatigue of Metals*, Andrzej Weronski and Tadeusz Hejwowski
75. *Classical and Modern Mechanisms for Engineers and Inventors*, Preben W. Jensen
76. *Handbook of Electronic Package Design*, edited by Michael Pecht
77. *Shock-Wave and High-Strain-Rate Phenomena in Materials*, edited by Marc A. Meyers, Lawrence E. Murr, and Karl P. Staudhammer
78. *Industrial Refrigeration: Principles, Design and Applications*, P. C. Koelet
79. *Applied Combustion*, Eugene L. Keating
80. *Engine Oils and Automotive Lubrication*, edited by Wilfried J. Bartz
81. *Mechanism Analysis: Simplified and Graphical Techniques, Second Edition, Revised and Expanded*, Lyndon O. Barton
82. *Fundamental Fluid Mechanics for the Practicing Engineer*, James W. Murdock
83. *Fiber-Reinforced Composites: Materials, Manufacturing, and Design, Second Edition, Revised and Expanded*, P. K. Mallick
84. *Numerical Methods for Engineering Applications*, Edward R. Champion, Jr.
85. *Turbomachinery: Basic Theory and Applications, Second Edition, Revised and Expanded*, Earl Logan, Jr.
86. *Vibrations of Shells and Plates: Second Edition, Revised and Expanded*, Werner Soedel
87. *Steam Plant Calculations Manual: Second Edition, Revised and Ex panded*, V. Ganapathy
88. *Industrial Noise Control: Fundamentals and Applications, Second Edition, Revised and Expanded*, Lewis H. Bell and Douglas H. Bell
89. *Finite Elements: Their Design and Performance*, Richard H. MacNeal
90. *Mechanical Properties of Polymers and Composites: Second Edition, Revised and Expanded*, Lawrence E. Nielsen and Robert F. Landel
91. *Mechanical Wear Prediction and Prevention*, Raymond G. Bayer

92. *Mechanical Power Transmission Components*, edited by David W. South and Jon R. Mancuso
93. *Handbook of Turbomachinery*, edited by Earl Logan, Jr.
94. *Engineering Documentation Control Practices and Procedures*, Ray E. Monahan
95. *Refractory Linings Thermomechanical Design and Applications*, Charles A. Schacht
96. *Geometric Dimensioning and Tolerancing: Applications and Techniques for Use in Design, Manufacturing, and Inspection*, James D. Meadows
97. *An Introduction to the Design and Behavior of Bolted Joints: Third Edition, Revised and Expanded*, John H. Bickford
98. *Shaft Alignment Handbook: Second Edition, Revised and Expanded*, John Piotrowski
99. *Computer-Aided Design of Polymer-Matrix Composite Structures*, edited by Suong Van Hoa
100. *Friction Science and Technology*, Peter J. Blau
101. *Introduction to Plastics and Composites: Mechanical Properties and Engineering Applications*, Edward Miller
102. *Practical Fracture Mechanics in Design*, Alexander Blake
103. *Pump Characteristics and Applications*, Michael W. Volk
104. *Optical Principles and Technology for Engineers*, James E. Stewart
105. *Optimizing the Shape of Mechanical Elements and Structures*, A. A. Seireg and Jorge Rodriguez
106. *Kinematics and Dynamics of Machinery*, Vladimír Stejskal and Michael Valášek
107. *Shaft Seals for Dynamic Applications*, Les Horve
108. *Reliability-Based Mechanical Design*, edited by Thomas A. Cruse
109. *Mechanical Fastening, Joining, and Assembly*, James A. Speck
110. *Turbomachinery Fluid Dynamics and Heat Transfer*, edited by Chunill Hah
111. *High-Vacuum Technology: A Practical Guide, Second Edition, Revised and Expanded*, Marsbed H. Hablanian
112. *Geometric Dimensioning and Tolerancing: Workbook and Answerbook*, James D. Meadows
113. *Handbook of Materials Selection for Engineering Applications*, edited by G. T. Murray
114. *Handbook of Thermoplastic Piping System Design*, Thomas Sixsmith and Reinhard Hanselka
115. *Practical Guide to Finite Elements: A Solid Mechanics Approach*, Steven M. Lepi
116. *Applied Computational Fluid Dynamics*, edited by Vijay K. Garg
117. *Fluid Sealing Technology*, Heinz K. Muller and Bernard S. Nau
118. *Friction and Lubrication in Mechanical Design*, A. A. Seireg
119. *Influence Functions and Matrices*, Yuri A. Melnikov
120. *Mechanical Analysis of Electronic Packaging Systems*, Stephen A. McKeown
121. *Couplings and Joints: Design, Selection, and Application, Second Edition, Revised and Expanded*, Jon R. Mancuso
122. *Thermodynamics: Processes and Applications*, Earl Logan, Jr.
123. *Gear Noise and Vibration*, J. Derek Smith
124. *Practical Fluid Mechanics for Engineering Applications*, John J. Bloomer
125. *Handbook of Hydraulic Fluid Technology*, edited by George E. Totten
126. *Heat Exchanger Design Handbook*, T. Kuppan

127. *Designing for Product Sound Quality*, Richard H. Lyon
128. *Probability Applications in Mechanical Design*, Franklin E. Fisher and Joy R. Fisher
129. *Nickel Alloys*, edited by Ulrich Heubner
130. *Rotating Machinery Vibration: Problem Analysis and Troubleshooting*, Maurice L. Adams, Jr.
131. *Formulas for Dynamic Analysis*, Ronald L. Huston and C. Q. Liu
132. *Handbook of Machinery Dynamics*, Lynn L. Faulkner and Earl Logan, Jr.
133. *Rapid Prototyping Technology: Selection and Application*, Kenneth G. Cooper
134. *Reciprocating Machinery Dynamics: Design and Analysis*, Abdulla S. Rangwala
135. *Maintenance Excellence: Optimizing Equipment Life-Cycle Decisions*, edited by John D. Campbell and Andrew K. S. Jardine
136. *Practical Guide to Industrial Boiler Systems*, Ralph L. Vandagriff
137. *Lubrication Fundamentals: Second Edition, Revised and Expanded*, D. M. Pirro and A. A. Wessol
138. *Mechanical Life Cycle Handbook: Good Environmental Design and Manufacturing*, edited by Mahendra S. Hundal
139. *Micromachining of Engineering Materials*, edited by Joseph McGeough

Additional Volumes in Preparation

Reliability Verification, Testing, and Analysis in Engineering Design, Gary S. Wasserman

Control Strategies for Dynamic Systems, John H. Lumkes, Jr.

Practical Guide to Pressure Vessel Manufacturing, Sunil Pullarcot

Nondestructive Evaluation, edited by Peter J. Shull and Michael Ehrlich

Diesel Engine Engineering: Dynamics, Design, and Control, Andrei Makartchouk

Bearing Design in Machinery, Avraham Harnoy

Handbook of Machine Tool Analysis, Ioan D. Marinescu, Constantin Ispas, and Dan Boboc

Implementing Concurrent Engineering in Small Companies, Susan Carlson Skalak

Mechanical Properties of Engineering Materials, Wole Soboyejo

Mechanical Engineering Software

Spring Design with an IBM PC, Al Dietrich

Mechanical Design Failure Analysis: With Failure Analysis System Software for the IBM PC, David G. Ullman

Micromachining of Engineering Materials

edited by
Joseph McGeough
The University of Edinburgh
Edinburgh, Scotland

CRC Press
Taylor & Francis Group
Boca Raton London New York

CRC Press is an imprint of the
Taylor & Francis Group, an **informa** business

FIRST INDIAN REPRINT, 2015

CRC Press
Taylor & Francis Group
6000 Broken Sound Parkway NW, Suite 300
Boca Raton, FL 33487-2742

Visit the Taylor & Francis Web site at
http://www.taylorandfrancis.com

and the CRC Press Web site at
http://www.crcpress.com

Printed and bound in India by Nutech Print Services

Preface

The present need for smallness of parts stems mainly from two requirements: greater compactness in the utilization of space and portability. The mechanical and electrical devices that make up these items need to be produced in ever-decreasing sizes, with tightly specified dimensions and accuracies. Although these miniature devices may be manufactured by various procedures, their shaping through removal of material constitutes a major means of production. Established and recently developed methods of machining continue to be investigated for the shaping of such parts to specified small dimensions. The term "micromachining" has thus emerged and is generally used to define the practice of material removal for the production of parts having dimensions that lie between 1 and 999 μm, although an upper limit of 500 μm has recently been considered to set the border between micro- and macromachining. This book is concerned with the technology of micromachining of materials utilized in engineering practice.

The advances in material removal that have led to micromachining are the subject of the first chapter. The smallness of micromachined parts can place strict limits on variations or errors in the dimensions. Precise measurement of the parts produced in micromachining is central to the progress of this technology. Methods of measurement available for micromachining are therefore discussed in Chapter 2. Further advancements in the technology can be aided through a theoretical understanding of micromachining. Such considerations are embodied within the treatments of many of the micromachining processes discussed in this book. The diversity of these processes renders difficult a generalized theoretical analysis of micromachining; however, the methods of molecular dynamics are becoming increasingly attractive for studies of micromachining, especially as the technology advances toward the shaping of parts in the nanometric range. The foundations of molecular dynamics that are needed for theoretical treatments of micromachining therefore form the basis of Chapter 3. A series of micromachining processes are then discussed. First, methods are presented that rely on mechanical means for material removal, by grinding (Chap. 4) and diamond turning (Chap. 5); growing interest in the ultrasonic technique is reflected in Chapter 6. Limitations in these mechanical methods can arise, usually as a result of the shapes or accuracies needed or the types of engineering material to be tackled. Their drawbacks have spurred investigation of alternative methods. To that end, electrodischarge and laser micromachining are studied in Chapters 7 and 8, respectively. These two methods rely on vaporization of the workpiece material, by, respectively, spark discharges in a liquid dielectric and photon light energy. The well-known electrochemical machining process has also been adapted for micromachining purposes, and in Chapter 9 its principles, which rely on both mask-based and maskless anodic dissolution processes, are explained. With the inexorable transition from micro- to nano-machining, processes that rely on the interaction of atomic or subatomic particles with engineering materials for shaping of components will continue to attract attention. The place of ion beam technology, in which charged atoms, or ions, are fired at a workpiece surface in order to remove atoms or groups of atoms, is investigated in Chapter 10. Likewise, Chapter 11 is devoted to machining in which a controlled fine-focused beam of electrons causes material removal, through the vaporization resulting from translation of kinetic to thermal energy. Electron beam technology is the foundation of many aspects of high-resolution lithography. This topic is studied in the final chapter of the book.

The diversity of the subject of micromachining has required that specialists in each of its main fields should prepare the chapters of this book. The international interest in the subject is evident, with authors coming from six countries on four continents. I am grateful to them all, for the benefit of their advice and expertise, and their patience in supplying me with their specialist chapters, and in many cases for lengthy subsequent dialogue.

Owing to the diverse character of the subject, a single notation for the book has been difficult to achieve. For ease of working, therefore, a list of principal symbols and their meaning is included in the appropriate chapters.

At Edinburgh University I am indebted to my colleagues Dr. Yaxin Tan, Gregory Skrabalak, and Martin Schopf for their assistance in checking many of the details of the chapters, and to Dr. G. O. Goudie for his cheerful liaison with individual authors. Avril Davies has also been most helpful in my dialogue with the authors and in my final preparation of the manuscript. Derek Jardine assisted with diagrams. The advice of Tim Jolly was much valued. At the publisher, Marcel Dekker, Inc., I have been appreciative of the support of Rita Lazazzaro and Eric Stannard, as the book has developed from its draft outline form through various stages of its production.

My greatest thanks have to be reserved for my wife, Brenda, and Andrew, Elizabeth, and Simon McGeough for their steadfast support and interest throughout the preparation of the book.

Joseph McGeough

Contents

Contributors

David M. Allen *Cranfield University, Bedford, England*

Y. Benkirane *Ecole Nationale Supérieure d'Arts et Métiers, Paris, France*

Madhav Datta* *IBM Corporation, Yorktown Heights, New York*

Kai Cheng *Glasgow Caledonian University, Glasgow, Scotland*

John Corbett *Cranfield University, Bedford, England*

* *Current affiliation*: Intel Corp., Hillsboro, Oregon.

H. El-Hofy *Alexandria University, Alexandria, Egypt*

A. B. M. Khairy *Alexandria University, Alexandria, Egypt*

D. Kremer *Swiss Federal Institute of Technology, Lausanne, Switzerland*

D. Macintyre *University of Glasgow, Glasgow, Scotland*

T. Masuzawa *Tokyo University, Tokyo, Japan*

Joseph McGeough *The University of Edinburgh, Edinburgh, Scotland*

Johan Meijer *University of Twente, Enschede, The Netherlands*

Shoichi Shimada *Osaka University, Osaka, Japan*

S. Thoms *University of Glasgow, Glasgow, Scotland*

Mohammad A. Younes *Alexandria University, Alexandria, Egypt*

Micromachining of Engineering Materials

1

Introduction

H. El-Hofy and A. B. M. Khairy
Alexandria University, Alexandria, Egypt
T. Masuzawa
Tokyo University, Tokyo, Japan
Joseph McGeough
The University of Edinburgh, Edinburgh, Scotland

1.1 ADVANCES IN MACHINING TECHNOLOGY

The shaping of materials for practical needs has been part of human activity since prehistoric days. Fragments of stone, bone, and wood were first used as tools by humans. In due time they learned to assemble several pieces into a device which could then be used more effectively for shaping materials than was feasible with a single-piece tool.

Progress in machining technology stems from those early days. Tools for drilling and cutting of metals are known to have been used in Ancient Egypt about 4000 BC. From the eras of the early civilizations up to the Middle Ages hand tools continued to be developed, in the production of basic utensils for everyday use and in the construction of major artifacts such as ships and carriages. Elementary mechanically driven devices were duly devised, and by the 17th century instruments for machining were widespread. More sophisticated tools for machining powered by water, steam, and electricity arose over the 18th and 19th centuries [1]. These and other advances continued into the 20th century, and gave rise to more adequate definitions of the practice of machining: the removal of a specified amount of material from a workpiece in order to produce a required space economically and accurately. In this practice, the operator was given a drawing of the finished part. The operator determined the machining strategy, arranged the machine, selected tooling, feeds, and speeds, and manipulated the machine controls to produce the given part, which then had to be inspected. Under these procedures, the accuracy and surface quality of products were not always satisfactory. Further developments of machine tools were necessary to improve accuracy and productivity. Templates, copying techniques, cam mechanisms, and indexing devices were introduced as a result of which the amount of labor needed in machining was also reduced.

The application of numerical control (NC) to machine tools in the 1950s opened the way for computer (CNC) and direct (DNC) numerically controlled machining. With the advent of robotics (1980s) enhanced accuracy in machining, uniformity in the items produced, and flexibility in manufacturing practices was achieved [2]. Machining processes and their associated machine tools have continued to benefit from the major strides made in the electronics and computer industries (1980s to 1990s). Ingenious machine designs and instruments have enabled the highly accurate machining of complex shapes and their measurement. Notable achievements in machining include techniques such as precision jig boring and grinding, and superfinishing. The 1990s witnessed further trends with increasing demands for components of dimensions on the order of micrometers (μm). Much of this interest has stemmed from the microelectronics and optical industries in which components have to be produced to ever-decreasing size or to finer accuracies. Requirements for microengineered products are also recognized for other applications. For instance, a small bearing in a watch supports a tiny spindle of microgear, and the size of the bearing hole must be smaller than the size of the microspindle. The stylus of the scanning transmission microscope provides a further example: it must gather geometrical information concerning the individual surfaces of atoms. Thus the manufacture of parts capable of dealing with atomic and molecular dimensions is becoming increasingly significant. Taniguchi [3] and then McKeown [4] and their coworkers have considered how equipment to achieve very fine shapes to high accuracy improved over the 20th century (Fig. 1). They classified machining in three divisions:

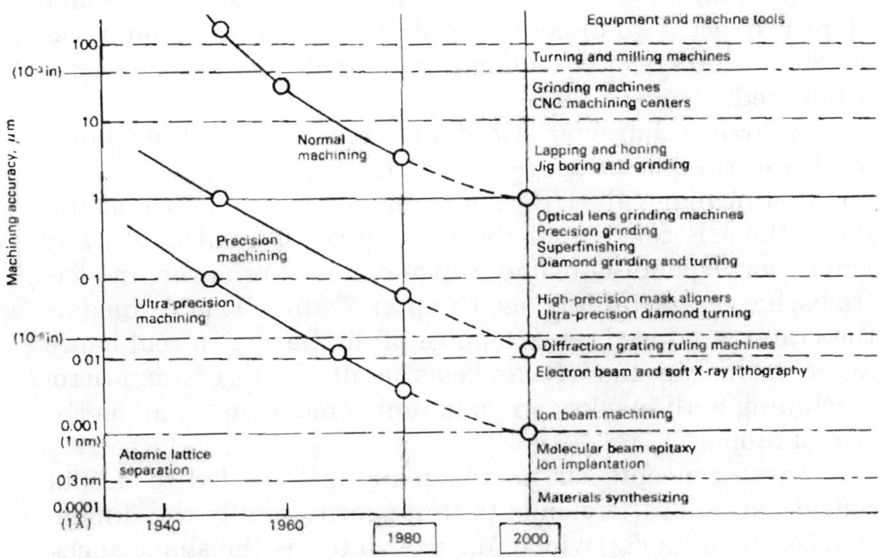

Figure 1 Machining accuracies. (After Refs. 3 and 4.)

normal, precision, and ultraprecision. The term "micromachining" is now associated with the qualities of precision and ultraprecision.

1.2 CHARACTERIZATION OF MICROMACHINING

Through the 1990s, Masuzawa drew attention to the need for characterization of micromachining, and in particular to means for direct determination of the shape of its products. (See, for example, [5].) He defines some basic groups of micromachining processes. The first set utilizes fixed and controlled tools, which can specify the profiles of three-dimensional shapes by a well-defined tool surface and path. These methods remove material in amounts as small as tens of nm, which is acceptable for many applications of micromachining. For finer precision and especially to atomic levels, the second set of micromachining processes employs masks to specify the shape of the product. Two-dimensional shapes are the main outcome; severe limitations occur when three-dimensional products are attempted.

A further characteristic of micromachining is the volume or size of the part removed from the workpiece, termed the "small unit removal" (UR). For example, in mechanical operations, the UR consists of the feed pitch, and cut depth and length corresponding to one chip of material removed; in electrodischarge machining (see Chapter 7) the UR is defined as the crater produced by one pulse of discharge. In tool-based micromachining, the UR can be as small as tens of nm. Micromachining with masks can yield unit removal as small as the size of atoms.

These concepts for micromachining have led to deeper consideration of the elements that specify finally the dimensions of its products, which Masuzawa terms the shape specification elements (SSE) [6]. The cutting tools used in mechanical turning, and the light beam in laser machining are examples of SSE. Indeed, most conventional machining processes are employed with tools as the SSE, and they can be readily applied to micromachining by miniaturization of their tools. Thus, turning, drilling, and milling have proved to be applicable to micromachining of shapes in the range of μm. The key lies in the production of microtools and in the maintenance of coordinate accuracy through tool-making, "chucking," and machining. The development of wire electrodischarge grinding has advanced significantly the technology of microtool production. Tool-making "on-the-machine" has proved to be a major benefit in the maintenance of accuracy. This aspect is considered further below.

1.3 TOOL-MAKING "ON-THE-MACHINE"

Since the tools for micromachining are often very small, the transfer of coordinate information from the stages of tool-making to actual machining must receive careful attention.

Three typical system configurations are shown in Figure 2. In system A, the tool coordinates are common for both tool-making and micromachining. Consequently, the highest accuracy should be expected. As tool-making and micromachining

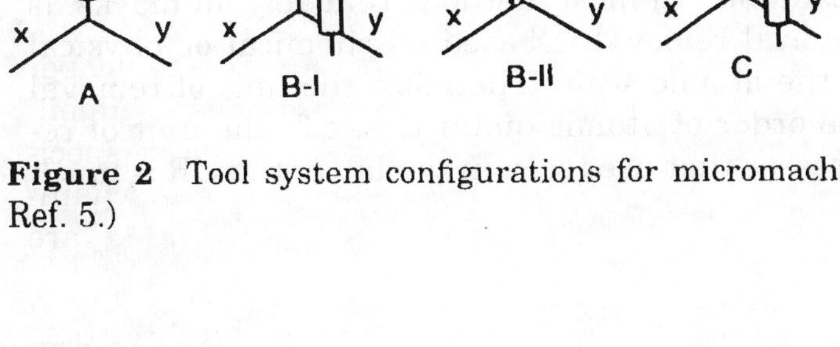

Figure 2 Tool system configurations for micromachining. (After Ref. 5.)

cannot be performed simultaneously this system leads to low productivity.

System B is most conventionally used in established machining processes. However, it is often difficult to place reference planes on the microtool. The result may be "off-centering," and/or tilting, as shown, respectively, in Figures 2 B-I, and B-II. Tool storage is possible between the stages of tool-making and micromachining, which is beneficial to productivity.

The tool and spindle are handled as a set in System C. Since sets can be stored in the same manner as the tools in System B, the productivity is again higher than that of system A. However, positioning error and the cost of extra spindles are the drawbacks of this system.

When a microproduct is removed from the machine, and placed in a parts container for subsequent assembly, it loses most of its coordinate information. Assembly is then made difficult. This difficulty can be overcome by assembly "on-the-machine." Since the space required for micromachining per se is usually small, the table can be divided into several parts for the modules on which tool-making, machining, product storage, and assembly take place. As the coordinates are common for these modules, a microproduct may then be more readily assembled.

Finally shape specification elements for processes in which masks are used have to be considered. Since the mask is specified in only two dimensions, these fabrication methods are usually applied to the production of thin or shallow shapes. Photofabrication of this type has advanced considerably reduced pattern sizes in semiconductor devices, with dimensions below the μm range already achieved and nanometric sizes becoming attainable. In most methods that rely on masks as the SSE, material removal is based on chemical or physical reactions on the atomic scale. Therefore the unit of removal can be on the order of atomic quantities; c.f., the unit of removal in processes that use tools: in the latter the UR is much higher, owing to the different mechanisms of material removal.

Mask-based processes offer certain main advantages:

1. Since the UR can be very small, production from the μm to nm range is possible.
2. Batch processes are possible enabling high production rates.

Mass production is a major issue in micromachining technology, as a large number of small parts are a common requirement. As noted above, mask-based processes such as photofabrication are capable of mass production. Most other processes have not yet been developed sufficiently to produce thousands of parts needed for microproducts. Modification of existing micromachining methods and new concepts for mass production based on them are major targets for future research and development.

1.4 MICROMACHINEABILITY OF MATERIALS

Until recently the main materials considered in micromachining have been metals and silicon for, respectively, tool- and mask-based processes. As industrial requirements for micromachining continue to grow, other substances such as ceramics, plastic, glass, and biological materials have been receiving attention. To that end, the micromachineability of engineering materials needs to be considered.

Established methods of micromachining by turning, drilling, milling, and grinding have already been applied to materials including copper and aluminum alloys, gold, silver, nickel, and polymethylmetacrylate (PMMA) plastics. The traditional measures of machineability, surface integrity, tool service life, and power consumption, remain valid for micromachining, as do qualitative indicators such as type of chip produced. Conditions of chip production for conventional material removal processes are highly affected by molecular scale phenomena. The forces arising are very small, and the ratio of normal to tangential component forces is high, especially at very small depths of cut. Simulation of such nanometric levels of chip removal may be undertaken by the methods of molecular dynamics, which are explained in Chapter 3, an example being given in Figure 3. These nanometric effects cause a considerable rise in specific cutting energy (or force), as may be noted from Figure 4. The length of the shear plane as an indication of material machineability also needs careful consideration; eventually there are three shear zones expected: primary, secondary, and an "apparent" shear region that is likely to form the clearance pocket between the bottom of the tool and the machined surface. The apparent shear zone is brought

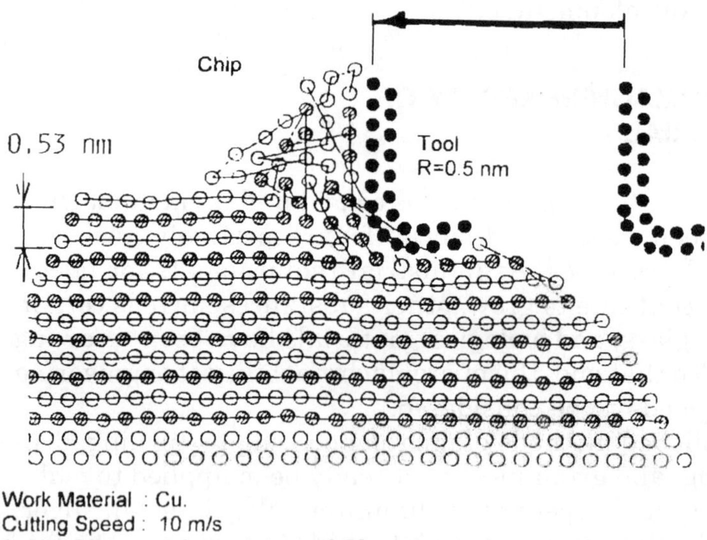

Work Material : Cu.
Cutting Speed : 10 m/s
Depth of cut : 2.2 atomic layer (0.53 nm)

Figure 3 Molecular dynamic simulation of nanometric chip removal. (After Ref. 14.)

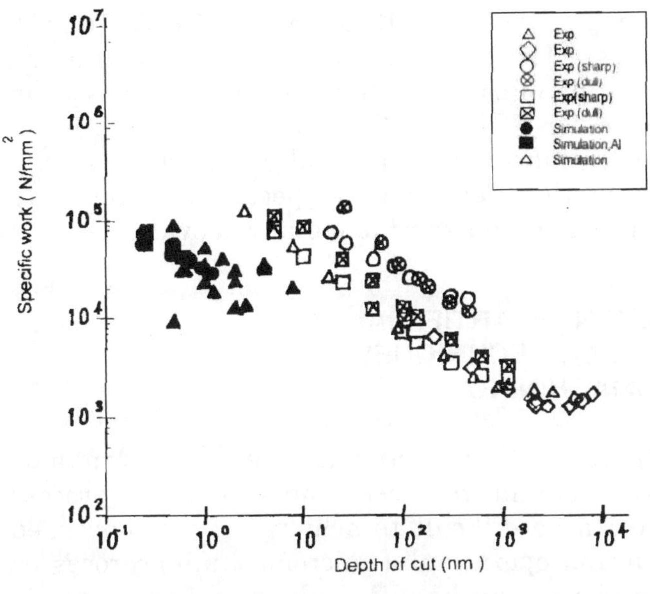

Figure 4 Dependence of specific cutting energy on depth of cut. (After Ref. 14.)

about by elastic recovery of work material and/or tool flank wear. Experimental evidence shows that pronounced dissipation of mechanical energy occurs due to plastic deformation of monolayers of the work part accompanied by "skidding" of the tool cutting edge. In effect, a new measure of "micromachineability" for conventional methods may be needed.

A micromachineability rating for unconventional processes may be less difficult to establish. These all rely on removal of micro amounts of materials either by mechanical means (e.g., ultrasonics), thermal erosion (electrodischarge, laser, electron beam), anodic dissolution (ECM), or ion impact (ion beam machining; see, e.g., [7]). Their principles of machining are reasonably well understood, and little significant difference occurs between material removal on the micro- and macro-scales. Many studies have been reported on machineability rating systems for ECM [8–10], EDM [10, 11], and laser processes [11, 12]. For example, Khairy drew on techniques for rating of conventional machining processes in establishing those for ECM [8]. He proposed that ECM machineability has to include dimensional and surface quality and power consumption, dependent, respectively, on the equilibrium machining gap and specific metal removal rate.

1.5 APPLICATION OF ARTIFICIAL INTELLIGENCE TECHNIQUES IN MICROMACHINING

It is evident therefore that micromachining often demands configurations of tool and workpiece and selection of process conditions that can be difficult to achieve. Those concerned with the design and operation of micromachining processes have limited proficiency in handling all the different procedures necessary and have to seek advice from experts and/or refer to a limited amount of literature. Such information is seldom available, as the bulk of expertise, based on practice, is not readily obtainable in technical documents or databanks.

The use of computerized fuzzy logic, artificial neural networks, and knowledge-based systems to increase the productivity and quality of micromachined components is a promising approach to overcoming difficulties. Fuzzy controllers, which use qualitative rules of discrete fuzzy classes, instead of continuous variables, to optimize micromachining performance have already been applied to electrodischarge machining albeit with limited success. As reported by Kruth [13], hybrid fuzzy control systems have been developed that can accommodate a large band of critical variables in electrodischarge micromachining, including those discussed above:

1. Effects of undesirable relative motion between tool tip and workpiece caused by vibration and/or wear;
2. Imperfect fit between mating parts in assembly, such

as machine spindle, part fixture, motion-drive unit, misalignment of tool;
3. Thermal deformation of the entire machining system over prolonged times of machining;
4. Deterioration of the condition of the machining zone due to pollution by debris and short-circuits.

Such approaches for unconventional processes were likewise anticipated for ultraprecision traditional methods by Ikawa and colleagues who emphasized the need for a machining strategy based on computer software to control the performances of all subsystems [14]. These requirements for micromachining were stressed more recently by Fujino et al. in their development of a multipurpose microprocessing machine [15].

Knowledge-based systems (KBS) have not yet been fully utilized in micromachining, although some work has been reported for electrodischarge (EDM) and electrochemical micromachining (ECM). Khairy's findings on ECM can be applied to the more general case of micromachining [16]. He shows that for the latter, optimization of tool design and operating conditions often requires complex computations and judgments, and that decisions reached are not necessarily unique; that is, different tool geometries and operating choices could be viable. He points out that any KBS for nonconventional micromachining depends greatly on the expertise of the personnel involved. A proper methodology for tool design becomes difficult to establish, owing to the complexity of the hydrodynamics of the flowing electrolyte and the variation in the process variables during machining. Khairy explains that the key outcome of tool design linked to selection of machining variables has to be the tool geometry that provides the required shape of the workpiece. A KBS has to recognize the following problems. First, the entire process of selecting the machining parameters, determination of the tool shape, finding the interdependences between process variables, and formulation of an appropriate plan can be difficult. Second, knowledge of micromachining technology is scarce and is restricted frequently to in-house use. Third, the multiplicity of boundary limits in analytical tool design and finding, for example, a realistic mode of material removal may lead to a number of alternative choices, if only an algorithmic strategy is adopted.

Such aspects make the selection of micromachining operating parameters a prime candidate for a KBS approach. Limited KBS have already been developed for highly specialized micromachining operations [17,18].

For micromachining in general the procedures for a knowledge base therefore become clear. The KBS should then be developed to contain as much expertise as possible that is relevant to the process. Most of this expertise will be available in technical reports, reference books, and case studies [19]. A protocol analysis, however, can be used to gather information on a specific micromachining process in readily formatted records of information which can thereafter be expressed as facts about concepts, practices, and rules for handling the facts.

KBS of this type can embody a number of working modules: information input, machining procedure, tool design, and economics; each module requires certain types of knowledge. Thus the KBS should be able to provide the necessary support required to guide successfully the micromachining of the part required.

REFERENCES

[1] W.L. Sims, The contribution of the machine tool industry to pioneering in new technologies, MA dissertation, University of Leicester (1988).

[2] L.T.C. Rolt, *Tools for the Job*, Her Majesty's Stationery Office, London (1986).

[3] N. Taniguchi, Current status in and future trends of ultraprecision machining and ultrafine materials processing, *Annals of the CIRP*, (32) 2, 1–8 (1983).

[4] P.A. McKeown, High precision engineering and the British economy. In *Proc. I. Mech. E., 200* (76), 1–8 (1986).

[5] T. Masuzawa and H.K. Toenshoff, Three-dimensional micromachining by machine tools. *Annals of the CIRP*, (46) 2, 621–628 (1997).

[6] T. Masuzawa, What will micromachining be? Private communication (1998).

[7] J.A. McGeough, W.J. McCarthy, C.B. Wilson, Electrical methods of machining. In Machine Tools, *Encyclopaedia Britannica* Vol. 28 712–736 (1987).

[8] A.B. Khairy, Towards a machinability system for ECM. In *Proc. CAPE-3 Conf.* (Editor J.A. McGeough), Edinburgh, UK, pp. 343–350 (1989).

[9] G. Bellows, ECM machinability and data rating. ASTME Paper MR 67, pp. 1–10 (1967).

[10] S. Ebeid, Computerised machinability data-base system as applied to ECM and EDM processes. In *Proc. 2nd AME Conf., MTC*, Cairo, 130–143 (1986).

[11] R. Snoeys, F. Staelens, and W. Dekeyser, Current trends in nonconventional metal removal processes. *Annals of the CIRP*, (35)2/1, 467–480 (1986).

[12] P. Hoffmann et al., Recent developments in laser system technology for cutting and welding applications. In *Proc. ISEM-XI*, 785–800 (1995).

[13] J.P. Kruth, Advances in physical and chemical machining. *Advancement in Intelligent Production*, E. Usui (Editor), Sept. 123–138 (1994).

[14] N. Ikawa et al., Ultraprecision metal cutting, the past, the present and the future. *Annals of the CIRP*, (42)2, 587–594 (1991).

[15] M. Fujino et al., Development of multi-purpose microprocessing machine. In *Proc. ISEM-XI*, April 613–620 (1995).

[16] A.B. Khairy, A knowledge-based system for electrochemical machining procedure. *J. Materials Processing Technology*, (58) 121–130 (1996).

[17] A. De Silva and J.A. McGeough, The production of full die

shapes by electrochemical arc/electrochemical machining. In *Proc. ISEM-X*, 107–110 (1989).

[18] F. Olsen, Investigations on optimising the laser cutting process. In *Proc. ASME Conf. on Lasers in Materials Processing*, Los Angeles, 64–80 (1983).

[19] T. Moriwaki and E. Shamoto, Ultraprecision diamond turning of stainless steel by applying ultrasonic vibration. *Annals of the CIRP*, (40)1, 559–562 (1991).

2

Measurement Techniques in Micromachining

Mohammad A. Younes
Alexandria University, Alexandria, Egypt

INTRODUCTION

The last two decades have shown an ever-increasing interest in higher precision and miniaturization in a wide range of manufacturing activities. These growing trends have led to new requirements in machining, positioning control, and metrology down to nanometer tolerances. Recent developments in silicon micromachining have made possible the fabrication of micromechanical elements of sizes typically ranging from 0.1 to 100 μm [1–4]. Slots and apertures for some applications such as color TV, electron gun masks, and jet-engine turbines are made as small as 5 μm. Microcircuit elements of 0.5 to 1 μm are commonly manufactured using X-ray or electron-beam lithography [5]. In order to assess and control the quality of micromachined parts it has been necessary to develop new measuring techniques, capable of effectively and accurately measuring the dimensions, geometry, profile, and surface roughness of microholes, slots, very thin films, microspheres, steps, and grooves of different configurations in micromachined parts. These parts and features can be either checked for configuration and completeness, or measured to determine actual sizes. Inspection and measurement of these features raise the demand for special equipment some of which depends on entirely new principles.

2.1 CLASSIFICATION OF MEASURING SYSTEMS

In addition to high-resolution calipers and coordinate measuring machines, equipment used for measurement of micromachined parts includes high resolution microscopes, laser-based surface followers, scanning electron microscopes (SEM), interferometers, profilometers and scanning probes (e.g., scanning tunneling microscopes STM), and scanning force microscopes (SFM). The practical use of almost all these methods depends on the development of high-precision scanning tables as well as high-resolution linear transducers. Measurements are carried out offline, in a metrology laboratory, as well as online, or in-process while the parts are being fabricated. In most measuring applications, noncontact methods are eventually used. Measuring systems rely, in their function, upon different principles and apply several technological methods. The systems used for dimensional measurement and topographic inspection can, however, be classified into two categories.

2.1.1 Category 1

In this category the size of an inspected feature is determined by measuring the distance between its edges (Fig. 1a). Accordingly, the system consists of three main parts: a sensor, a pre-

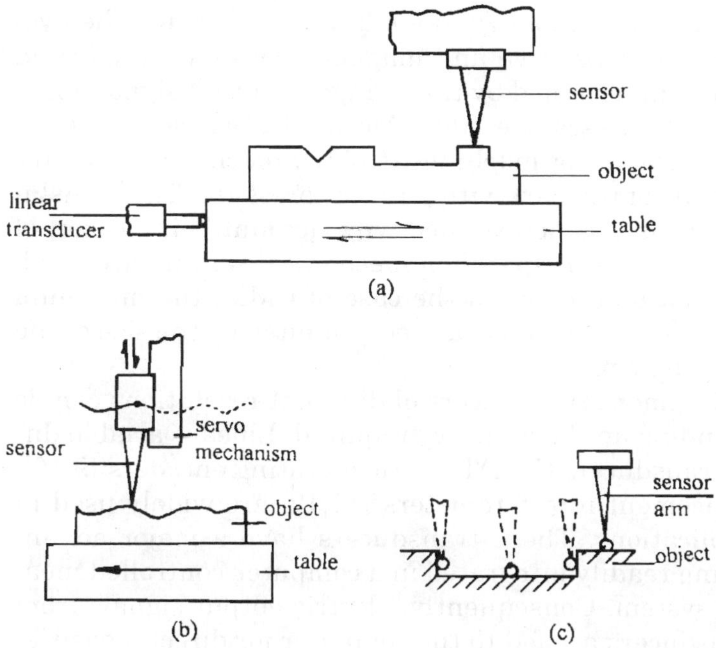

Figure 1 Different configurations of measuring instruments.

cision table, and a displacement transducer. By such an arrangement, the sensor determines the exact position of the feature edges while the precision table moves the object or the sensor for edge location. The displacement transducer can then measure the distance moved between edges and indicate the size of the feature in the specific direction.

Sensors can be mechanical, magnetic, capacitive, . . . and in many instances optical. Tables of stable and precise movement have recently been developed (e.g., [6]). For short travels of submicrometer and nanometer resolution levels, piezoelectric-driven stages are recommended [7,8].

In optical sensors, the position of the measured edge is realized by the change in the reflected pattern as a light beam crosses an edge. In optical microscopes, the edge position is determined by a stationary index line placed inside the eyepiece unit. For capacitive and magnetic sensors the position of the edge is determined by the change in output signal noted as the sensor crosses the edge. Mechanical sensors (e.g., in coordinate measuring machines (CMM)) touch the inside (or outside) walls of the part with a preset pressure. The translation of the sensor is determined with account being taken of the size of the sensor tip. When mechanical sensors are used, which are not stationary in the case of CMM, the minimum internal dimension to be measured is limited by the size of the sensor (stylus) tip.

Displacement transducers of different resolutions can be used depending on the accuracy required. Linear variable differential transducers (LVDT), optical grating encoders [9,10], and displacement interferometers [11,12] are widely used in many applications. These transducers have a major advantage of being readily integrated in a computer-controlled measurement system. Consequently, electric output signals from these transducers are fed to the computer for direct measurement and/or control. The above-mentioned category of measuring instruments is used to measure sizes of object features other than the height except for the case of CMM.

2.1.2 Category 2

In order to measure height, profile, or surface topography another category is used. This can be classified into two main types, whole field contouring and single profile methods. Whole field contouring includes interferometric and holographic techniques. Single profile (SP) methods include mechanical stylus instruments, optical profile followers (OPF), scanning tunneling microscopy, scanning electron microscopy, and atomic force microscopy (AFM).

Single Profile and Height Measuring Methods

Here the sensor is forced to follow the profile of the inspected surface based on specific criteria (Fig. 1b). Height variations can be recorded provided they are small enough to maintain the validity of the working principle. The sensor can be of the contact-type as in most CMM machines and stylus-type roughness meters or it may be noncontact utilizing several principles. CMM machines can accurately measure step height provided there is sufficient room for the insertion of the sensor tip (Fig. 1c). The resolution depends on the operating principle, and values as small as 1 Å are characteristic of some of these instruments (e.g., SEM, STM, AFM). The working principles of some of these systems that have potential use in the measurement of micromachined parts are illustrated in the following sections. The range of their application, accuracy, and resolution limits are also examined.

Whole Field Contouring

In whole field contouring (WFC), a contour image (interferogram) of the inspected object surface is recorded (Fig. 2) by use of several interferometric or holographic arrangements. The contour images are analyzed by means of appropriate computer algorithms, and surface height ordinates are accordingly determined. In this case, the resolution depends on both the arrangement used as well as the algorithm adopted. Some interference microscopes [13] have vertical resolution on the order of 0.1 nm, with maximum vertical step height of 100 µm. Besides roughness such methods can also measure and evaluate height of microsteps and grooves.

2.2 MICROSCOPES

Microscopes are widely used for the inspection and measurement of tiny object features. Basically, a microscope produces a magnified virtual image of the inspected object. Three types of microscopes, namely, optical, electron, and interference are used. The principle and application of the first two types are discussed in the following section, while that for the third type is presented in Section 2.4.2.

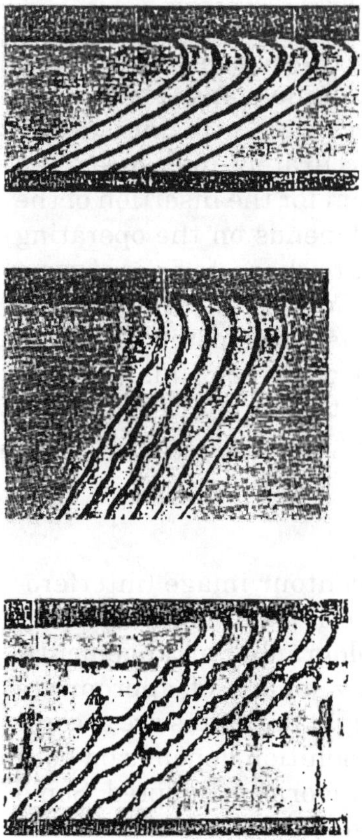

Figure 2 Typical interferograms.

2.2.1 Optical Microscopes

An optical microscope consists basically of two lenses (Fig. 3), a high-power short focal length objective, and a low-power, longer focal length, eyepiece. In practice the objective and the eyepiece are not single lenses. To reduce the effects of aberrations, each is assembled from two or more lenses. Generally a microscope is equipped with several objectives to provide different magnifications, which can be as high as 1000x [14]. Optical microscopes can be used as stand-alone inspection instru-

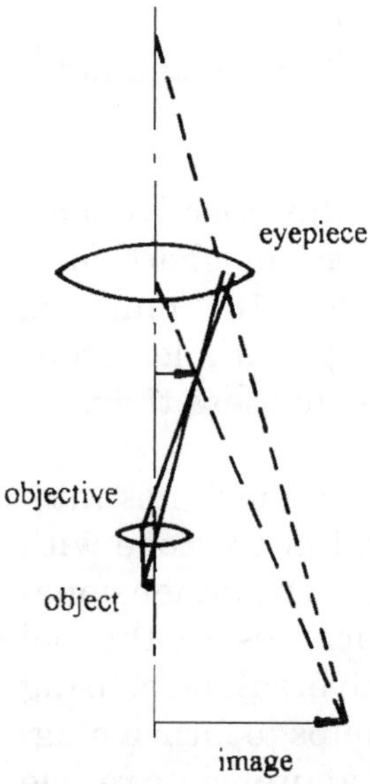

Figure 3 The principle of the optical microscope.

ments that are commonly employed for the visual inspection of printed circuit boards. Modern microscopes are equipped with video or CCD cameras where the field of view is observed on a cathode ray tube (CRT) monitor. Such advanced systems are presently fitted on production lines to monitor the quality of microfeatures of manufactured parts.

The resolving power of an optical microscope a is given by the Abbe equation [14]

$$a = \frac{1.22\lambda}{2n\,\sin(i)} \tag{1}$$

where

a: distance between two points on the surface of an object that can be seen separated in the image plane
λ: effective wavelength of illumination used

n: refractive index of objective medium
i: suspended angle of lens which depends on its diameter and focal length

(n sin(i)) is the numerical aperture (NA) of the objective lens, which is higher for a high-power lens having short focal length. For white light illumination, $\lambda = 5.6 \times 10^{-4}$ mm and, assuming NA = 1, then, a = 0.27 μm. An optical microscope with NA = 0.6 can effectively detect two points less than 0.5 μm apart.

Many commercially available measuring machines integrate the high resolving power of an optical microscope with a high-resolution x–y stage to measure different dimensional features of a product. Examples of these machines are the tool maker's microscopes (TMM), and the universal measuring machines (UMM) (Fig. 4). The microscope helps to form a magnified image of the inspected workpiece. An appropriate reticle with cross-lines fixed inside the eyepiece is used to mark the ends of the dimensions to be measured. The dimension size is

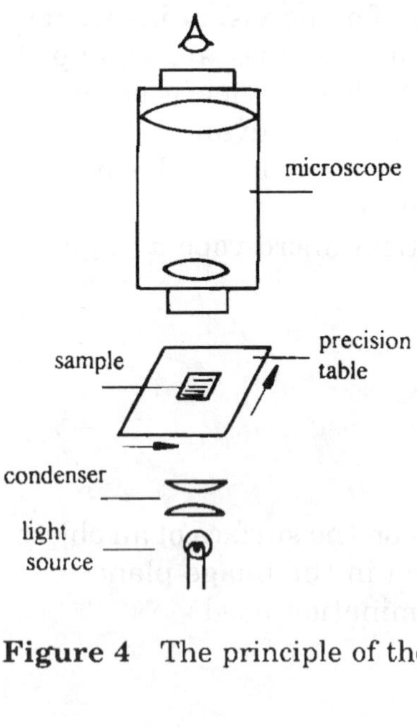

Figure 4 The principle of the tool maker's microscope (TMM).

determined by the distance moved with the x–y stage between the two ends. An accuracy level of 1 μm is commonly available with Abbe's metroscopes. Better resolutions are achieved from interferometric displacement measurement [11]. Indeed, dimensions of holes, slots, and other features of an object less than 200 μm can easily be measured with accuracy better than 1 μm. TMMs are used for dimensional measurement of both internal and external part features.

2.2.2 Electron Microscopes

In electron microscopes the inspected surface is interrogated by a focused beam of electrons. The beam is collimated and then focused by means of coils that generate a radial magnetic field to control the shape of the electron beam. Focusing is achieved by varying the focal length of the objective lens coil through the control of the coil current. The effective resolution is almost 10^5 that of an optical microscope [15]. In this case the wave length λ is given by the De Broglie formula [16],

$$\lambda = \frac{h}{mv} \tag{2}$$

where

h: Plank's constant
m: mass of electron
v: electron velocity

As an electron of mass m and charge e pass through a potential difference V, its kinetic energy is

$$\frac{1}{2}mv^2 = eV \tag{3}$$

and

$$\lambda = \frac{h}{(2meV)^{1/2}} \tag{4}$$

It is obvious that the wavelength λ depends on the potential difference V. For $V = 60$ kV, λ is about 0.05 Å. From Abbe's equation (1), the resolving power a for the electron microscope can be on the order of 2.4 Å for $\lambda = 0.05$ Å.

Two main types of electron microscopes are available, namely, transmission electron (TEM) and scanning electron (SEM) types. In most engineering applications SEM is used. Figure 5(a) shows a schematic diagram of the main components of an SEM microscope. The electron gun generates a stream of electrons (electron beam) that is collimated by a coil (lens). The objective (lens) focuses the electron beam onto the surface of the specimen. In order to scan the specimen surface with the focused electron beam (electron probe), a beam deflecting unit is used.

When an electron beam impinges on the surface of the specimen different phenomena are observed [17]. Some elec-

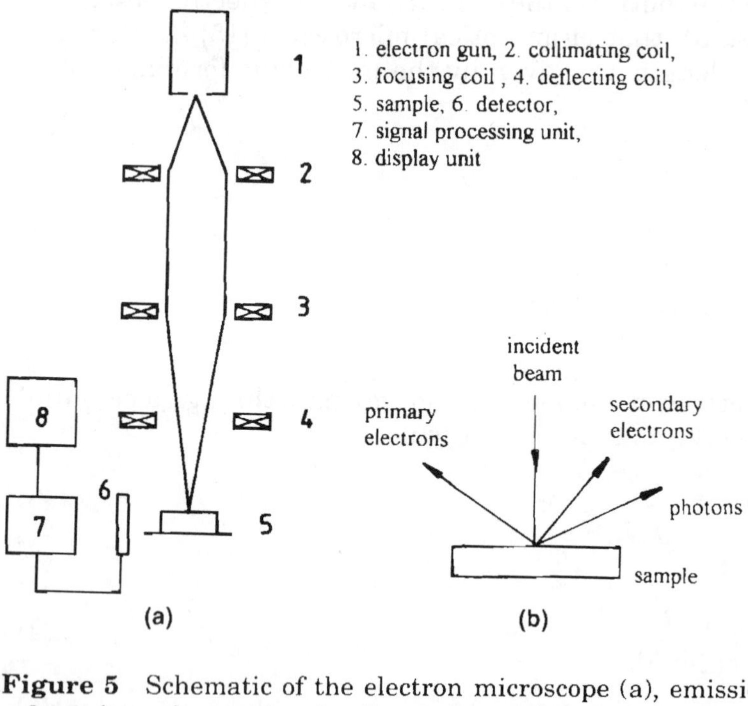

Figure 5 Schematic of the electron microscope (a), emissions resulting from electron bombardment (b), and typical micrographs (c) and (d).

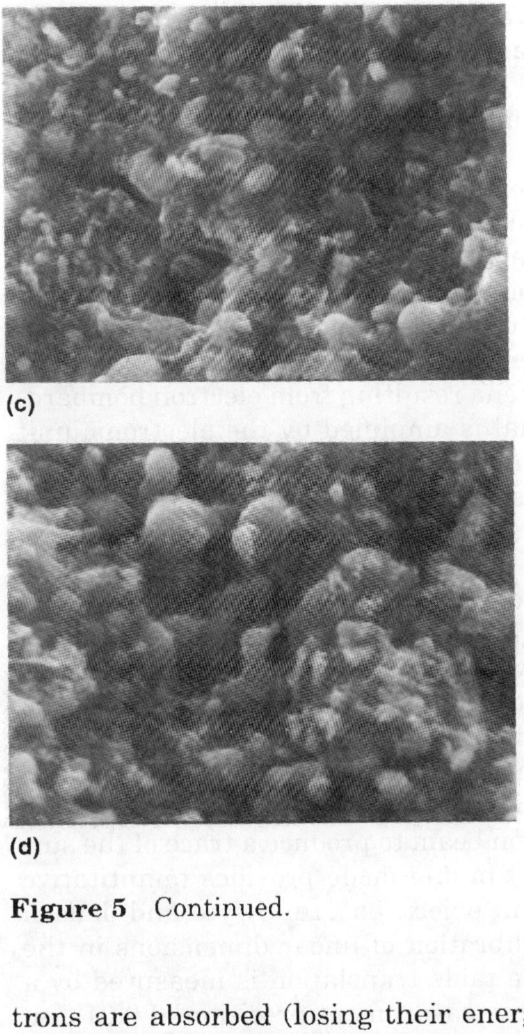

(c)

(d)

Figure 5 Continued.

trons are absorbed (losing their energy on collision). Others are back-scattered (reflected) either elastically or inelastically and these are called primary electrons. In elastic reflection, electrons do not lose any of their energy but change their direction. For inelastic reflection, electrons interact with specimen atoms and lose some of their energy before deflecting back out of the specimen surface. Electrons that penetrate inside the specimen interact with the material atoms resulting in the ejection of secondary electrons. Secondary electrons formed near the surface may escape producing secondary electron emits. In addition to primary scattered electrons and secondary emitted ones, X-ray and even light photons are also produced as a result of electron bombardment (Fig. 5b). The amount and ratio of back-scattered electrons, secondary emission, and other radiation depend on the beam energy, specimen geometry, and substrate atomic number. Since the atomic number and beam energy are practically constant, specimen geometry is therefore the main controlling factor.

A detector is used to collect the emission from the specimen surface (Fig. 5a). Several types of detectors are available for the different phenomena resulting from electron bombardment. The detector signal is amplified by the electronic unit and used to modulate the brightness of a cathode ray tube. The CRT is adjusted to scan synchronously with the electron probe. Variations in the recorded brightness produce the highly magnified image of the scanned object. Figure 5(c, d) shows typical SEM micrographs that reveal surface details in the submicrometer range.

SEM are used in two modes. In the conventional raster-scanning mode the deflection coils (Fig. 5a) move the electron beam across the stationary object to produce a two-dimensional image of the surface. In the second mode the specimen is fixed to a precision table which scans the specimen under a stationary focused electron beam to produce a trace of the surface relief. This intensity-profile mode provides quantitative linear measurement of an object feature. Swyte and Jensen [5] used SEM for the calibration of linear dimensions in the range 0.1 to 100 μm. The table translation is measured by a laser interferometer with measurement precision of 0.01 μm. Electron detector and interferometer signals are fed to a computer that analyzes the electron intensity with respect to position profile to obtain the linear dimension for a specific feature. A typical example of inspected objects is a microscopic chromium metal line deposited on a glass substrate by means of a photolithography technique [5]. In such an application, the line is 0.5-μm wide, and 1-μm thick, with an edge slope of 70°.

Asai et al. [18] used raster-scanning to obtain a sectional curve of diamond tool edges. A newly developed SEM featuring two secondary electron detectors was used to measure the cutting edge radius which is on the order of 45 nm [19,20]. The image created by the difference signal of the two detectors emphasizes the convexity and concavity of the sample surface.

As well as profile recording, SEM is used for roughness measurement. Based on the principle that the back-scattered electron signal is proportional to the slope of the surface in the direction of scanning, the surface profile can be obtained. By integrating the signal, Sato and Ohmori [19] detected a roughness profile of surfaces having slope in a specific direction. Three-dimensional surface topography of the specimen can be obtained by integrating scans covering the entire image, at resolutions on the order of 0.001 μm. To measure the roughness of surfaces having slope in an arbitrary direction, Sato and Ohmori [20] proposed a method to detect the normal of the object surface by comparing the intensity of the back-scattered electron signal of the specimen with that of a standard ball. The surface topography is processed from the measured data of the normal.

2.3 LASER-BASED SYSTEMS

In laser-based systems optical phenomena, observed as laser light scans across an engineering surface, are applied for online and in-process inspection of surface features. These phenomena include diffraction, reflection, refraction, scattering, and others. Some methods that depend on such phenomena are explained below.

2.3.1 Diffraction Method

In micromachining, slots of dimensions on the order of few nanometers up to 200 μm are produced. In the range of 50 μm or higher, these dimensions can readily be measured, in the laboratory with high magnification microscopes; however, for online measurement other techniques are adopted.

Younes [21] used a focused laser beam for online measurement of fine surface grooves having different cross-sections (Fig. 6). The principle of his technique depends on the condition that the reflected pattern from a flat surface will be modulated by the presence of a microsurface groove. A flat smooth surface reflects the beam into a single spot (Fig. 6b); however, as the focused beam crosses the edge of a groove, the reflected pattern is divided into two parts (Fig. 6c). Scanning of this field, therefore, produces a double-peaked signal (Fig. 6d), the relative amplitudes of which change as the surface moves against the focused spot. The slot width can be evaluated from the recorded signal of a photosensitive device (PSD) as the sample scans beneath the stationary laser spot. Moreover, the recorded signal also represents an approximation of the slot profile (cross-section). Figure 6(a) shows the arrangement used, while the PSD signals recorded as the focused spot scanned three different grooves are shown in Figure 6(e). Inspected surfaces should be smooth, since a speckle pattern limits the measurement accuracy.

The double-peaked pattern can be explained by the Pekrinck and Kennedy model [22]. Levy et al. [23] observed a similar pattern when inspecting the height of a submicrometer step. The method described above is noncontact and can be used online provided the specimen travel is precisely controlled. The proposed system is sensitive to changes in groove width; however, variations in depth are less detectable especially for high depth-to-width ratios, or steep sides. Although different incidence angles can be used, normal incidence results in more accurate slot description. The resolution of measurement is directly proportional to the sampling rate of the data acquisition unit R and inversely proportional to the table speed v. For $v = 300$ mm/min, and $R = 10$ kHz, the resolution can be on the order of 0.5 μm. A linear photodetector array can be used to indicate the start and finish of the double-peaked pattern, while a linear transducer measures the distance covered during this event.

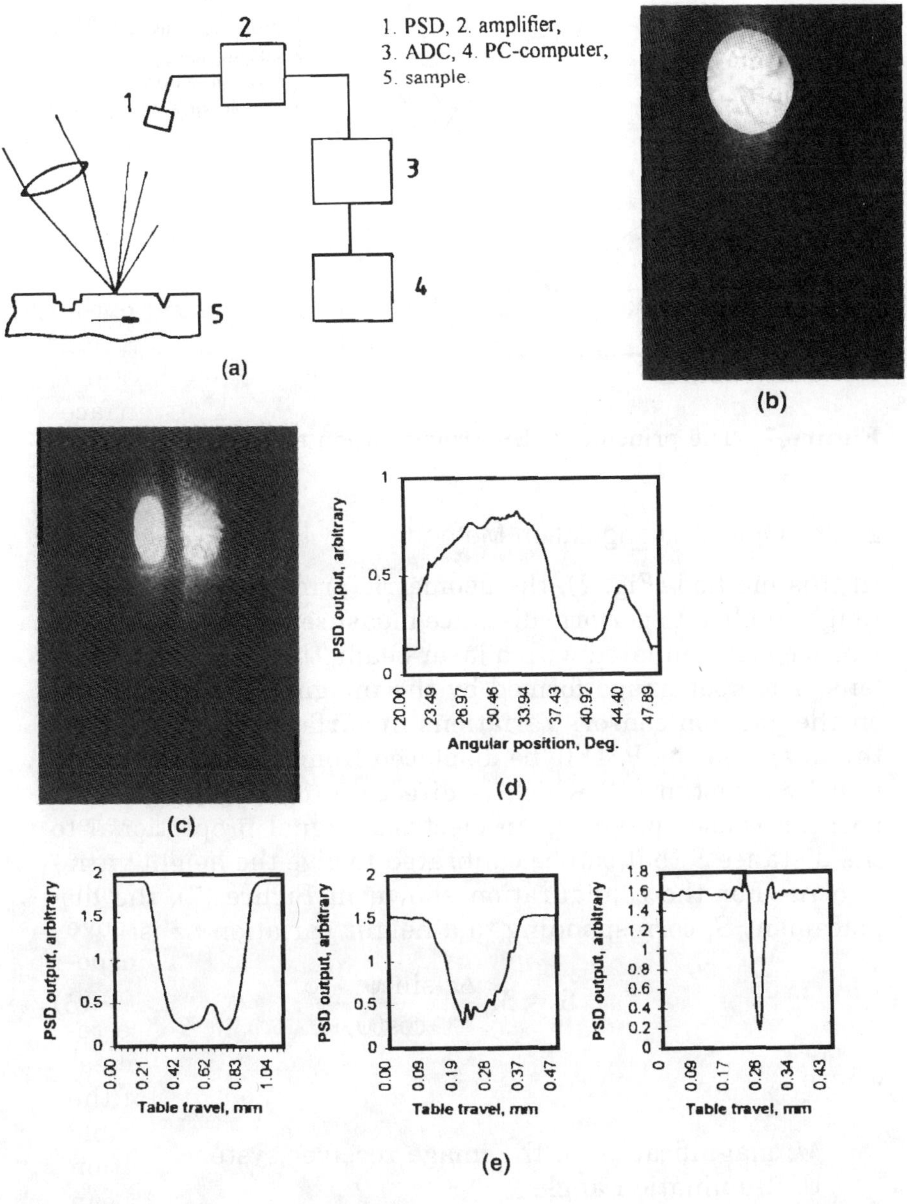

Figure 6 The principle of online dimensional measurement using diffraction method: (a) setup; (b) reflection from a flat surface; (c) diffraction by a microgroove; (d) intensity distribution of the diffraction pattern in (c); (e) PSD signal when scanning typical grooves.

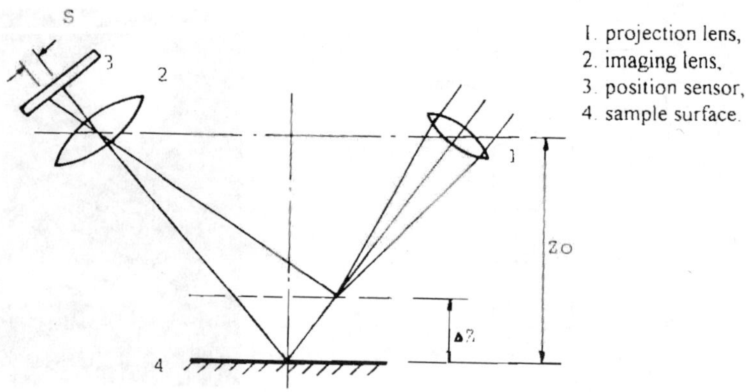

Figure 7 The principle of the triangulation method.

2.3.2 Optical Triangulation Method

In this method (Fig. 7), the geometric principle of triangulation is applied to perform distance measurements. The sample surface is illuminated with a laser beam, through a projection lens. The spot image formed by the imaging lens is received on the position sensor. Variations in surface height Δz cause the image on the PSD to be displaced from its reference position by a distance S which is directly proportional to Δz. A position sensor produces an electrical signal proportional to the distance S that can be calibrated to give the height variation Δz. For the configuration shown in Figure (7), the displacement S, corresponding to a height variation Δz is:

$$S = M \frac{\Delta z \sin(\psi + \phi)}{\cos(\psi)} \tag{5}$$

where

M: magnification in the image receiver system
ψ: illumination angle
ϕ: imaging angle
Δz: height deviation from a reference level

Some applications use normal illumination ($\psi = 0$), for which Equation (5) reduces to

$$S = M \cdot \Delta z \cdot \sin(\eta) \qquad (6)$$

where η is the angle between the illumination and imaging directions.

The height resolution and range of this method depend on the wavelength of the illumination λ, the numerical aperture (NA) of the imaging lens, and the geometrical configuration. Using inclined illumination and normal imaging, Costa [24] achieved height resolution of 0.49 μm ($\lambda = 0.6328$ μm, NA $= 0.6$, inclination angle $= 65°$); however, it can be as high as a few microns. The range of height variation detected can be as small as few micrometers or as large as several millimeters. The application ranges from the measurement of surface topography [24,25], in-process measurement and control [26], range-finding [27], and as noncontact probes on coordinate measuring machines [28]. A critical analysis of errors evolved in triangulation-based systems is given by Kilgus and Svetkoff [29].

2.3.3 Optical Followers

The optical follower is a measuring instrument that scans an object surface with a focused laser beam. If the beam is initially focused on a specific surface point, height variations as the object moves set the beam out of focus. The out-of-focus condition is detected by a sensory unit that activates a servomechanism to bring the beam back in focus. The vertical displacement required to bring the beam back in focus is recorded against the linear displacement of the object. This record represents the profile of the surface along the scanned line. Figure 8 shows the main components of an optical follower. The laser beam is filtered using the lens-pinhole spatial filter, then collimated and focused on the sample surface. The focused spot is imaged on the surface of the position sensor by use of the beam splitter and the imaging lens. The detector signal is fed to the

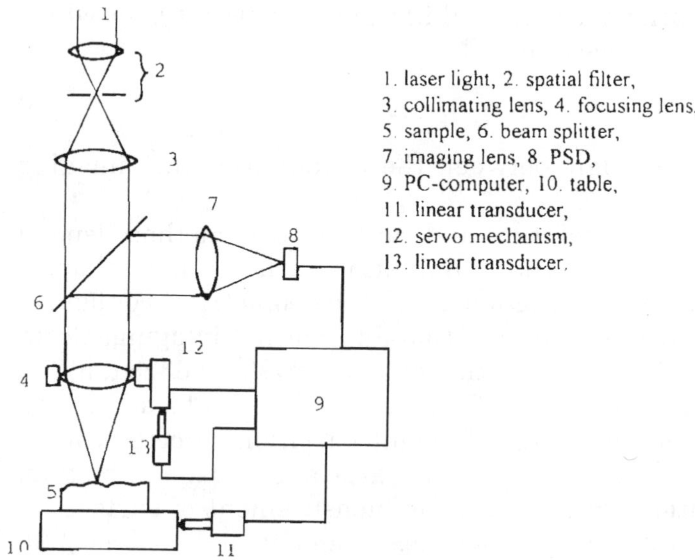

Figure 8 The main components of an optical follower.

personal computer, where the focusing condition is determined. As the table moves, height variations in the sample surface set the beam out of focus. Accordingly, the servomechanism is activated to bring the spot back in focus by moving the focusing lens vertically. Two linear transducers are used: the first measures the horizontal displacement of the table, while the other measures the vertical displacement of the lens. A record of the vertical lens displacement against the horizontal table displacement produces a profile trace of the sample surface along the scanned line. By use of an x–y table multiple traces, and consequently, three-dimensional maps of surface profiles can be constructed. The accuracy of height measurement depends on two main factors, the resolution of the displacement transducer, and the precision of the sensory unit. On the other hand, spatial resolution depends on the precision of the stage moving the object. Several techniques are used to detect the focusing condition in optical followers [4,30–32]. They include defect-of-focus and astigmatic methods, some basic principles of which are discussed below.

Defect-of-Focus Principle

That the position of the focus image along the beam axis is strongly dependent on the distance between the sample surface and the objective lens (Fig. 9a) is the basic criterion used for focusing detection. For a small variation in sample height z, and for an objective magnification M, the change in the position of focus image along the beam axis is given by [4]:

$$S = 2M^2z \tag{7}$$

For a magnification, M = 40x, a height change of ± 1 μm results in a change in focus position along the beam, $S = 3.2$

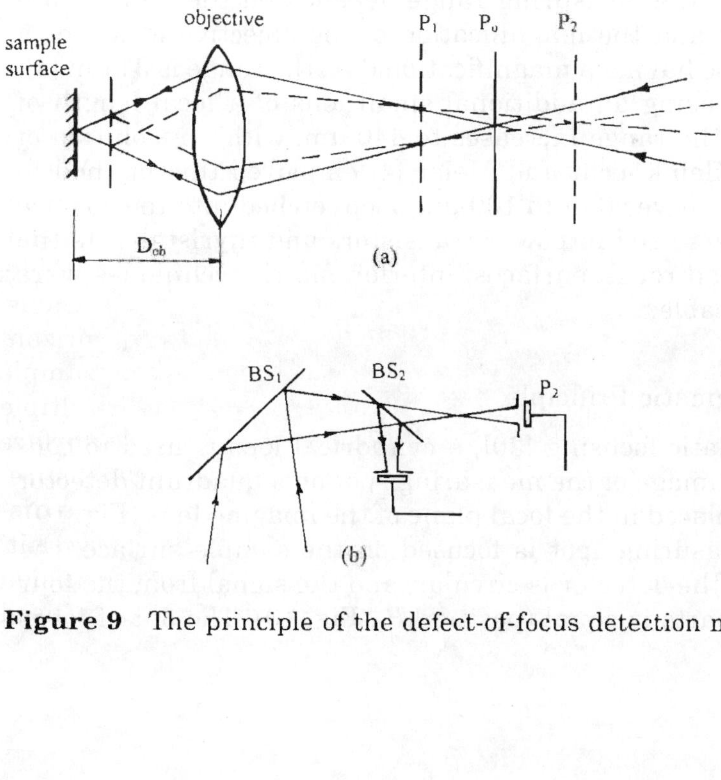

Figure 9 The principle of the defect-of-focus detection method.

mm. If P_o is the plane where the image of the in-focus spot is formed, two planes, P_1, and P_2, at equal distances from P_o, are determined and two equal aperture detectors are placed, one at each plane. The signal of the second detector P_2, is subtracted from that of the first one. If the difference between the two detector signals $(P_2 - P_1) = 0$, the laser beam is focused on the sample surface; however, if $(P_2 - P_1) \neq 0$, then the beam is out of focus. In this case the servomechanism is activated to bring the beam back in focus. The distance moved by the objective lens to bring the beam back in focus is measured by use of a displacement transducer. A record of this displacement against the sample movement represents the surface profile along the scanline. For practical reasons two beam splitters BS1, and BS2, are used as shown in Figure 9(b).

If the height variation reaches a certain limit, the focus image goes to infinity and beyond that the beam becomes divergent leading to $(P_2 - P_1) < 0$, for $z < 0$ causing runaway. Therefore, the measuring range depends on the optical configuration and the magnification of the objective lens. For a microscope having a magnification 64x, the range is 31 μm [4]; however, using an additional small lens of a focal length of 150 mm, the range increases to 440 μm, with a resolution of ±1 μm. Kleinknecht and Meier [4] employed this method to measure grooves (30- to 100-μm deep) etched into the surface of Si wafers used in power transistors and thyristors. As the grooves had rough surfaces, interferometric techniques were not applicable.

Astigmatic Principle

In astigmatic focusing [30], a cylindrical lens is used to converge the image of the measuring spot on a quadrant detector which is placed in the focal plane of the imaging lens (Fig. 10). If the measuring spot is focused on the sample surface, the image on the detector is circular, and the signal from the four detector parts is equal (position P_0, Fig. 10). The out-of-focus

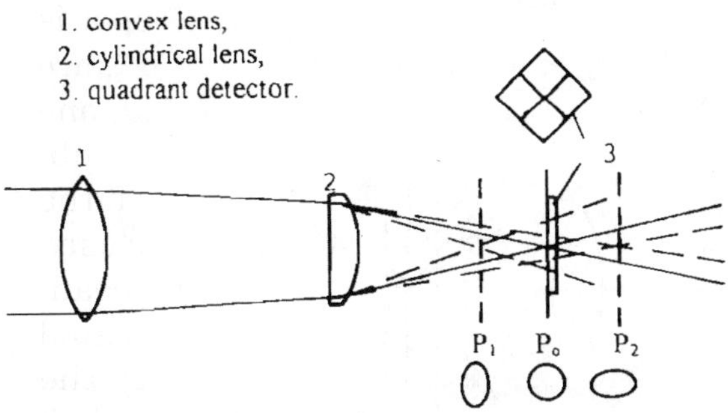

Figure 10 The principle of astigmatic focus detection.

condition caused by surface height variations changes the circular shape of the image into elliptical form (positions P_1 and P_2, Fig. 10). The direction of the major axis of that ellipse depends on the position of the surface point relative to the reference position corresponding to *Po*. A difference signal can be calibrated to produce a surface profile record as the surface scans beneath the illumination spot.

The difference signal is linear in a short range around the focus and depends on the power and numerical aperture of the objective lens [30]. For an objective (60x, NA = 0.8), the range is ±0.5 μm; however for (40x, NA = 0.65), it becomes ±1.5 μm. The resolution can be as small as 2 nm. Results obtained by this method showed good agreement with stylus instruments; however, optical diffraction at the edges of a step on the surface resulted in small crests in the record.

Two-Beam Method

Shimokohbe et al. [31] used a two-beam system to detect surface relief and inclination. The arrangement shown in Figure 11 applies two focused laser beams, B_1, and B_2, fired sequentially one after the other to intersect at the sample surface. The objective lens forms an image of the spots on the surface

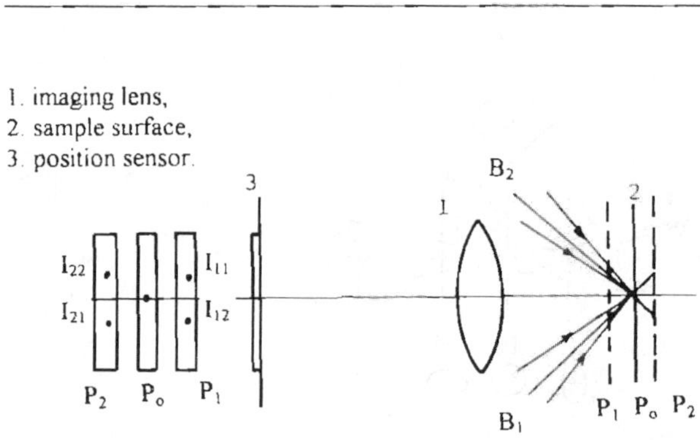

Figure 11 The principle of the two-beam method.

of a position sensor. At the reference position P_0, the two spots coincide. Height deviation from the reference position results in two intersection points between the laser beams and the sample surface. The distances between the spot images P_1, and P_2 on the detector, and the reference image P_0, depend on the value of surface deviation z. The direction of surface deviation is easily determined by the relative positions of beams on the detector. This method produces good results in the case of smooth surfaces. For relatively rough surfaces, the presence of speckle pattern caused by diffraction limits the accuracy. For smooth surfaces, measurement uncertainty is on the order of ± 0.2 μm.

2.4 INTERFERENCE METHODS

Interferometry is the field in which light interference phenomena are exploited for the purposes of measurement and inspection. An interference fringe pattern results when two wavefronts of monochromatic coherent light having a phase relationship combine. In most cases the two wavefronts originate from a single source. Several techniques are used to produce two waves from a single source [33]. One of the two waves is considered as a reference wave, while the other is the object wave which is modulated in phase by the object surface. As the two waves recombine a fringe pattern is produced that depends directly on the phase modulation of the object wave.

To produce two wavefronts from a single source the wavefront or the amplitude can be divided. In some interference microscopes, a birefringent material is used to split the electric and the magnetic components of the electromagnetic light wave thus generating the reference and object wavefronts.

2.4.1 Interference Comparators

Interference comparators are used for two major applications: to produce contour interferograms of inspected surfaces or as displacement transducers. Evaluation of contour interference patterns renders valuable information regarding surface topography, surface roughness, as well as dimensions of micro-surface features [34,35]. Interferometric transducers are also used to measure the translation of highly precise tables down to the nanometer resolution [11,36].

Many widely used interferometers are based on the Michelson principle (Fig. 12a). Coherent light from the monochromatic source 1, is collimated by lens 2, then divided into two waves by the beam splitter 3. The reference wave is reflected back by the reference plane 4, while the object type is reflected after being modulated by the object surface 5. The two waves recombine to form a fringe pattern by using imaging unit 6. The fringe pattern (Fig. 12b) is determined according to the phase relationship between the object and the reference waves, which in turn depends on the shape and orientation of the object surface with respect to that of the reference.

If the object and reference surfaces are normal to each other, the pattern produced represents contour fringes; however, for inclined surfaces, the fringes are rows of profiles. Parallel equidistant straight fringes indicate a plane surface in-

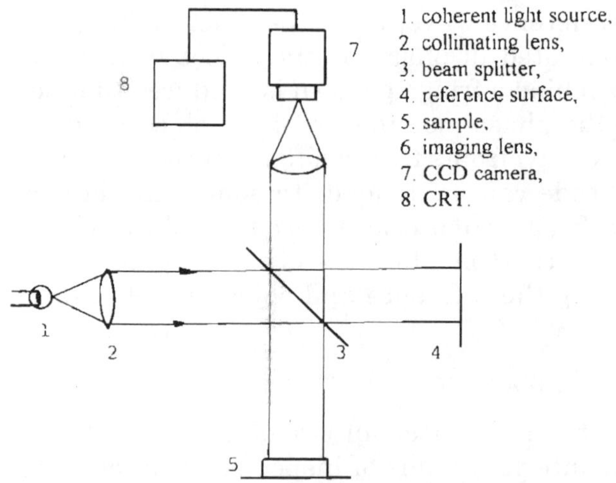

(a)

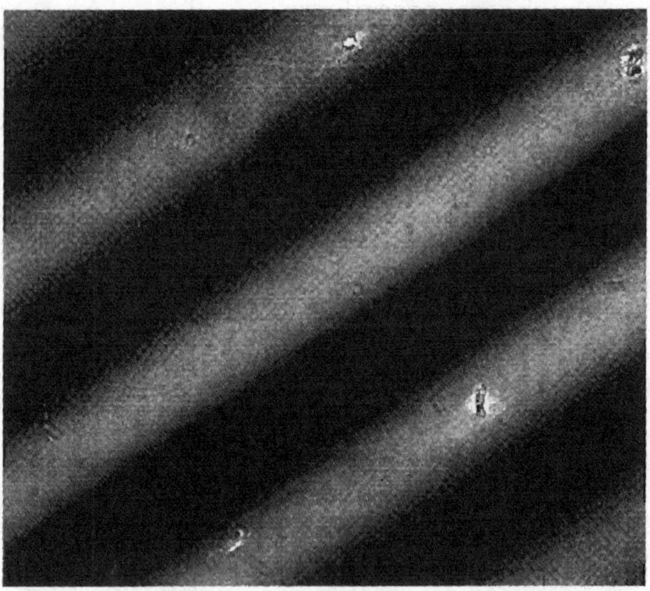

(b)

Figure 12 The principle of the Michelson interferometer (a) and a typical interferogram (fringe pattern) (b).

clined to the reference plane. Any out-of-flatness in the sample surface alters the obtained fringe pattern. In effect, the fringe pattern represents contour lines of equidistant points from the reference plane. The contour step in this case equals $\lambda/2$, where λ is the wavelength of the light used.

Personal computers (PCs) have recently been used for the analysis of interferograms. CCD and video cameras are used to capture the fringe pattern, which is then analyzed, by means of dedicated software. Different surface topography parameters are evaluated from the interference pattern. The image of the fringe pattern is transferred electronically to digital values representing the grey levels or intensity that are treated mathematically to produce contour maps and surface profiles as well as surface parameters. Dimensions of microsurface grooves or steps can also be evaluated from the interferograms.

2.4.2 Interference Microscopes

Interference microscopes provide higher resolution than their optical counterparts. They form magnified images of the inspected part surface modulated by interference contour fringes. These represent the microtopography of the inspected part surface and provide invaluable information about microsurface features such as form, profile, and roughness as well as dimensions of grooves, slots, and scratches. In many applications, calibrated reticles can be placed inside the eyepiece unit to measure dimensions of surface features. Several arrangements are adopted to produce the interferograms of examined parts. In all cases light is split into two wavefronts, one going to a reference plane and the other to the inspected surface. After reflection, wavefronts recombine undergoing constructive and destructive interference, producing the interferograms with dark and bright fringe pattern.

Figure 13(a) shows Tolansky's arrangement for interference microscopy [37]. By this arrangement, it is possible to measure microsteps of 500-μm height with an accuracy on the

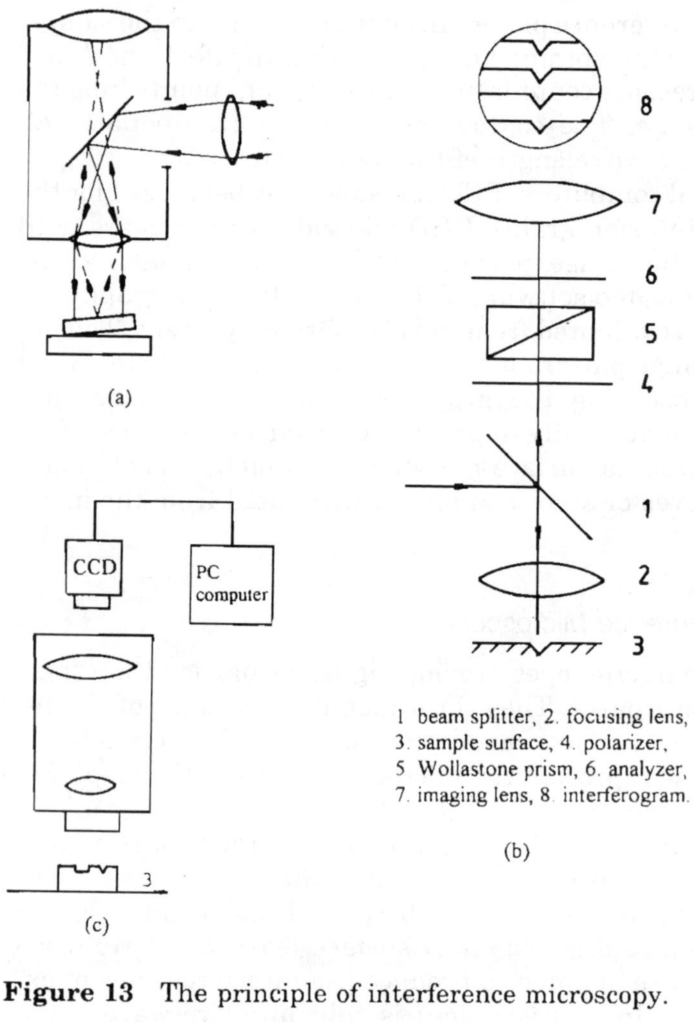

Figure 13 The principle of interference microscopy.

order of ±3 μm. In the Tolansky interferometer the wavefront is divided into two types, the reference and the object waves. In other types of interferometers, the two waves are formed by separating the electric and the magnetic components of the electromagnetic wave of light [38]. In the interference microscope (Fig. 13b) a Wollastone birefringence prism 5 is placed in the space between the objective 2 and eyepiece 7. A polarizer plate 4 is placed in front of the Wollastone prism while an analyzer 6 is in the back. The semireflecting plate 1 directs the light onto the surface of the inspected object 3. With this arrangement the reflected wavefront imaged by the objective 2 is divided into two wavefronts representing the electric and the magnetic fields. These wavefronts interfere with each other after passing through the analyzer 6. The resulting interferogram shows minute variations in the inspected surface. This procedure is useful for examining fine surface structures (e.g., grooves, slots, and scratches) as well as surface roughness. Commercially available interference microscopes can measure height variations as small as 0.1 nm (see, e.g., [13]). Moreover, roughness values in the sub-Angstrom range can also be evaluated. Like optical microscopes, interference microscopes are now equipped with CCD cameras that are interfaced to PCs (Fig. 13c).

A major limitation of interferometry is the need for fairly reflective surfaces. Apart from this limitation, it provides a powerful tool that produces a whole field image revealing microfeatures of inspected surfaces. The resulting interferogram can be used for direct visual inspection, or numerical evaluation of dimensions and surface roughness.

2.5 SURFACE PROFILERS

Surface profilers are instruments that use a fine stylus or tip to trace the fine details of an engineering surface. Height variations along the traced line modulate the force interacting between tip and surface or the tunnel current passing between them when they are very close to each other. By monitoring the deflection of the tip caused by the tip-surface force or the tunnel current, it is possible to produce a profile record of the traced line. Multiple line traces generate three-dimensional records of the inspect surfaces.

2.5.1 Stylus Instruments

Mechanical stylus instruments are most widely used for surface topography assessment. They have a wide range of height resolutions; however, their spatial resolution is limited by the size of the stylus tip used. As shown in Figure 14(a), an arm (lever) carrying a microtip (stylus) represents the sensor that scans across the inspected surface. Height variations along the scanned line change the force between the stylus tip and the surface points, and consequently lever deflection. An appropriate transducer transforms the lever deflection into an electrical signal proportional to height variations. The signal is then amplified, digitized, and analyzed to evaluate surface parameters.

The stylus movement is measured relative to a datum that should conform with the nominal shape of the measured

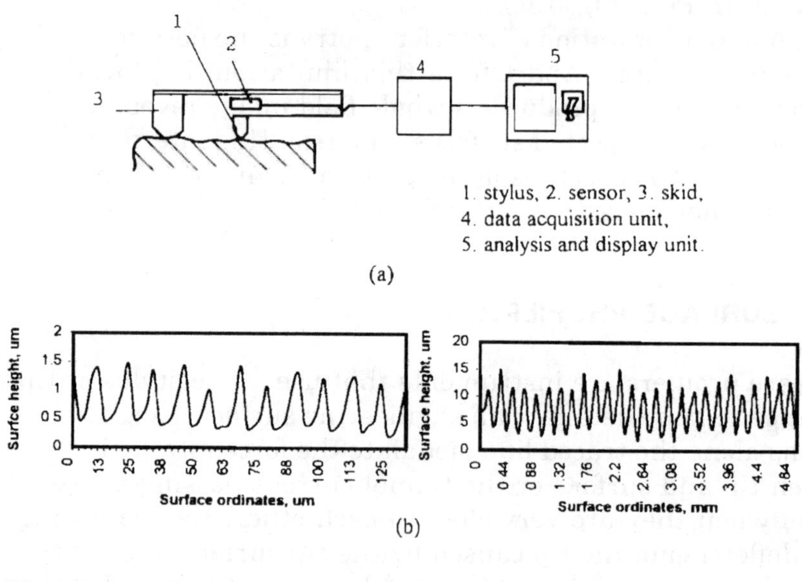

Figure 14 The principle of the stylus instrument (a) and typical profile traces (b).

surface [39]. While a straight-line datum has been used, a simple skid arrangement (Fig. 14a) is most common. Generally the stylus is a conical or pyramid diamond with a flat or rounded tip. The pyramid is normally 90° with a 2- to 2.5-μm flat, while the cone is 60° with 12.5-μm radius. Tips of 2.5 μm are also in use [40]. Figure 14(b) shows typical stylus traces of engineering surfaces.

The tip radius should be smaller than the radius of curvature of the bottom of surface valleys, otherwise the profile is commonly modulated by the stylus tip. A measuring force must be applied to ensure contact between stylus tip and surface points. Typical force levels are about 70 mg giving a pressure of 2500 Nmm^{-2} [40]. Such pressure is less than the yield strength of most materials. However, for soft materials or coatings this pressure may exceed the yield strength and cause surface damage.

The large development of computer science and technology has greatly increased the capability and spatial resolution of stylus instruments. Very accurate and precise linear transducers have been developed for scanning the inspected samples. Despite noise limitations, high resolutions of more than 0.1 nm have been achieved [41]. A reproducibility of 1.5 nm rms over a 40-mm traverse range was reported by Lindsey et al. [42]. Song and Vorberger [43] showed experimentally that lateral resolution on the order of 0.05 to 0.15 μm can also be obtained. A fine stylus, of small loading and high magnification, in the lateral direction has been used to measure fine surface structures. Razor blade styli with tip width 0.05 to 0.15 μm and stylus load of 0.6 to 1.2×10^{-6} N were used, a piezostage providing very slow and stable stylus motion in which horizontal magnification up to 50,000x or greater was obtainable.

Stylus instruments provide profile records of traced lines. Three-dimensional surface profile maps can also be plotted by use of scanning tables. Since profile signals can be digitized, surface geometry is analyzed and different surface parameters valuated using PCs interfaced with the measuring instrument. In addition to the assessment of surface roughness, stylus instruments can be used to measure microsurface grooves with depth values on the order of several microns. A major disadvantage of stylus instruments is that they rely on contact-type methods. They are relatively slow and the measuring pressure may, in some cases, damage the inspected surface.

2.5.2 Scanning Tunnelling Microscopes

Scanning tunnelling microscopes (STM) are another family of stylus-type instruments where the stylus does not contact the inspected surface. STM instruments are capable of lateral resolution sufficient to resolve protruding atoms on reconstructed surfaces. The vertical resolution is as small as 0.02 μm.

These instruments operate in vacuum, air, oil, liquid nitrogen, and water, giving images which are direct topographs of inspected surfaces [7]. In addition to the promising advantages of STMs in microtopographic mapping of highly finished surfaces, other applications such as microlithography, micromachining, polymer science, and biotechnology are also being studied. The usefulness of STMs for the analysis of diamond-turned surfaces, ruled grating replicas, X-ray reflecting optics, and optical discs has recently been demonstrated [2].

Although the concept of tunnelling in solid-state physics first appeared in the late 1920s [45], the first successful tunnelling experiment with an externally and reproducibly adjustable vacuum gap was reported by G. Binnig et al. of the IBM Zurich Research Laboratory in 1982 [46]. The principle of scanning tunnelling microscopy is demonstrated in Figure 15(a), in which a stylus of very sharp tip (ultimately one atom) is brought very close to the inspected surface (<1 nm apart). At such closely adjacent distances, free electrons from the conductive sample surface atoms tunnel to the conductive stylus tip, producing a very weak current. In the one-dimensional case, the tunnel resistance and, consequently, the tunnelling

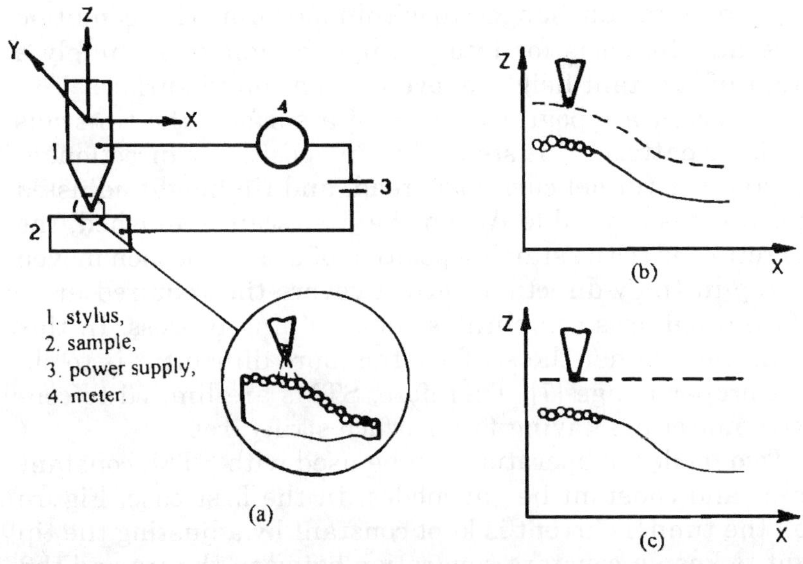

Figure 15 Scanning tunneling microscopy basic principle (a), constant current mode (b), and constant height mode (c).

current at low voltage and temperature is exponentially dependent on the tip-sample separation d [46,47]:

$$I \propto \exp(-2kd) \tag{8}$$

where

I: tunnelling current
d: distance between tip and surface
k: constant

For vacuum tunnelling, $k = h^{-1}\sqrt{2m\,\phi}$ where h is Planck's constant, m is electron mass, and ϕ is effective local work function, for a work function of 4 eV, k = 1.0 Å^{-1}. The current decreases by an order of magnitude when the distance d is increased by 1.0 μm. If the current is kept constant within ±2%, the gap remains unchanged to within 0.01 μm. This condition represents the basis for interpreting the image as simply a contour of constant height above the measured surface.

To record a topographic map of a surface, the tip scans in a raster pattern. It is stepped in the positive x-direction; at each step the tunnel current is read, and tip height adjusted to get the desired value. When the first scan is finished, the tip is returned to the starting position of this scan, then moved one step in the y-direction until it covers the required area. Surface roughness complicates the scanning process. In this regard, the rougher the surface, the more difficult it is to obtain a proper image [7]. Therefore, STMs are limited to conductive materials having fine surface structures.

Two modes of operation can be used with STM, constant current and constant height modes. In the first case, Figure 15(b), the tunnel current is kept constant by adjusting the tip height to keep a constant separation between the tip and the surface. The displacement of the servomechanism required to bring the tip to the constant current (separation) position is then recorded as the z-ordinate. In this case, the scanning speed is determined by the response of the feedback circuit, which maintains the average current constant. The constant current mode can be used for surfaces that are not atomically flat (i.e., $\gtrsim$ 10-nm peak-to-valley height).

In the constant height mode (Fig. 15c) the tip scans the surface while its vertical position, relative to a mean reference plane, and the current are kept constant, the voltage variation being monitored. Under such circumstances, the scanning speed depends on how fast the feedback circuit responds to achieve constant current by adjusting the voltage. Scanning is faster; however, the tip may be damaged unless the surface is tolerably smooth.

Special attention should be paid to vibration conditions, since the performance of an STM microscope can be enhanced by use of proper isolation [7]. Fine resolutions, required for scanning the tip and measuring height variations, are realized by using piezoelectric elements for x, y, and z translations.

Controlled voltage is applied to the piezoelectric crystal in specified directions. Consequently, the crystal contracts or elongates according to the sign of the electric field. This effect is linear and precisely controlled [49]. Separate piezoelectric elements for each axis translation have been used [46]. On the other hand, a single piezoelectric tube that provides translation in the three axes has been described by Binnig and Smith [50].

STM microscopes have many potential applications in measurement and fabrication of micromachined parts. They are successfully used for the measurement of surface topography as low as the atomic scale. Yang and Talke [51] used STM to investigate surface roughness of magnetic recording disks, using line graphs and aerial images at sampling intervals of 125 and 5 nm. Fine diffraction gratings, 2000 lines/mm, were also examined using STM to reveal detailed surface topography [52]. Besides measurement of surface topography STMs were used to modify surface structure and manufacture parts bits on highly oriented pyrolytic graphite [53].

2.5.3 Atomic Force Microscope

Atomic force microscopes (AFM) are noncontact profilometers which use a very fine stylus fixed to the end of a thin cantilever. They trace actual profiles of highly smooth and flat surfaces (Fig. 16a). The profile is recorded by making the stylus follow the profile of a constant force between stylus tip and surface points. As the fine stylus tip is brought close to the surface (30 to 150 μm) attraction forces between atoms are generated. As the tip comes closer to the surface the atomic force increases. Such a force is balanced by the plastic force generated by bending the cantilever. Therefore the stationary cantilever bend is a direct measure of the atomic force [56]. Most probes use capacitive sensors to determine cantilever deflection, and hence the atomic force [8,54,55]. A feedback loop is usually used to keep the force at a specified value by maintaining a constant tip-sample spacing. The signal, from the

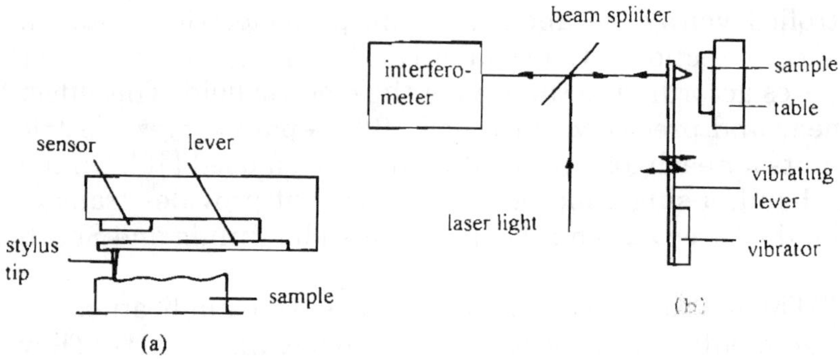

Figure 16 The principle of atomic force microscopy with capacitive sensor (a), and with vibrating tip (b).

capacitive sensor, is fed to a servomechanism that controls the tip-sample spacing.

Another arrangement (Fig. 16b) uses a fine tip fixed to the end of a vibrating lever. Changing the tip-sample spacing leads to proportional change in the attraction force, which modifies the compliance of the lever. The vibration amplitude is also affected by the shift in the lever resonance [56]. The tip vibration amplitude as a function of the frequency ω is given by

$$A = \frac{A_0(\omega/\omega_0)}{[1 + Q^2(\omega/\omega_0 - \omega_0/\omega)^2]^{1/2}} \tag{9}$$

where

- A: amplitude of vibration
- A_0: amplitude at resonance
- ω: tip frequency
- ω_0: tip resonance frequency, $\omega_0 = c_1 \sqrt{k}$, c_1 is a function of lever mass, and k is the spring constant
- Q: quality factor ($Q \gg 1$), Q is a measure of system damping, and $Q = (1/c)\sqrt{km}$

where

c: damping factor
m: lever mass

The larger the value of Q, the smaller is the damping. As the tip approaches the sample surface, interatomic attraction forces reduce the spring constant of the lever by the atomic force derivative f'. The resonance frequency then becomes $\omega_0' = c_1 \sqrt{k - f'}$. Thus by measurement of the shift in the lever resonance $\Delta\omega = \omega_0' - \omega$, the force derivative can be found [56]. The change in the resonance frequency changes the amplitude of vibration. This means that, as the tip approaches the surface, the attraction force rises leading to a proportional change in both lever resonance frequency and vibration amplitude. By monitoring these values as the tip scans across the surface, it is possible to trace the surface profile.

The amplitude of vibration can be measured optically [39], or by use of an interferometer [56]. A signal representing the amplitude of vibration is usually used in a feedback loop to maintain the tip at a specific distance from the surface as the tip scans across a determined path. This technique has been applied to measure surface roughness of ultrafine surfaces. Micro- and sub-micrometer surface features can also be measured. Line profiles and 3-D maps of surfaces can be recorded and presented. V-shaped grooves on silicon wafer and steps of height 50 μm have been mapped and measured [56]. Profiles of a photoresist grating (0.1-μm line width and 0.09-μm thickness) have been recorded by the same technique.

2.6 MEASUREMENT OF MICROHOLES AND SLOTS

2.6.1 Optical Methods

Small holes produced by micromachining (typically 200 μm or less) can be measured with commercially available microscopes. Tool Makers' microscopes are provided with linear transducers to measure displacement in the x- and y-directions with resolutions on the order of 1 μm [57]. These measurements are digitally displayed, and in many cases, can be interfaced with a computer. Very small patterns of semiconductor devices can successfully be inspected for geometrical characteristics. Shallow holes with large diameter-to-depth ratio are readily measurable with microscopes. However, deep holes of nozzles, receptacles, or dies are not suitable for microscopes. By means of microscopes it is possible to measure the size and shape of the two entrances of the hole, while the inside geometry is not accessible.

Coordinate measuring machines can measure the size and geometry of holes having a minimum diameter of 1 mm [58]. The main reason for such a limitation is related to the size of the stylus or feeler. To maintain an adequate level of accuracy in displacement measurement the feeler must maintain a certain rigidity, hence, its dimensions must not lie below certain limits.

New techniques using very thin feelers have recently been developed to measure holes. One method uses a very thin vibrating feeler and senses the electrical contact between the feeler and the internal wall of the hole [59]. Another method takes the deflection of the feeler into consideration and compensates for such deflections [60]. The principles of these two methods are described below.

2.6.2 Vibroscanning Method

Masuzawa et al. [59] developed a new method for measurement of microhole geometry which is called vibroscanning (VS). It is a nondestructive technique based on sensing electrical contact between the surface of the hole and a vibrating feeler which has a small raised part (tip) at its end (Fig. 17a). As the vibrating feeler gradually approaches the surface of the hole, the tip starts to touch the surface intermittently.

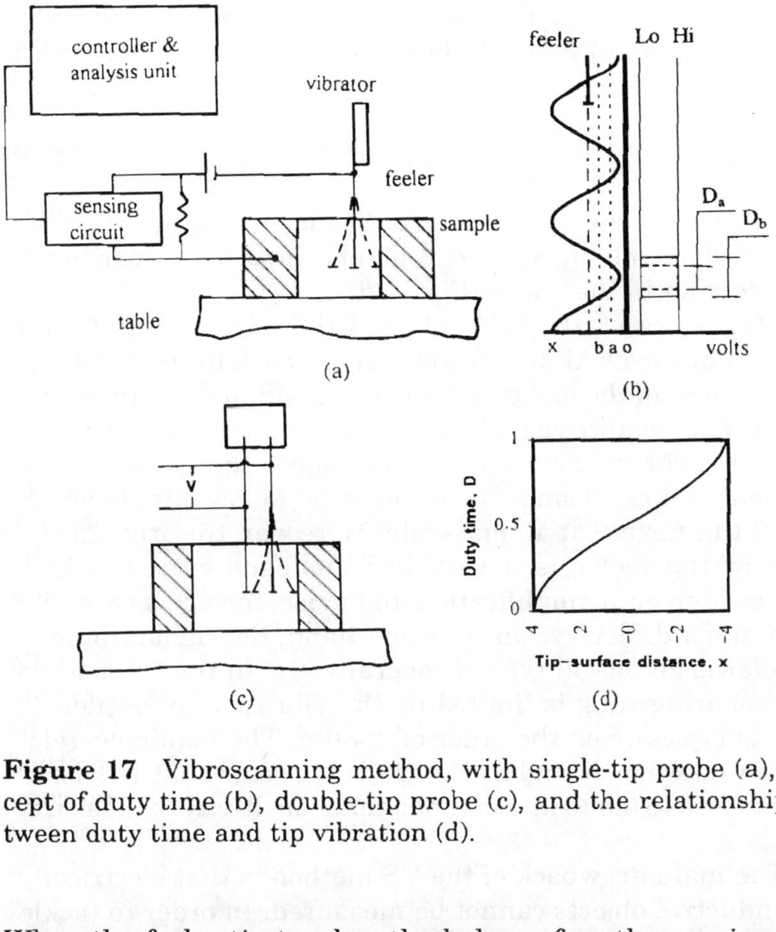

Figure 17 Vibroscanning method, with single-tip probe (a), concept of duty time (b), double-tip probe (c), and the relationship between duty time and tip vibration (d).

When the feeler tip touches the hole surface the sensing loop closes and the voltage goes high to a level Hi. The voltage stays high for a short period called the duty time D, which depends on the distance between the hole surface and the neutral axis of feeler vibration x (Fig. 17b). If the surface is at position o relative to the neutral axis of the feeler, no contact occurs and the voltage remains at the low level L_0. As the feeler is moved close to the hole surface, position a, the duty time will be D_a. As the feeler comes closer to the surface, position b, the duty time increases to D_b. The relationship between D and x depends on the amplitude of vibration A, and can be expressed as

$$D = |\arccos(x / A)|/\pi(-A < x < +A) \qquad (10)$$

From knowing the vibration amplitude A, the distance between feeler and surface x can be obtained by measuring D and referring to the curve (Fig. 17d).

Masuzawa et al. [59] explained the advantages of their VS technique over AFM and STM techniques. In the AFM, the displacement of the feeler end must be magnified by an optical system. This requires a wide open space around the feeler end. Although STM can be used for small holes, the speed of measurement is limited since the feeler must follow minute variations of the surface at atomic scale. Moreover, the original signal from the feeler is a very low analogue current which requires high gain amplification and noise, in such a case, can pose a major difficulty. On the other hand, the signal from the VS feeler is an on–off type of several volts. In the VS method the measuring rang is limited by the vibration amplitude A, which is typically on the order of ± 4 μm. The nonlinear relationship between D and x (Fig. 17d) suggests that D should be kept close to 0.5 to attain consistent sensitivity of measurement.

The main drawback of the VS method is that electrically nonconductive objects cannot be measured. In order to tackle this problem, Masuzawa et al. [3] developed a twin probe (Fig. 17c) that can be used to measure holes in electrically nonconductive objects. The probe has two elements connected to the terminals of the scanning loop. As the front vibrating element touches the surface of the hole it actually contacts the tip of the back element, and the sensing loop is closed. A silicon-based twin probe was fabricated which is 300-μm long, 20-μm thick, and 15-μm wide [3]. Holes having 125-μm diameter were measured with an estimated accuracy of ± 0.5 μm. However, measurement of very deep holes requires low frequency of feeler vibration and, consequently, a long measurement time. Electrically nonconductive dust or inclusion inside the hole may cause runaway and damage of the feeler.

2.6.3 Elastic Transmission Method

Zhang and Yang [60] introduced a new principle that allows the use of a very thin stylus to measure blind holes 200 μm in diameter and 700 μm in depth on a 3-D measuring machine with an accuracy better than 1 μm. The proposed elastic transmissing principle takes into account the deflection of the stylus due to the measuring force. The effective sensitivity of the probe is determined accordingly and the error caused by the deflection is compensated for automatically. The elastic transmission principle can be explained using the widely known parallel strip mechanism shown in Figure 18(a), where the cantilever stylus (1) is treated as a spring with constant k_1. As the stylus tip (M) is moved a small distance δ_M, by the measuring force, the point N will be displaced a corresponding distance δ_N depending on the mechanical response of the system. If the spring constant for the parallel strips (2) is k_2 and

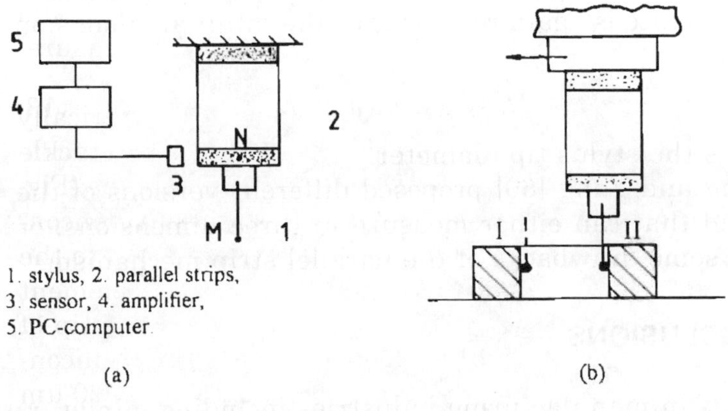

Figure 18 Elastic transmission probe (a) and measurement procedure (b).

the mass of the moving parts is m, the displacement of point M, δ_M is related to δ_N of point N by

$$m\delta_N'' + c\delta_N' + (k_1 + k_2)\delta_N = k_1\delta_M \tag{11}$$

where c is the damping factor.

In a static process

$$\delta_N = k\delta_M = \frac{k_1}{k_1 + k_2}\,\delta_M \tag{12}$$

where k is the static transfer coefficient.

The displacement sensor 3 (Fig. 18(a)) measures the actual displacement of point N by means of an appropriate transducer. The corresponding displacement of stylus tip M is calculated from Equation (12), based on an assumed static process. The read-out unit shows the displacement of the stylus tip directly. The stylus is first inserted inside the hole by use of a microscope. The measuring head is then moved in the negative x-direction until it touches the hole surface on the left-hand side and deflects a preset amount, position I in Figure 18(b). At this position the x-coordinate of the 3-D is recorded, x_1. The measuring head is then moved in the positive x-direction to touch the other side of the hole, position II, in Figure 18(b), and deflects the same amount as in position I. The x-coordinate is again recorded, x_2; the internal dimension will be

$$D_x = |x_2 - x_1| + d_t, \tag{13}$$

where d_t is the stylus tip diameter.

Zhang and Yang [60] proposed different versions of the probe head that can either measure in three dimensions, or overcome some drawbacks of the parallel strip mechanism.

2.7 CONCLUSIONS

New trends in manufacturing industries including miniaturization and large-scale integration have enhanced the development of measurement science and technology. Miniaturization calls for new measuring techniques to evaluate microfeatures of tiny parts at high resolution. In addition, computer-integrated, high-speed, noncontact techniques are required for applications including in-process and online inspection. In the future, accurate and intelligent measurement systems will therefore be an integral part of industrial manufacturing lines.

Advances in the field of measurement are expected to be integrated in manufacturing systems along four main lines: probe and sensor performance, precision and resolution of translation mechanisms, accuracy and response of linear and angular transducers, together with computer interfaces and software.

Probes and sensors are being greatly improved to achieve higher resolutions and better accuracy. Fine details of micromachined parts, including features that are difficult to reach, need to be inspected at an appropriate resolution. AFM, STM, and electron microscopes provide high resolutions; however, the measuring time is relatively long, which limits the application of such systems for online inspection especially for large-scale production.

Precision tables that can provide very accurate fine motion for both workpiece and sensor at reasonable speed are another challenge facing the development of measuring instruments. Piezoelectric actuators are highly precise; however, their speed and range of travel are limited.

Linear transducers are used in measuring systems to determine the displacement of axes. High-response long-range transducers are necessary for the realization of precise measurement. Research is being undertaken to improve transducer response, resolution, and range of measurement and to reduce environmental impacts on accuracy.

Integration of accurate sensors, precision tables, and high-resolution transducers in a measuring system is only possible through computer control. Table movement, displacement measurement, analysis of sensor signals, and activation of feedback systems will all be controlled by computers. Great improvements are therefore expected in computer hardware and software, as well as interfacing technology.

NOTATION AND SYMBOLS

- a: distance between two points on the surface of an object that can be seen separated in the image plan, μm
- λ: effective wavelength of illumination used, μm
- n: refractive index of objective medium
- i: suspended angle of lens which depends on its diameter and focal length
- h: Planck's constant, 6.626 × 10–34 Js
- m: mass of electron, 9.110×10^{-31} kg
- v: electron velocity, 2.99×10^{8} m/s
- V: potential difference, V
- e: electron charge, 1.602×10^{-19} C
- M: magnification in the image receiver system
- Ψ: illumination angle, deg.
- ϕ: imaging angle, deg.
- η: angle between illumination and imaging directions, deg.
- Δz: height deviation from a reference level, μm
- df: change in the focus position along the optic axis, μm
- I: tunneling current, A
- d: distance between tip and surface, μm
- k: constant, $Å^{-1}$
- ϕ: effective work function, eV
- A: amplitude of vibration, μm
- A_0: amplitude at resonance, μm
- ω: tip frequency, Hz
- ω_0: tip resonance frequency, Hz
- SEM: scanning electron microscope
- STM: scanning tunneling microscope
- SFM: scanning force microscope

AFM: atomic force microscope
CMM: coordinate measuring machine
LVDT: linear variable differential transducer
OPF: optical profile follower
WFC: whole field contouring
NA: numerical aperture
TMM: tool maker's microscope
UMM: universal measuring machine
VS: vibroscanning
PSD: position sensitive device
CRT: cathode ray tube

REFERENCES

[1] P.A. McKeown, The role of precision engineering in manufacturing of the future, *Annals of the CIRP—The International Institution for Production Engineering Research*, (36)2, 495–501 (1987).

[2] P. Dario, M.C. Carrozza, N. Croce, M. Montesi, and M. Cocco, Non-traditional technologies for microfabrication. *J. Micromech. Microeng.*, (5) 64–71 (1995).

[3] T. Masuzawa, B.J. Kim, C. Berguard, and M. Fujino, Twin-probe vibroscanning method for dimensional measurement of microholes. *Annals of the CIRP*, (46)1, 437–440 (1997).

[4] H.P. Kleinknecht and H. Meier, Optical profilometer for monitoring surface contours of Si power devices. *Proc. SPIE—The International Society for Optical Engineering*, (398), 266–273 (1983).

[5] D.A. Swyte and S.W. Jensen, Electron microscope-based system for accurate micro-dimensional measurement. *Prec. Eng.* (3)1, 11–15 (1981).

[6] M. Holmes, D. Trumpet, and R. Hocken, Atomic-scale precision motion control stage (the Angstrom stage). *Annals of the CIRP*, (44)1, 455–460 (1995).

[7] P. Hansma and J. Tersoff, Scanning tunneling microscopy. *J. Appl. Phys.*, (61)2, R1-R23 (1987).

[8] L.P. Howard and S.T. Smith, A metrological constant force stylus profiler. *Rev. Sci. Instrum.* (65)4, 892–902 (1994).

[9] S.J. Martin, *Numerical Control of Machine Tools*. Hodder and Stoughton, London (1986).

[10] ACU-RITE Microscale, ACU-RITE Inc., New York, USA.

[11] C.R. Steinmetz, Sub-micron position measurement and control on precision machine tools with laser interferometry. *Prec. Eng.* (12)1, 12–24 (1990).

[12] C. Schonenberger and F. Alvarado, A differential interferometer for force microscopy. *Rev. Sci. Instrum.*, (60)10, 3131–3134 (1989).

[13] New View 100 3D Imaging Surface Structure Analyzer, Zygo Corp., Middlefield, CT.

[14] F.A. Jenkins and A. White, *Fundamentals of Optics*, 3rd ed. McGraw-Hill, New York (1957).

[15] S. Wischnitzer, *Introduction to Electron Microscopy*, 2nd ed., Pergamon, New York (1970).

[16] Olive C. Wells, Alan Boyde, Eric Lifshin, and Alex Rozanwich, *Scanning Electron Microscopy*, McGraw-Hill, New York (1974).

[17] P.R. Thornton, *Scanning Electron Microscopy*, Chapman and Hall, London (1968).

[18] S. Asai, Y. Tagushi, K. Horio, T. Kasai, and A. Kobayashi, Measuring the very small cutting-edge radius for diamond tool using a new kind of SEM having two detectors. *Annals of the CIRP*, (39)1, 85–88 (1990).

[19] H. Sato, and M. Ohmori, Surface roughness measurement by scanning electron microscope. *Annals of the CIRP*, (31)1, 457–462 (1982).

[20] H. Sato and M. Ohmori, Measurement of surface shape by

scanning electron microscope using detection of normal. *Annals of the CIRP*, (35)1, 365–368 (1986).

[21] M.A. Younes, On line form detection of macro surface grooves using laser light. In *Proc. 6th Int. Conf. Prod. Eng. Design and Control PEDAC'97*, Vol. 1, Alexandria, Egypt, 9–18 (1997).

[22] B.J. Pekrinck and J. Kennedy, Optical method for fatigue crack detection. *Appl. Opt.*, (19)18, 3224–3229 (1980).

[23] D. Levy, L. Singher, J. Shamir, and Y. Leviatan, Step height determination by a focused Gaussian beam. *Opt. Eng.*, (34)11, 3303–3313 (1995).

[24] M.F.M. Costa, Surface inspection by an optical triangulation method. *Opt. Eng.*, (35)9, 2743–2747 (1996).

[25] M. Zahidi, M. Assoul, J. Mignot, and B. Ballaton, A fast 2D/3D optical profilometer for wide range topographical measurement. *Wear*, (165), 197–203 (1993).

[26] Stephen D. Murphy (Editor), *In-Process Measurement and Control*. Marcel Dekker, New York (1990).

[27] J. Lewandowski and L. Dejardins, Absolute and relative range measurement with a single optical system. *Proc. SPIE—The International Society for Optical Engineering*, (2599) 44–53 (1995).

[28] K.H. Goh, N. Philips, and R. Bell, The applicability of a laser triangulation probe to non-contact inspection. *Int. J. Prod. Res.*, (24)6, 1331–1348 (1986).

[29] D.B.T. Kilgus and D.J. Svetkoff, Imaging geometry and error sensitivity in triangulation-based optical receivers. *Proc. SPIE—The International Society for Optical Engineering*, (2599) 106–119 (1995).

[30] K. Mitsui, In-process sensors for surface roughness and their applications. *Prec. Eng.* (8)4, 212–220 (1986).

[31] A. Shimokohbe, H. Osada, and K. Gotoh, An optical non-contact probe. *Prec. Eng.* (7)4, 195–200 (1985).

[32] M.J. Downs, W.H. McGivern, and H.J. Ferguson, Optical sys-

tem for measuring the profile of super-smooth surfaces, *Prec. Eng.* (7)4, 211–215 (1985).

[33] D. Malacara (Editor), *Physical Optics and Light Measurement*. Academic, London, 1–45 (1988).

[34] J.F.W. Galyer and C.R. Shotbolt, *Metrology for Engineers*. Cassell, London (1969).

[35] R.K. Jain, *Engineering Metrology*, 13th ed. Khanna, Delhi (1991).

[36] C.V. Collete and A.D. Hope, *Engineering Measurement*, 2nd ed. English Language Book Society and Pitman, London (1983).

[37] S. Tolansky, *Multiple-Beam Interference Microscopy of Metals*. Academic, London and New York (1970).

[38] Nikon Metrological Microscope "OPTIPHOT," Nikon Corp., Japan.

[39] D.J. Whitehouse, *Handbook of Surface Metrology*. Institute of Physics, Bristol and Philadelphia (1994).

[40] R.E. Reason, The measurement of surface texture. In *Modern Workshop Technology*, part 2, Macmillan, Bristol (1970).

[41] J.M. Bennett and J.H. Dancy, Stylus profiling instrument for measuring statistical properties of smooth optical surface. *Appl. Opt.* (20), 1785–1802 (1981).

[42] K. Lindsey, S.T. Smith, and C.J. Robbe, Sub-nanometer surface texture and profile measurement with NANO-SURF2, *Annals of the CIRP—The International Institution for Production Engineering*, (37)1, 519–522 (1988).

[43] J.F. Song and T.V. Vorberger, Stylus profiling at high resolution and low force. *Appl. Opt.*, (30)1, 42–50 (1991).

[44] I.H. Musselman, P.A. Peterson, and P.E. Russell, Fabrication of tips with controlled geometry for scanning tunneling microscopy. *Prec. Eng.*, (12)1, 3–6 (1990).

[45] R.H. Fowler and L. Nordheim, *Proc. R. Soc. London A119*, 173–177 (1928).

[46] G. Binnig, H. Rohrer, C. Gerber, and E. Weibel, Tunneling through a controllable vacuum gap, *Appl. Phys. Lett.* (40)2, 178–180 (1982).

[47] G. Binnig, H. Rohrer, C. Gerber, and E. Weibel, Surface study by scanning tunneling microscopy, *Phys. Rev. Lett.*, (49)57, 57–61 (1982).

[48] R. Piner and Reifenberger, Computer control of the tunnel barrier width for the scanning tunneling microscope. *Rev. Sci. Instrum.*, (60)10, 3123–3127 (1989).

[49] B.I. Bleaney and B. Bleaney, *Electricity and Magnetism*, 3rd ed. Oxford University Press, Oxford (1983).

[50] G. Binnig and D.P.E. Smith, Single-tube three-dimensional scanner for scanning tunneling microscopy. *Rev. Sci. Instrum.*, (57)8, 1688–1689 (1989).

[51] M. Yang, and F.E. Talke, Surface roughness investigation of magnetic recording disks using STM and profilometry measurements. *Wear*, (170) 15–24 (1993).

[52] H. Kaizuka, Application of capacitor insertion method to scanning tunneling microscopes. *Rev. Sci. Instrum.* (60)10, 3119–3122 (1989).

[53] J.A. Miller and R.J. Hocken, Scanning tunneling microscopy bit making on highly oriented pyrolytic graphite: Initial results. *J. Appl. Phys.*, (68)2, 905–907 (1990).

[54] Y. Xu, S.T. Smith, and P.D. Atherton, A metrological scanning force microscope. *Prec. Eng.* (19)1, 46–55 (1996).

[55] J. Bay, S. Bouwstra, E. Laegsgaard, and O. Hansen, Micromachined AFM transducer with differential capacitive read out. *J. Micromech. Microeng.*, (5)2, 161–165 (1995).

[56] Y. Martin, C.C. Williams, and H.K. Wickramasinghe, Atomic force microscope-force mapping and profiling on a sub 100-Å scale. *J. Appl. Phys.*, (61)10, 4723–4729 (1987).

[57] Mitutoyo "linear scale" digital length measuring system. Mitutoyo, Japan.

[58] J.A. Bosch, *Coordinate Measuring Machines and Systems*. Marcel Dekker, New York (1995).

[59] T. Masuzawa, Y. Hamasaki, and M. Fujino, Vibroscanning method for nondestructive measurement of small holes. *Annals of the CIRP*, (42)1, 589–592 (1993).

[60] G.X. Zhang and S.M. Yang, A 3D probe for measuring small blind holes. *Annals of the CIRPl*, (44)1, 461–464 (1995).

3

Molecular Dynamics Simulation of the Atomic Processes in Microcutting

Shoichi Shimada
Osaka University, Osaka, Japan

3.1 INTRODUCTION

In conventional metal cutting, machining accuracy is mainly governed by the inherent performance of the machine tool, account having to be taken of characteristics such as error motion, its stiffness, and dynamic conditions. As a result, machining allowances can be as high as hundreds of μm. However, in ultraprecision cutting, such as single point diamond turning (SPDT) (see Chapter 5), the machining allowance and chip thickness can be reduced, respectively, to less than approximately 10s of μm and 1 μm. Single point diamond turning is a well-known ultraprecision cutting method where a fine single point diamond tool mounted on a specially designed precision machine tool is used. Metal surfaces can be finished under precisely controlled machine and environmental conditions. Accuracies of, respectively, 10 and 1 nm have been attained in practice and under experimental conditions (aided by advanced control techniques) [1–3].

Under the highly precise motion of such a machine tool, the primary factor affecting machining accuracy is the controllability or repeatability of the thickness of cut, that is, the undeformed chip thickness effectively removed at the cutting edge. Experiments with a specially prepared fine diamond cutting tool used on such a machine tool have confirmed that very fine chips, of undeformed thickness as small as 1 nm, can be

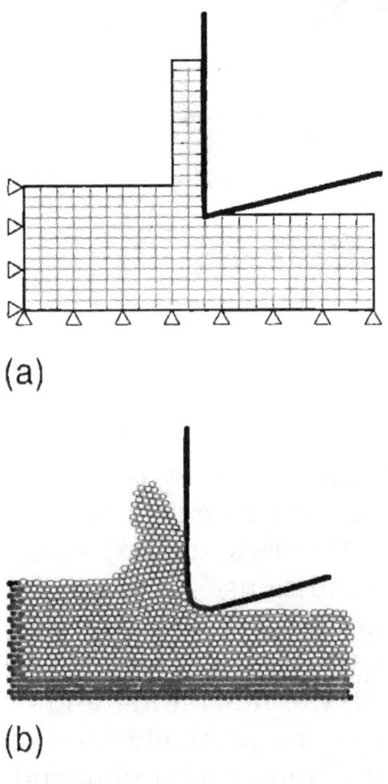

Figure 1 Models for analysis of nanometric cutting process: (a) continuum; (b) atomic.

removed in the turning of some highly machineable work materials [4].

However, the accuracy attainable, and mechanisms such as chip formation and surface generation in microcutting are still not well understood, owing to limitations in availability of experimental and measurement techniques, and in analytic methods for studying such machining conditions. Furthermore, microcutting that occurs in a small region which contains only a few layers of molecules can consequently be atomistic, or discrete in nature, rather than continuous, as is assumed in conventional continuum mechanics; see Figure 1. In studies of such atomistic processes, which are difficult to investigate experimentally, computer simulation by molecular techniques is useful. Several such analyses of microcutting have been developed [5–13].

In this chapter, the principle of molecular dynamics (MD) simulation of micromachining and the procedures used to determine the accuracies attainable are described. As noted above, the case of diamond machining is used to illustrate the technique, although molecular dynamics is now being increasingly used in studies of other methods of micromachining.

3.2 PRINCIPLES OF MOLECULAR DYNAMICS

The MD method provides an accurate means of analysis of deformation and/or fracture of an atomic lattice model of a workpiece and its adjacent tool. Figure 2 shows schematically a computer simulation of the deformation of a workpiece being machined with a diamond cutting edge. Every atom shown in Figure 2 is in motion and interacts with neighboring atoms in a manner that may be determined from the interatomic potential function.

The resultant interatomic force $\boldsymbol{F}_i$ to a specified workpiece atom i, at position $\boldsymbol{r}_i$ from the neighboring workpiece and tool atoms may be calculated from Equation (1) below, where

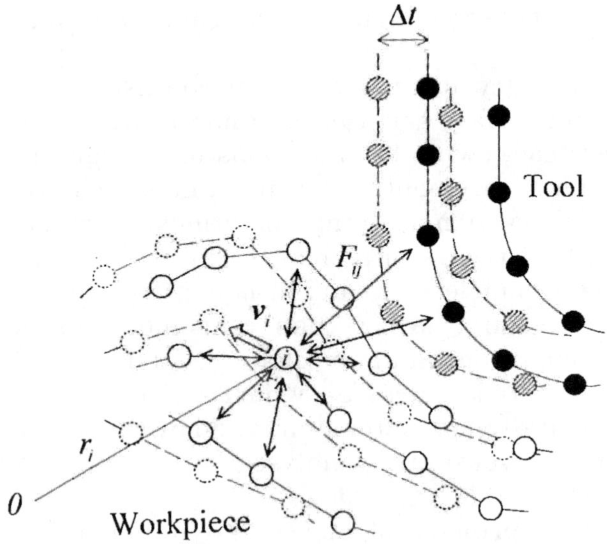

Figure 2 Principle of MD simulation.

$\phi(\boldsymbol{r}_{ij})$ is the interatomic potential function between atoms i and j.

The equation of motion of atom i is obtained from Equation (2), where m and v_i are, respectively, its atomic mass and velocity. Thus we write:

$$F_i = \sum_j \frac{d\phi(r_{ij})}{dr} \tag{1}$$

$$\frac{d^2 r_i}{dt^2} = \frac{dv_i}{dt} = \frac{F_i}{m}. \tag{2}$$

In the simulation, the tool is fed in a stepwise procedure into the workpiece, at time intervals of Δt, that are shorter than the period of lattice vibration. As indicated in Figure 3, the position $\boldsymbol{r}_i(t + \Delta t)$ and velocity $v_i(t + \Delta t)$ of the atom i after the tool has been fed in the period of Δt are calculated by means of Equations (3) and (4), finite difference numerical

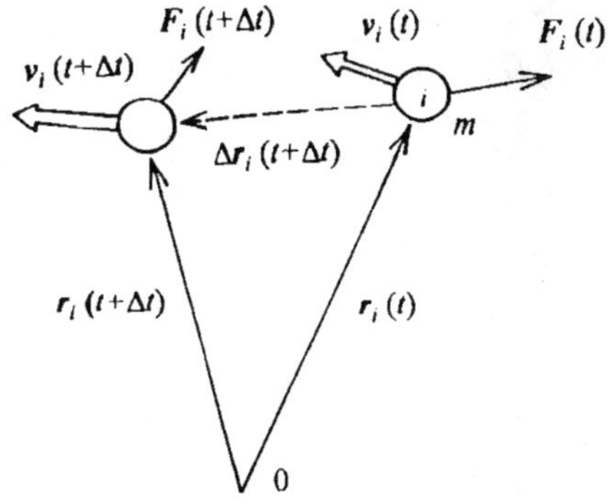

Figure 3 Schematic diagram of numerical calculation in MD simulation.

methods being employed. This calculation is then repeated in order to describe the motion of individual atoms. The behavior of the collection of atoms that comprise the tool/workpiece model can then be analyzed through synthesis of the movement of the individual atoms.

$$r_i(t + \Delta t) = r_i(t) + \Delta t v_i(t) + \frac{(\Delta t)^2}{2m} F_i(t) \tag{3}$$

$$v_i(t + \Delta t) = v_i(t) + \frac{\Delta t}{2m} (F_i(t + \Delta t) + F_i(t)). \tag{4}$$

3.3 MD SIMULATION OF MICROMACHINING

The {111} plane of copper and aluminum, subjected to a hypothetical Morse-potential is assumed to undergo orthogonal cutting by the {111} plane of a diamond edge, as shown in Figure 4 [8,10]. The three variables in the Morse potential function, D, a, and r_0 are determined from the cohesive energy, elastic modulus, and equilibrium interatomic distance. The elastic modulus and equilibrium interatomic distance are assumed to be derived from the average value of those for the

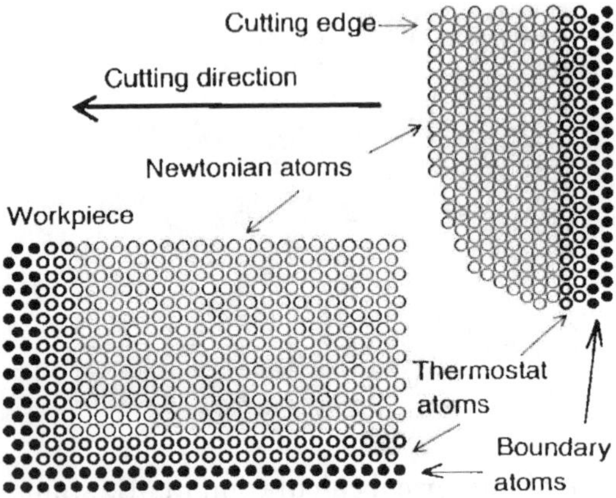

Figure 4 Atomic model of microcutting for MD simulation.

metal workpiece and diamond tool, while the cohesive energy is estimated from experimentally determined values for the separation energy of the junction of metal and diamond in an ultrahigh vacuum [14].

The model is composed of the boundary, thermostat, and Newtonian atoms. The static boundary atoms are fixed to the rigid bases of the workpiece and cutting edge. Those atoms in the two adjacent layers are thermostat atoms which act to stabilize the temperature of the model (see Fig. 4). In these layers, as described in Equation (6), the velocities of atoms v_i are scaled to be v_i' for each computation time-step, so that these layers are considered to remain on average at room temperature, T (293K). In Equation (6), N and k denote, respectively, the number of atoms in these layers of average constant room temperature, and Boltzmann's constant. The Newtonian atoms are in motion following the equation of motion described in Equation (2).

The interatomic distance, which is given as a variable in the Morse potential function, is assumed to be the same as the initial spacing between atoms. The initial velocity of each atom is considered statistically to follow a Maxwell–Boltzmann distribution related to the initial bulk temperature (of 293 K). Computational time-steps of 1 and 10 fs are typically employed, depending on the purpose of the analysis. Micromachining at cutting speeds of 20 and 200 m/s was simulated (the latter value being used to reduce time of computation).

$$\phi(r_{ij}) = D[\exp\{-2a(r_{ij} - r_0)\} - 2a \exp\{-a(r_{ij} - r_0)\}] \quad (5)$$

$$v_i' = v_i \sqrt{\frac{NkT}{\sum_i \frac{m_i v_i^2}{2}}}. \quad (6)$$

In general, thermal conduction in metals cannot be analyzed correctly in MD simulation, as the simulation used does not take electron behavior into account.

As thermal conduction in metals is mainly governed by electron mobility, in practice metals exhibit considerably higher thermal conductivities, and consequently smaller thermal gradients than those estimated by MD simulation based on lattice vibration. Thus for a more realistic simulation of micromachining, the gradient in the thermal field should be scaled, or adjusted, to coincide with that calculated from continuum thermal conduction theory [13], as indicated in Figure 5, since chip formation is considerably affected by the thermal conditions in the shear zone.

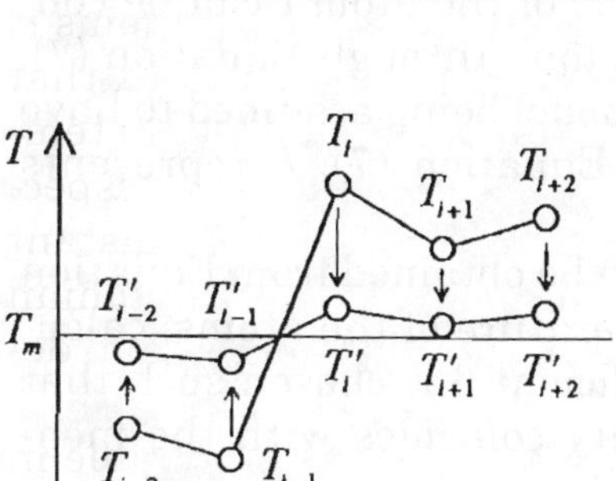

Figure 5 Scaling of equivalent temperature of atoms in MD model.

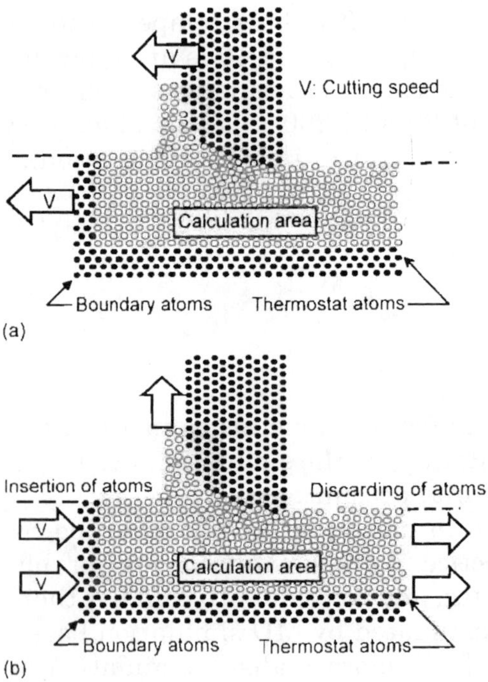

Figure 6 Schematic illustration of translational calculation area method; (a) principle; (b) practical calculation.

The equivalent temperature T_i of the atom i can be converted from the velocity v_i of the atom, through Equation (7), the atom in the two-dimensional model being assumed to have a twofold degree of freedom (in Equation (7), k represents Boltzmann's constant).

The scaled temperature T_i' can be obtained from Equation (8), where T_m is the average temperature of the atoms, calculated from Equation (9). The coefficient β is chosen such that the simulated thermal conductivity coincides with the measured macroscopic value.

$$T_i = \frac{mv_i2}{2k} \quad (7)$$

$$T_i' = T_i - (T_i - T_m)\beta \tag{8}$$

$$T_m = \frac{1}{N}\sum_i T_i. \tag{9}$$

For analysis of the micromachining phenomena under steady-state chip removal, the translational calculation area method, by which the cutting distance can be infinitely extended, is used in the simulations as illustrated in Figure 6 [15].

In order to limit the area of analysis of the behavior of individual atoms by the MD method, the boundary and thermostat layer areas are considered to move at the same speed as the cutting edge, as illustrated in Figure 6(a). The procedures for calculation are shown in Figure 6(b). New atoms are successively inserted (from the left-hand side in this figure) into the area of calculation, in which the cutting edge is fixed. Atoms leaving the calculation area are discarded from the right-hand or top-side, as given in Figure 6(b).

3.4 ATOMISTIC FORCES OF CHIP FORMATION AND SURFACE GENERATION

Simulation of the forces on the diamond atoms at the tool surface enables calculation of the cutting force as the resultant. Figure 7 presents a comparison between experimental and MD simulation of the force per unit width of the cutting edge as a function of nominal undeformed chip thickness, for a copper workpiece [10]. In experiments, where relatively larger undeformed chip thicknesses of 20 to 450 nm were employed, the force of cutting was measured by a quartz force transducer with a resolution of 0.01 N. Cutting forces at very small undeformed chip thickness, less than 100 nm, may be calculated from classical simple shear plane models on the assumption that the shearing stress is the same as the theoretical one.

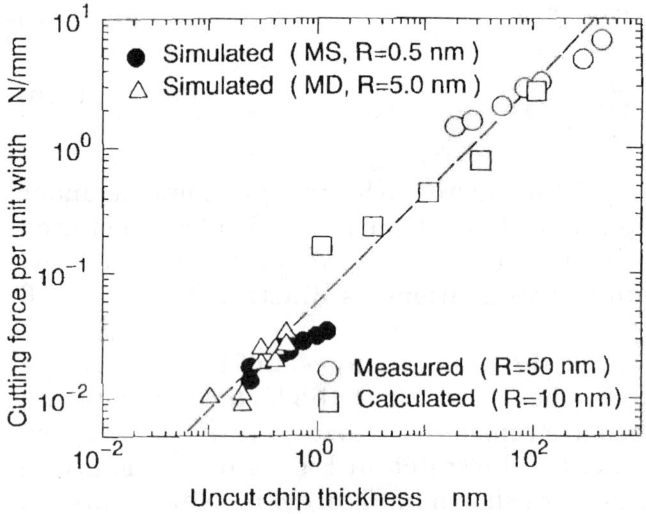

Figure 7 Size effect in cutting force per unit width of cutting edge as a function of undeformed chip thickness.

The experimental effects of magnitude of cutting force on chip morphology can be approximately predicted in these simulation studies. This correlation suggests that MD simulation is a useful tool for analysis of microcutting phenomena, even to nanometric conditions.

Figure 8 shows an example of the deformation behavior of the workpiece atoms at the leading edge of the cutting tool. The lattice of the workpiece is noted to become deformed, or buckled, due to the ploughing effect of the cutting edge. When the strain energy stored in the deformed lattice exceeds a specified level, the atoms begin to rearrange, so that the lattice strain is relaxed.

However, since the stored strain energy is usually not sufficient to provide for complete or flawless rearrangement, some dislocations are generated in the workpiece lattice. As the cutting edge advances, many dislocations are successively

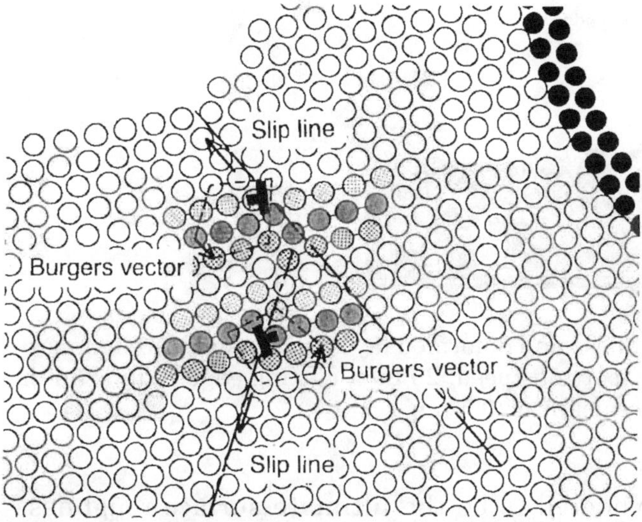

Figure 8 Generation of dislocations in workpiece due to ploughing of cutting edge.

generated in the workpiece at its interface with the tool in a fashion similar to that described above. Some dislocations move into the shear zone, "disappearing" from the free surface. By the disappearance of a dislocation, a chip is formed for the length corresponding to an atomic layer. This effect corresponds to the elemental process of morphological chip formation. As a result of successive generation and disappearance of the dislocations, a stable chip removal process can be envisaged.

The other dislocations penetrate into the workpiece under the cutting edge. After the cutting edge has passed, these dislocations begin to move back, before finally disappearing from the workpiece surface, as a consequence of the relaxation of the lattice, as the workpiece "springs back." As a result of this relaxation, atomic-size steps are formed on the workpiece surface (see Fig. 9). The height of these steps on the work sur-

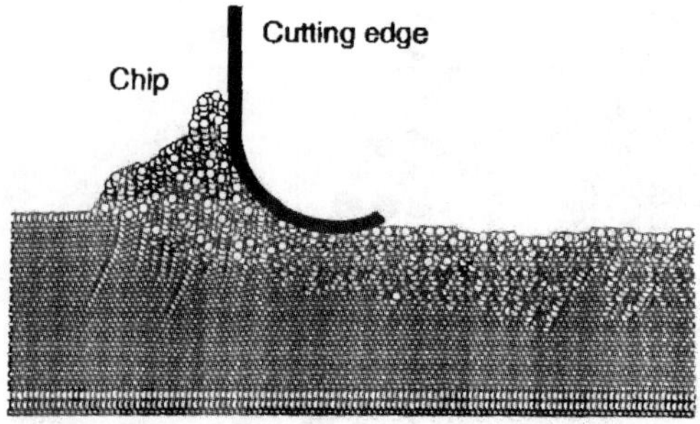

Figure 9 Roughness and deformed layer on work surface in microcutting of copper (cutting speed: 20 m/s, cutting edge radius: 5 nm, undeformed chip thickness: 0.5 nm).

face can be considered to be the ultimate surface roughness attainable in microcutting [9].

3.5 ULTIMATE ACCURACY ATTAINABLE IN MICROCUTTING

Surface generation due to material removal from a workpiece by a tool is a basic outcome of metal cutting. The motion needed between tool and workpiece is provided by a machine tool, on which both are mounted. Efficient metal cutting therefore relies greatly on the accuracy with which the characteristics of machine tool movement are transferred to the tool edge as it cuts into the work surface. In addition to machine tool performance, two primary factors affect the accuracy achievable in micromachining: the minimum thickness of an undeformed chip (MTC), removed at the cutting edge from a workpiece under a specified cutting condition [16,17]; and "transfer fidelity," the extent to which the profile of the cutting edge is reproduced on the workpiece surface. The surface roughness, achieved when the workpiece is cut under perfect machine motion, may be regarded as representative of the highest level of transfer fidelity.

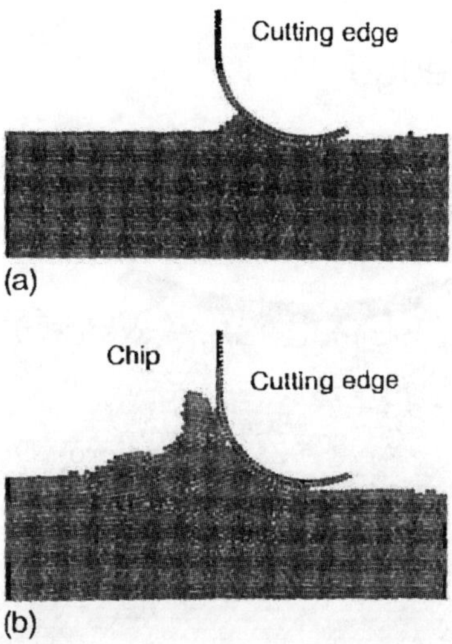

Figure 10 Chip formation in copper microcutting (cutting speed: 200 m/s, cutting edge radius: 5 nm). (a) $t_n = 0.2$ nm; (b) $t_n = 0.3$ nm.

Figure 10 illustrates chip formation for copper undergoing micromachining with a cutting edge of radius 5 nm. Chip formation cannot be observed for an undeformed thickness t_n smaller than 0.2 nm. However, the initial stage of its formation becomes apparent for undeformed chip thickness greater than 0.3 nm. The MTC for these conditions is therefore 0.3 nm. For a variety of edge radii in the microcutting of copper MD simulation indicates a MTC of 0.05 to 0.1 times the radius of the cutting edge. As the edge radius of a fine diamond cutting tool is typically 20 to 30 nm, the MTC in practice can be deduced to be about 1 to 3 nm, which is consistent with experimental findings [4]. The MTC varies with the material. For example, for aluminum, a larger MTC, about 0.1 to 0.2 of the cutting edge radius is estimated from MD simulation [9].

Figure 11 illustrates the differences in deformation behavior in the shear zone, when copper and aluminum are cut with a diamond cutting edge of 10 nm. The size of the deformed region for aluminum is noted to be considerably larger.

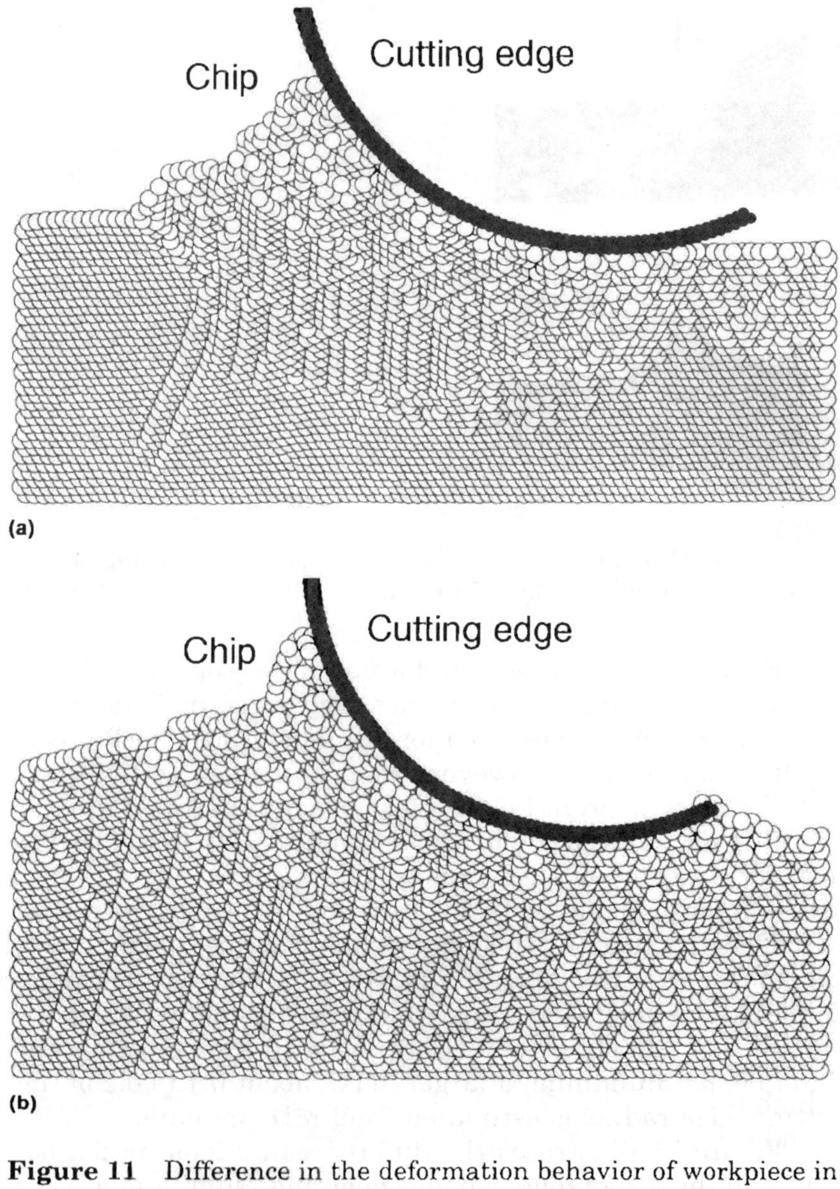

Figure 11 Difference in the deformation behavior of workpiece in microcutting of: (a) copper and (b) aluminum. (Cutting speed: 200 m/s, cutting edge radius: 10 nm, undeformed chip thickness: 1 nm.)

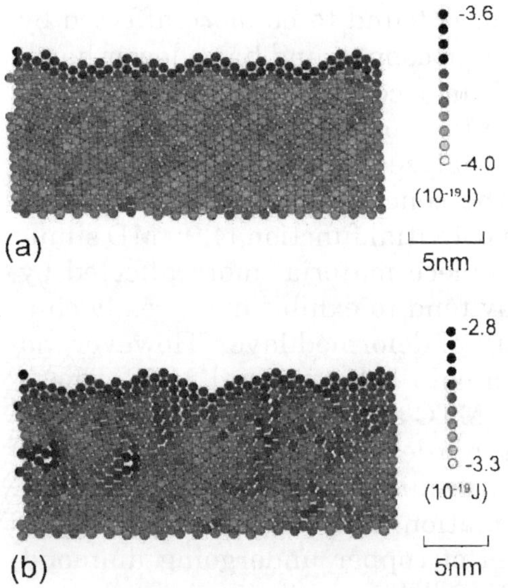

Figure 12 Difference in the quality of work surface in microcutting of (a) copper and (b) aluminum. (Cutting speed: 200 m/s, cutting edge radius: 10 nm, undeformed chip thickness: 1 nm.)

Figure 12 illustrates the difference in the quality of work surface expressed by potential energy distribution of atoms, when copper and aluminum are cut at the same cutting condition as shown in Figure 11. Atoms which are distorted or disordered show relatively higher potential energy than relaxed ones. Simulation predicts a transfer fidelity of the cutting edge profile to the work surface of 0.5 to 1.0 nm and 1.5 to 2.0 nm, respectively, for copper and aluminum [9,11]. No residual strain is evident on the copper work surface. Conversely, the aluminum exhibits vacancies and many atoms possessing high potential energy on its surface [12]; its surface roughness and residual strain are therefore greater than those of copper.

MTC and transfer fidelity are both influenced by the physical relationship occurring between the tool and workpiece at their interface, and/or the hardness of the work material. For example, aluminum is found to be more affected by diamond as a tool material than copper, and has a lower hardness than the latter. The former condition is sometimes described as "affinity." These effects of conditions arising at the tool–work interface can be analyzed by changing the process variables associated with the cohesive energy for the tool–work interface in the Morse potential function [4,9]. MD simulations indicate that a workpiece material more affected by the tool at the interface may tend to exhibit more ready chip formation and a larger depth of deformed layer. However, no significant change occurs in chip behavior and MTC. These results suggest that smaller MTC and higher transfer fidelity in micromachining occur on harder workpiece materials.

Grain boundary effects in micromachining may also be analyzed through MD simulation. For example, simulation based on a two-grain model of copper undergoing diamond turning reveals that a distorted layer, albeit of nm dimensions, remains on the work surface of polycrystalline metals, owing to the trapping of dislocations at the grain boundaries [12,13].

3.6 FUTURE TRENDS IN ULTRAHIGH SPEED MACHINING

Ultrafast cutting to high precision remains a formidable task in machine tool development for micromachining. For example, while surface finish and cutting forces needed may be specified for high-speed cutting, physical conditions may arise that are difficult to anticipate, the effects of which on machining might be catastrophic. MD simulation provides a useful analysis of the phenomena that may arise.

To that end, Figure 13 shows examples of the MD simulation of material removal at high speeds [18]. In Figure 13(a), at a cutting speed of 200 m/s, the smaller depth for the deformed layer on the machined surface is noted, which may be

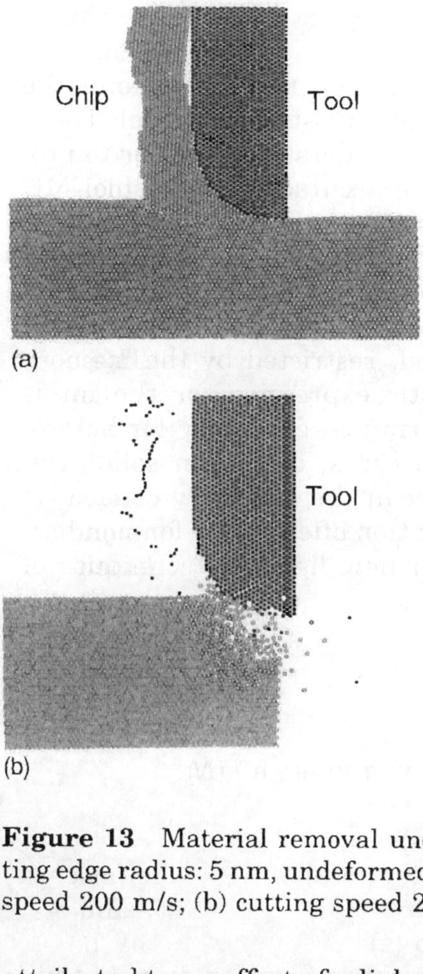

Figure 13 Material removal under ultrahigh cutting speed (cutting edge radius: 5 nm, undeformed chip thickness: 1 nm): (a) cutting speed 200 m/s; (b) cutting speed 2000 m/s.

attributed to an effect of adiabatic shearing in chip formation. When the cutting speed is increased tenfold, to 2000 m/s, a catastrophic dispersion of work material arises. When the cutting edge engages the workpiece, the associated kinetic energy, supplied to individual atoms, is considered to exceed the specific cohesive energy of the work material. These results suggest that the surface finish required may be linked to optimal speed of cutting.

3.7 CONCLUSIONS

As micromachining approaches nanometric dimensions, the mechanisms involve changes at atomic structure level. These atomistic mechanisms have to be understood in order to provide information on the ultimate accuracies attainable. MD simulation is proving to be a useful tool in such analyses. At present, the drawbacks of MD for these kinds of analyses lie partly with the relatively long times of computation needed (which can be on the order of days), and limitations in the dimensions of the model treated, restricted by the memory size of the computer. A realistic expression for the interatomic potential has to be determined especially for heterogenous materials. For practical cases, defects in solids can be difficult to model in MD, as can the boundary conditions needed. Nonetheless, MD simulation offers scope for theoretical treatments, which can shed new light on techniques of micromachining.

NOTATION AND SYMBOLS

F_i	resultant interatomic force to atom i (N)
r_i	position of atom i (m)
v_i	velocity of atom i (m/s)
v_i'	scaled velocity of atom i (m/s)
m	atomic mass (kg)
Δt	computation time-step (s)
$\varphi(r_{ij})$	interatomic potential function between atoms i and j (eV)
D	parameter of Morse potential function concerned with cohesive energy (eV)
a	parameter of Morse potential function concerned with elastic modulus (Å^{-1})
r_0	parameter of Morse potential function concerned with equilibrium interatomic distance (Å)
N	number of atoms in model or thermostat layer

k	Boltzmann's constant (J/K)
T	macroscopic temperature (K)
T_i	atomic temperature (K)
T_i	scaled atomic temperature (K)
T_m	average atomic temperature (K)
t_n	undeformed chip thickness (nm)

REFERENCES

[1] R.R. Donaldson and D.C. Thompson, Design and performance of a small precision CNC turning machine. *Annals of the CIRP*, (35) 1, 373–376 (1986).

[2] Y. Kami, M. Yabuya, and T. Shimizu, Research and development of ultra-precision positioning system. In *Proc. 4th Biennial Joint Warwick / Tokyo Nanotechnology Symposium*, Warwick (1994).

[3] H. Mizumoto, S. Arii, A. Yoshimoto, T. Shimizu, and N. Ikawa, A twist-roller friction drive for nanometer positioning. *Annals of the CIRP*, (44) 1, 523–526 (1996).

[4] N. Ikawa, S. Shimada, H. Tanaka, and G. Ohmori, An atomistic analysis of nanometric chip removal as affected by tool-work interaction in diamond turning. *Annals of the CIRP*, (40) 1, 551–554 (1991).

[5] J.F. Belak and I.F. Stowers, A molecular dynamics model of the orthogonal cutting process. In *Proc. ASPE (American Society for Precision Engineering) 1990 Annual Conference*, Rochester, 259–262 (1990).

[6] T. Inamura, H. Suzuki, and N. Takezawa, Cutting experiments in a computer using atomic model of a copper crystal and a diamond tool. *J. Japan, Society for Precision Engineering*, (56) 8, 1480–1486 (in Japanese), (1990).

[7] I.F. Stowers, J.F. Belak, D.A. Lucca, R. Komanduri, R.L. Rholar, T. Moriwaki, K. Okuda, N. Ikawa, S. Shimada, and H. Tanaka, Molecular dynamics simulation of the chip forming

process in single crystal copper and comparison with experimental data. In *Proc. 1991 ASPE Annual Meeting*, Santa Fe 100–103 (1991).

[8] S. Shimada, N. Ikawa, G. Ohmori, H. Tanaka, and J. Uchikoshi, Molecular dynamics analyses as compared with experimental results of micromachining. *Annals of the CIRP*, (41) 1, 117–120 (1992).

[9] S. Shimada, N. Ikawa, H. Tanaka, G. Ohmori, and J. Uchikoshi, Feasibility study on ultimate accuracy in microcutting using molecular dynamics simulation. *Annals of the CIRP*, (42) 1, 91–94 (1993).

[10] S. Shimada, N. Ikawa, H. Tanaka, G. Ohmori, and J. Uchikoshi, Molecular dynamics analysis of cutting force and chip formation process in microcutting. *J. Japan Society for Precision Engineering*, (59) 12, 2015–2021 (in Japanese) (1993).

[11] S. Shimada, N. Ikawa, H. Tanaka, and J. Uchikoshi, Structure of micromachined surface simulated by molecular dynamics analysis. *Annals of the CIRP*, (43) 1, 51–54 (1994).

[12] S. Shimada, R. Inoue, J. Uchikoshi, and N. Ikawa, Molecular dynamics analysis on microstructure of diamond turned surface. *Proc. SPIE*, (2576), 396–405 (1995).

[13] S. Shimada, N. Ikawa, R. Inoue, and J. Uchikoshi, Molecular dynamics modeling of thermal conduction in microcutting of metals. In *Proc. 1995 ASPE Annual Meeting*, Austin, 21–24 (1995).

[14] Y. Mori, K. Sugiyama, K. Endo, and H. Gotoh, Interaction force between solid surfaces (1st report)—An atomistic estimation of interfacial surface energy. *J. Japan Society for Precision Engineering*, (52) 10, 1795–1801 (1986).

[15] W.G. Hoover, C.G. Hoover, J.F. Belak, I.F. Stowers, and A.J. DeGroot, Molecular dynamics modeling applied to indentation and metal cutting problems. Thrust area report FY90, Lawrence Livermore National Laboratory, Livermore, 4-1–4-8 (1990).

[16] N. Ikawa and S. Shimada, Cutting tool for ultraprecision machining. In *Proc. 3rd Int. Conference On Production Engineering*, Kyoto, 357–364 (1977).

[17] N. Ikawa, S. Shimada, and H. Tanaka, Minimum thickness of cut in micromachining. *Nanotechnology*, (3), 6–9 (1992).

[18] S. Shimada, N. Ikawa, R. Inoue, and J. Uchikoshi, Molecular dynamics simulation of high speed metal cutting mechanism. In *Proc. JSPE 1996 Spring Meeting*, Tokyo 143–144 (in Japanese) (1996).

4

Abrasive Micromachining and Microgrinding

Kai Cheng
Glasgow Caledonian University, Glasgow, Scotland

4.1 INTRODUCTION

The rising demand for high and ultrahigh accuracy products is widening the application of abrasive machining into areas of micron scale with depth of cut of only a few 100 or 10 nanometers, that is, micromachining as generally understood. Abrasive micromachining and microgrinding (AMMG) techniques are of particular interest in the optical and electronic industries and are gaining new stature as manufacturers seek tighter tolerances and higher quality.

Minimal subsurface damage and high removal rates constitute major advantages in AMMG. These techniques are not particular useful in fabrication and manufacturing if the two advantages are not met. In fabricating optics, for example, large quantities of glass are removed and essentially all of the figuring is accomplished with pitch polishing laps. By use of microgrinding, however, long polishing times can be avoided, if almost all of the optical figuring is accomplished through the well-controlled and more suitable grinding process. As shown in Figure 1, the introduction of microgrinding can drastically reduce the optics production cycle and achieve an improved fabrication quality [1]. The main advantages to microgrinding before polishing is indicated are that it can greatly reduce the long and unpredictable polishing process and the amount of subsurface damage in a finished component or product. Furthermore, it can deterministically produce components with high accuracy of dimension and form.

For metals, AMMG processes produce controlled surface conditions involving size, finish, geometry, and metallurgical structure. A microfinished surface in particular is metallurgically free of fragmented, amorphous, or smeared metal from previous grinding operations. The processes may also partly

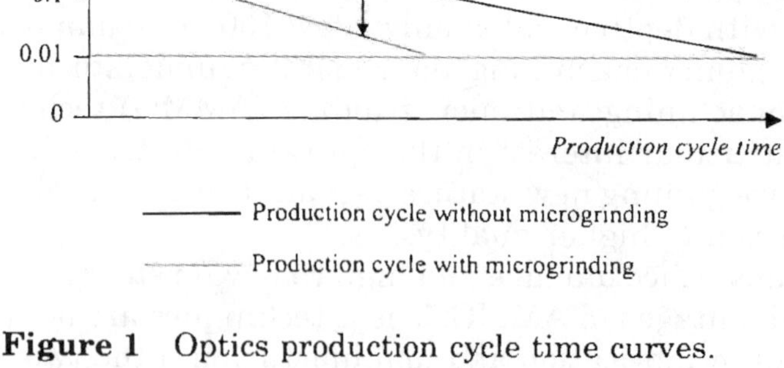

Figure 1 Optics production cycle time curves.

restore surface integrity by eliminating surface stresses and grinding thermal damage.

The present chapter is therefore concerned with the principles and process characteristics of AMMG, and their applications. Micromachining analyses and models are presented that account for the mechanisms and modes of actual material removal under various machining conditions. The machining rates, optimal process parameters, and tool characteristics are discussed for typical engineering materials such as silicon, glass, ceramics, germanium, and metals. Applications are described that illustrate how AMMG techniques can be utilized in the production of products particularly in the optical and electronic industries.

4.2 PRINCIPLES

In abrasive machining hard abrasive particles are used as the cutting medium. Although AMMG techniques have been being practiced for more than a decade, there are still no clear definitions of the relevant terminology used and well-defined boundaries on their types of applications. Nevertheless it is clear that both of the processes involve micrometer-sized abrasive grains interacting with the workpiece surface at that level of scale. The term “abrasive micromachining” is often associated with loose abrasive machining techniques such as lapping, polishing, and honing, while microgrinding is associated more with deterministic grinding processes.

4.2.1 Abrasive Micromachining Mechanisms

Material removal is a fracture process in which abrasive grains interact with the workpiece surface. An effectively single layer of abrasive grits, usually block-shaped, tumble on the workpiece surface under the effects of a downward force. The force is transferred into multiple microloads on grits that chip the workpiece substrate and cause material removal. The un-

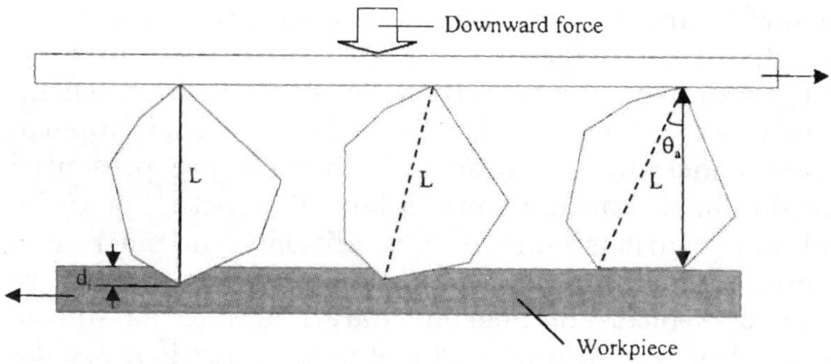

Figure 2 Penetration depth of an abrasive grit.

derlying removal mechanisms are complex as machining proceeds at such a micrometer scale; many physical and possible chemical factors are involved. However, an understanding of the underlying principles and relevant processing conditions is needed, especially as fresh potential industrial applications arise.

The depth of penetration of the grit tips into the material depends on the topography of the abrasive grits, and also the geometry and kinematic motion of the grits and workpiece. The penetration of an abrasive grit onto the workpiece surface can be envisaged as an indentation process as shown in Figure 2. The penetration depth (d) of the grit can be calculated from the following geometrical relationship [2].

$$d_i = L(1 - \cos\theta_a), \tag{1}$$

where

L: Grit size,
θ: Tumbling angle.

The tumbling angle θ_a can be determined from measurement of the nominal coefficient (g) of friction between the grit and workpiece material, defined as the ratio of drag-to-normal force; that is,

$$\mu = \frac{\ln(\cos\theta_a)}{\theta_a}. \tag{2}$$

The penetration depth (d_i) represents the machining interaction between an abrasive grit and the workpiece surface and directly affects the surface roughness of the micromachined surface. The depth also greatly depends on the hardness of the abrasive grits and the workpiece material, the grit size, uniformity of size of grits, and downward force of machining. The penetration results in three types of micromachining action on the workpiece surface, namely, brittle, ductile, and smeared mode as shown in Figure 3 [3].

Brittle Mode

A brittle mode micromachined surface consists of a disrupted uppermost layer and a subsurface compacted layer lying between the former layer and the bulk material. Brittle fracture occurs owing to the microindentations produced in the micromachining process. The fracture involves two principal crack systems: lateral cracks, which are responsible for material removal, and median cracks, for strength degradation. Subsurface damage extends into the bulk material with a magnitude nearly equal to the grit size. The magnitude and depth of permanent deformation accompanying the brittle process depend on the grit size, with smaller grits producing a higher overall percentage deformation caused by plastic flow. Brittle mode micromachining as described is solely a mechanical process.

Ductile Mode

A ductile mode micromachined surface consists of an upper compacted layer that sits above the bulk material. Little subsurface damage occurs. The ductile surface is the result of material removal due to shearing, where the material is planed off at a microscale level. The efficiency of ductile removal is strongly influenced by environmental effects such as lubrication and heat generated during the machining, which may al-

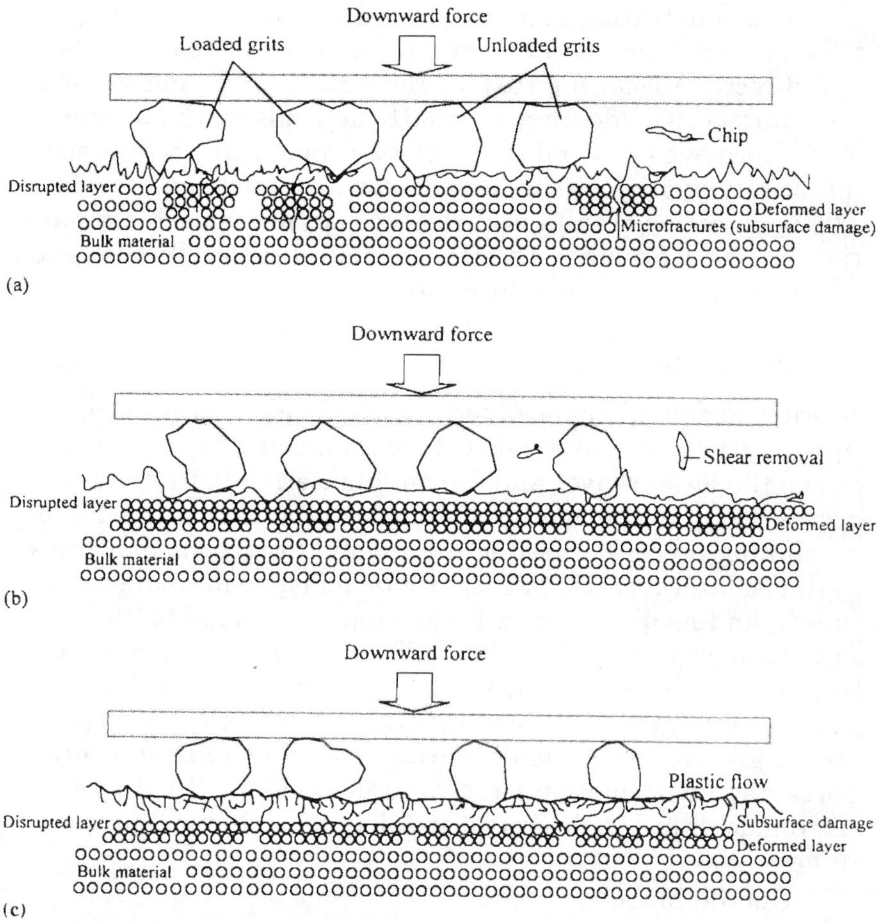

Figure 3 Three types of micromachining modes in abrasive micromachining: (a) brittle; (b) ductile (insufficient force for producing chips); (c) smeared (insufficient force for producing chips).

ter the fracture toughness of the workpiece surface. A higher degree of permanent deformation arises in this process. The magnitude and depth of the deformation are dependent on the grit size. Ductile mode micromachining is also a mechanical process but with extremely high surface stresses. Weighing measurement and acid etching confirm that the material is actually removed in the ductile mode and the surface stresses are concentrated in a thin layer. In the ductile mode, the energy is transferred into a permanent deformation of the workpiece material; in contrast, in the brittle mode, more machining energy is transferred to a fracture process.

Smeared Mode

A third mode of abrasive micromachining can be observed, in which little material is actually removed, but instead is smeared. As shown in Figure 3(c), the peaks of the fractured layer are plastically deformed and simply cover the disrupted layer. Etching or X-ray scanning such a surface can expose the damage that was produced in the previous abrasive machining process. This damage seems to occur when either very small abrasive grits (under 0.5 ~ 1 μm) are used, or the machining process is inhibited by the type of slurry employed. The phenomena are commonly seen in bonded abrasive machining when a tool dulls and sparks out [3].

4.2.2 Microgrinding Mechanisms

In microgrinding small size chips are usually involved. A major consequence of their smallness is that the normally ductile material in micromachining exhibits toughlike properties, similar to brittle substances such as glass and ceramics when the volume deformed is limited to a very small size [4]. Very small whiskers produced when a normally ductile metal was deposited electrolytically at a very small rate were found to be essentially defect-free. When these whiskers were tested in bending they were found to be completely brittle, to have unusually high elastic fracture strain, with the fracture stress approaching the theoretical strength of the material. This observation demonstrates that when the size of the deformation zone (resulting from the small size of abrasive grits) approaches the mean defect structure spacing, normally ductile materials behave as tough and very brittle. An understanding of this phenomenon is useful in studying the mechanisms of microgrinding processes.

The small undeformed chip thickness involved in microgrinding has a further consequence: the chip-forming model shifts from one involving concentrated shear to microextrusion, as shown in Figure 4. When the undeformed chip thick-

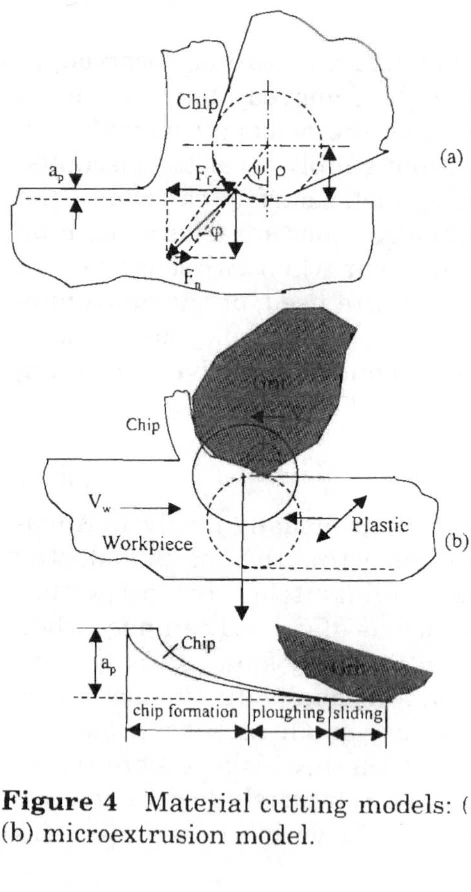

Figure 4 Material cutting models: (a) concentrated shear model; (b) microextrusion model.

ness (a_p) becomes less than the radius (ρ) at the cutting tool or grit tip, the effective rake angle of the tool or grit has such a large negative value that the model in Figure 4(b) replaces that in Figure 4(a).

For a sharp grit, the relationship between its tip radius and the nominal cutting depth at the chip forming point is:

$$a_p = \rho - \rho\cos\psi = \rho(1 - \cos\psi) \tag{3}$$

and as

$$\psi = 45^\circ - \varphi = 45^\circ - \arctan\left(\frac{F_f}{F_n}\right). \tag{4}$$

In metal cutting with diamond abrasive grits, the friction coefficient is [5]:

$$\frac{F_f}{F_n} = 0.1 \sim 0.8$$

so there is

$$\varphi = \arctan\left(\frac{F_f}{Fn}\right) = 5.7^\circ \sim 38.7^\circ.$$

Thus, the minimum cutting depth with chip formation is

$$\begin{aligned} a_{p(\min)} &= \rho(1 - \cos\psi) = \rho[1 - \cos(6.3^\circ \sim 39.3^\circ)] \\ &= (0.006 \sim 0.226)\rho. \end{aligned} \tag{5}$$

This conclusion shows that the chips can be formed even though a grit tip radius is bigger than the nominal cutting depth [5]. The chips are possibly smaller in size than those of abrasive grits. However, the chip formation mechanisms in grinding processes are much more complicated because of the multiplicity of cutting points and their irregular geometry varying during the machining.

The model in Figure 4(b) gives a more realistic illustration of grinding mechanisms involved within the grit–workpiece surface interaction. Grit grinding can be divided into three phases including sliding, ploughing, and cutting since the effective rake angles of abrasive grits are largely negative [6,7]. But for a grit the phases do not necessarily occur in the order of sliding–ploughing–cutting during one grinding cycle.

When the grit engages with the workpiece surface in microgrinding, the grit slides without removing any material on the surface due to the elastic deformation of the grit–surface interface or the possible wear-flat areas of the grit. This is termed the "sliding phase." As the stress between the grit and workpiece surface is increased beyond the elastic limit, plastic deformation known as the "ploughing phase" occurs. Ploughing is normally associated with side flow of material from the cutting path into ridges, but it can also include plastic deformation of the material passing under the grit tip edge. A chip is formed when the workpiece material can no longer withstand the shearing stress. The chip formation stage is the "cutting phase." From consideration of the specific energy expended, the cutting phase is the most efficient. Sliding and ploughing are inefficient, since the energy is wasted in deformation and friction with little contribution to material removal. Furthermore, a high temperature may result, producing an excessive rate of grit wear, and the workpiece surface may suffer subsurface damage.

4.3 MICROMACHINING RATES

Predominant parameters in abrasive micromachining include the grit size, grit size uniformity, and machining downward force or pressure [8]. The material removal rate (MRR) is given by

$$MRR = C_p \times P \times V \times \rho_d, \tag{6}$$

where

C_P: Denotes coefficient of proportionality (cm^2/N),
P: Total load normal to the workpiece surface (N),

P_d: Material density of the workpiece (g/cm^3),
V: Machining velocity (cm/s).

The equation indicates the MRR is directly proportional to machining velocity and total load applied on the grits. But the machining velocity and total load applied on the grits do not affect the magnitude of abrasion, that is, the chip size. Clearly, Equation (6) does not include the size and machining behavior of grits, which have significant effects in micromachining materials. Whereas the behavior of smaller abrasives has not been previously studied in detail, a comprehensive investigation is now needed for abrasive micromachining, especially in relation to the micromachining efficiency of the grits.

The micromachining efficiency of abrasive grits is determined from the rate at which they are capable of removing material. The MRR of the grits is also dependent on these aspects:

1. Grit wear
2. Micromachining cooling media
3. Disposal rate of the dull grits
4. Grit size.

The effects of the above-mentioned aspects of micromachining efficiency are now described in detail and case studies presented in which different abrasive grits are used in the micromachining of various engineering materials.

4.3.1 Grit Wear

Three main modes of wear in general occur in AMMG: attritious wear, grain fracture, and bond fracture [6]. Attritious wear involves dulling of abrasive grains and the growth of wear flats by rubbing against the workpiece. Grain fracture refers to removal of abrasive fragments by fracture within the grit, and bond fracture occurs on an abrasive tool by dislodging the abrasive from the binder. Another type of wear is binder erosion, which is likely to reduce the bond strength of a grinding wheel or tool and promote grit dislodgement, especially with resin and metal bond wheel or tools. Attritious wear amounts to a few percent of the total wear, but it is often the most important form of wear since it controls the grinding forces and hence also the rate of grain and bond fractures. With increasing tool hardness, there is relatively more grain fracture and less bond fracture, because of the greater bond strength allied to the greater probability of the abrasive grit undergoing fracture prior to being finally dislodged. Therefore, it is believed the wear in AMMG is mostly due to grain fracture.

The suitability of abrasive grits for micromachining particular materials depends to a large extent on their attritious wear resistance. Attritious wear and the dulling of the grits are both mechanical and chemical. Chemical effects are likely to be more significant when the abrasive is harder than the workpiece and any of its included phases. As an abrasive grit interacts with the workpiece at the elevated temperatures reached in the grinding zone, numerous chemical reactions may occur involving the abrasive, workpiece material, binder, atmosphere, and micromachining coolant or slurry in various combinations.

For instance, diamond grits, despite their extreme hardness, are not suitable for abrasive machining of most ferrous metals. This anomalous behavior is attributed to excessive attritious wear mainly by the degradation of diamond to graphite because the degradation appears to be more rapid in the presence of iron and other ferrous metals unsaturated in carbon, owing to their affinity for carbon. This phenomenon is the basis for the use of diamond grits in grinding some cast irons with high carbon content. Cubic boron nitride (CBN) grits, although softer than diamond grits, do not have affinity for carbon and thus wear less in working on most ferrous metals.

Another example is silicon carbide abrasives. They are harder than friable aluminum oxide abrasives, although they are usually less effective for micromachining most ferrous materials. This is because of the tendency for silicon carbon to react with, and adhere to, iron at elevated temperatures. The main chemical reaction appears to be the dissociation of silicon carbide, and this reaction could also promote attritious wear when micromachining titanium and other nonferrous materials [9]. Dissociation of silicon carbide at elevated machining temperatures could be driven by the affinity of silicon or carbon for the workpiece. Therefore, silicon carbide abrasives tend to work better than aluminum oxide on some ferrous metals with excessive carbon, but not on carbon-deficient ferrous metals which are also unsaturated in carbon. This case is analogous to that on applications of diamond abrasives discussed above.

4.3.2 Micromachining Cooling Media

AMMG are normally performed with the use of a fluid, which is generally considered to have two main roles, namely, cooling and lubrication. In microgrinding the fluids are commonly used as coolants, although their role as lubricants is often more important in machining metals. In AMMG, a coolant can have great effect as a cooling, lubricating, and working medium to support the processing operations.

The type of coolant has a significant effect on the efficiency of the micromachining process. Coolant media affect the process primarily by influencing the rate of wear of the grits [10]. In the following study, the effect of four different coolants has been investigated, namely, water, emulsified oil (Quaker 101), ethylene glycol, and triethanolamine. The effect of these coolants for micromachining glasses with diamond grits was observed with the aid of an optical microscope.

As shown in Figure 5, water is unsuitable as a coolant. After five minutes of micromachining with water, most of the diamond grits were mechanically removed and those remaining were severely abraded. About three hours of machining with ethylene glycol as a coolant was achieved before the diamond grits underwent noticeable abrasion wear. During a

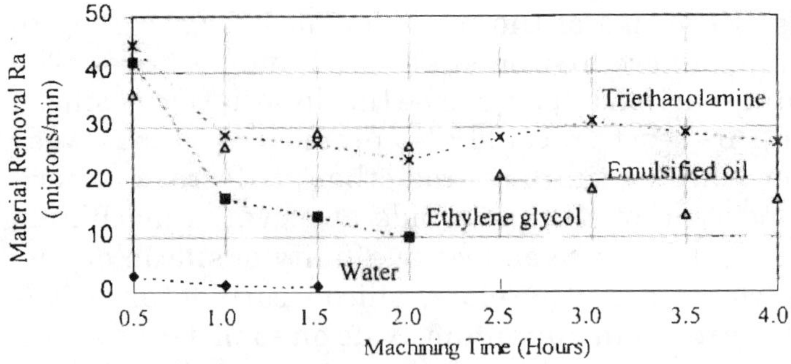

Figure 5 Effect of different coolants on the machining efficiency.

three-hour period of micromachining with emulsified oil, some of the grits were mechanically removed, but no typical flat spots developed on any working grits. The most satisfactory results were obtained with triethanolamine as a coolant, with which no significant wear on the grits was developed even after six hours of machining.

These observations may be related to process efficiency, by comparing machining rates of diamond grits on BK-7 glass for these four evaluated coolants over an extended time of micromachining. Figure 5 shows that triethanolamine and emulsified oil are more effective than ethylene glycol in prevention of tool dulling, and the results also confirm the unsuitability of water for working as a coolant or slurry medium.

4.3.3 Disposal Rate of Dull Grits

The disposal rate of dull abrasive grits is a complex process, and depends greatly on the bond matrix in an abrasive tool. Hard bond matrices exhibit a slow rate of disposal of worn grits, thus increasing the possible dulling effect on a tool or the workpiece machined. A softer bond would be more susceptible to wear, that is, disposal of dull grits. However, if a bond is too soft, it is found to be insufficiently firm to hold sufficiently exposed grits on the tool surface. Consequently only grits with very small exposure are used in machining, resulting in a low material removal rate.

A commonly used expression for performance of an abrasive tool in terms of grit and bonding matrix is the "G ratio," defined as the volumetric ratio of material removal to tool wear. A high G ratio indicates low tool consumption. High G ratio for an abrasive tool is desirable, but a more wear-resistant tool may result in higher forces and energy, thereby increasing the likelihood of thermal damage to the workpiece. Disposal rate of the dull grits of a tool is closely related to the G ratio of the tool. Low disposal rate of the dull grits on a tool may result in inferior self-sharpening, which can cause high machining forces and poor workpiece surface quality.

In addition to the disposal rate of the dull grits, their friability and size also play a significant part. The greater their friability, the more grits become susceptible to grain fracture prior to final dislodgement from the bond matrix, which has a large bearing in restricting the size of the wear-flat areas initially dressed onto the grits and their subsequent growth by attrition. Without this mode of self-sharpening, the abrasive grits would become extremely dull. Similarly, abrasive grits that are too tough are generally not suitable for precision grinding operations [6].

4.3.4 Grit Size

The size of abrasive grits greatly influences tool micromachining efficiency. In loose abrasive machining, grit size is generally considered to affect directly the material removal rate and the final surface roughness of the workpiece. However, when optical lens surfaces are micromachined with diamond grits, the relationship between these factors is more complex. Machining tests show that the glass removal rate does not always rise with increasing grit size. If the diamond grits are too large, they prevent direct contact between the peaks of the glass surface texture and the bond matrix material. Consequently, the dull grits are not discarded at the appropriate stage, and the abrasive tool loses its machining efficiency. If the grits are too small for a given surface roughness, they are removed together with the bond, and consequently no properly exposed grits occupy the working surface of the abrasive tool. The result is poor working efficiency. Figures 6(a) and (b) show how the performance of an abrasive tool with approximately similar bonding material hardness changes with different diamond grit size. In these experiments, an initially sharp tool with an average grit size of 12 μm was used on two workpiece surfaces of, respectively, 2.2 μm rms and 1.2 μm rms. In the former, higher roughness, case some self-sharpening effect occurred. In Figure 6(a) the removal rate from the workpiece after initially rising, stabilizes, with little further appreciable change. On micromachining workpieces with an initial surface roughness of 1.2 μm rms, with the 12-μm-grit initial sharp

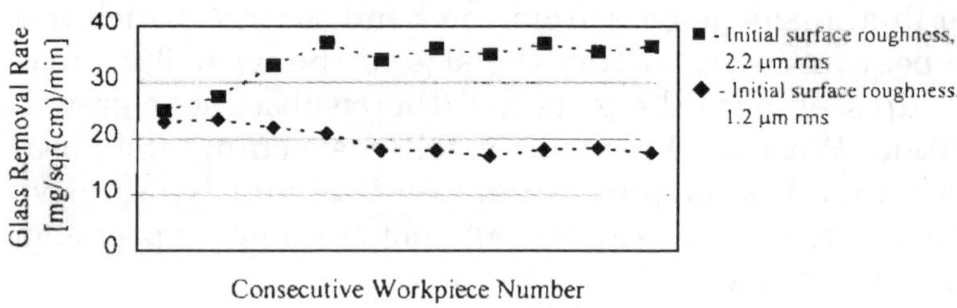

(a)

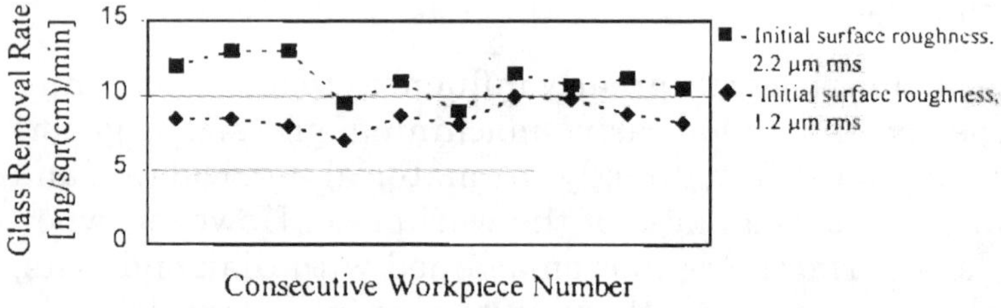

(b)

Figure 6 Variation in glass removal rate as function of number of workpieces machined: (a) tool bonding material hardness, Rockwell 62 B; grit size 12μm; (b) tool bonding material hardness, Rockwell 80 B; grit size 6.5 μm.

tool, a pronounced dulling process occurred even although the diamond grits' exposure was about 3.7 μm. The results in Figure 6(b) were obtained with an abrasive tool of 6.5-μm diamond grit size. The tool produced very low removal rates on workpieces with an initial surface roughness of 2.2 μm rms. However, workpieces with an initial surface roughness of 1.2 μm rms were ground at a slightly lower and constant rate, the initial sharpness of the tool being maintained. The 6.5-μm-grit tool with a diamond exposure of 2 μm was noted to micromachine more quickly than the smoother surfaces in comparison with the 12-μm-grit tool which had a diamond exposure of 3.7 μm. Thus for an initial smooth workpiece the tool with 6.5-μm grit is a more appropriate vehicle to use than that with 12-μm grit.

4.4 ACCURACY AND DIMENSIONAL CONTROL

Tighter tolerance, superior surface quality control, consistent part geometry, and lower cost are increasing requirements for machining magnetic and electronic materials such as ferrite ceramics, silicon and quartz wafers, and germanium crystals. The dimensional accuracy and surface quality of manufactured components or products can directly determine their performance. For instance, the frequency and quality of a resonator are determined by the geometry and surface quality of a quartz wafer: very thin wafers of uniform thickness and fine surface finish are needed to obtain high frequency with minimal frequency broadening [11]. Currently, the semiconductor industry is exploring the production of ultrahigh density one-GB DRAM which requires form and dimensional accuracy on the order of 0.1 μm over a substrate area of 20 mm by 20 mm. The manufacture of a component of such high accuracy from brittle materials like silicon and germanium is difficult to achieve by currently available techniques [12]. The production of the ground or lapped surface of a ferrite recording head for data storage poses similar difficulties in fabrication. The surface is found to have a shallow magnetically inactive layer which results from the residual stress produced in grinding or lapping. The surface of a ferrite head that is free of damage from machining is much needed for achieving better magnetic permeability [13]. Microgrinding and abrasive micromachining can be effective techniques to meet these needs, by precise control of the grain depth of cut, and avoidance of subsurface damage through ductile-regime grinding. Abrasive micromachining is mainly used for improving the functionally significant properties of the workpiece surface, combined with tighter dependable size control.

4.4.1 Control of Accuracy

Although the achievement of high precision workpieces or components requires careful attention to all aspects of machining processes, the key elements that particularly affect the workpiece accuracy may be identified:

1. Machine tools
2. Operational factors
3. Selection of abrasive tools
4. Workpiece materials.

Figure 7 illustrates major aspects associated with each of these four primary elements. Inadequate attention to any aspect can influence the workpiece machining accuracy in terms of its dimensions, tolerances, geometrical form, and surface quality. Their influence on the machining process through effects such as cutting, ploughing, and the frictional interaction with the resultant process variables of machining forces and power needs to be understood. These process interactions in turn influence eventual characteristics such as the production of components of required highly accurate dimensions, stringent geometrical shape, and superior surface texture and in-

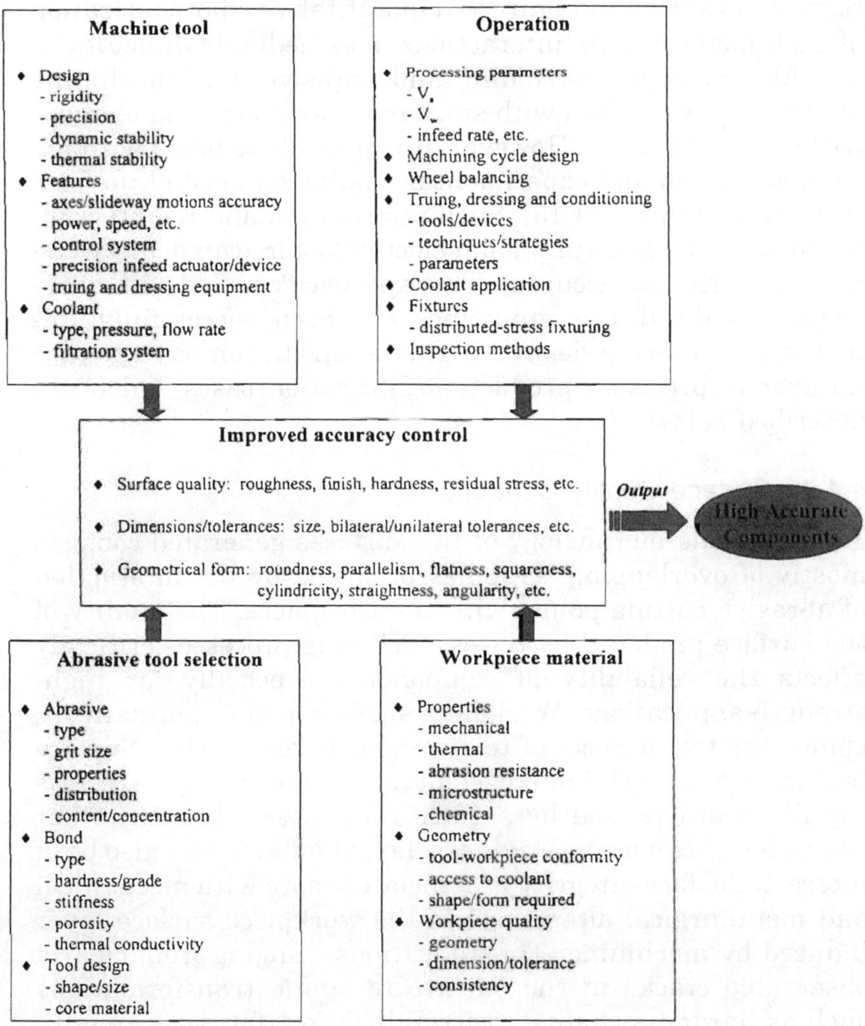

Figure 7 Selected features influencing microgrinding/abrasive micromachining processes.

tegrity. As shown in Figure 7, a quantitative characterization of each factor and its interactions is exceedingly difficult.

Moreover, microgrinding and abrasive micromachining processes have to deal with stochastic variations and changes within control limits. However, it is possible to begin with basic information and experimental results, on most of the process input factors [14,15]. Such information and results combined with the fine-tuning of selected factors can achieve the desired workpiece accuracy. This approach coupled with experience-based rules or guidelines has been successfully employed in achieving desired highly accurate components in a number of precision production/fabrication cases which are described below.

4.4.2 Surface Quality Control

The fine-scale morphology of the surfaces generated consists mostly of overlapping scratches produced by the interaction of abrasive cutting points with the workpiece. The quality of the surface produced by these machining processes critically affects the reliability of components especially for high-strength applications. Workpiece surface quality normally includes the two aspects of texture and integrity [16]. Surface texture refers to the microgeometry or topography, which is usually characterized by surface roughness, although other characteristics such as waviness, lay, and flaws may also be of interest. Surface integrity is associated more with mechanical and metallurgical alterations to the workpiece surface layer induced by machining. The alterations, ranging from clearly observable cracks in the surface to subtle transformations such as hardness change, recrystallization, fatigue strength, or residual stress in the underlying metallic structure, are caused mainly by the forces and excessive temperatures involved in the machining processes.

Surface Roughness

For a ground surface, arithmetic average roughness (R_a) and peak-to-valley roughness (R_t) are used to define quantitatively the surface profile: R_a is the average value of the vertical deviations from the nominal surface over a sampled surface length; R_t, also referred to as the total roughness, is defined as the difference in elevation between the highest peak and lowest valley in the sampled traversing length. The differences between R_t and R_a are considerably larger, which is the consequence of the broad distribution of peak heights and valley depths within the surface profile. For AMMG machined surfaces, the R_t roughness is typically 3 to 10 times R_a [17,18].

It is possible theoretically to predict an "ideal" surface roughness from an abrasive machining process by modeling how the abrasive cutting points on the rotating abrasive wheel kinematically interact with the workpiece. It is assumed that clean cutting, whereby the cutting edges remove all material encountered in their paths leaving behind the resulting cutting traces, generates the surface. From the geometry traced and the relationship between the cutting points and workpiece, the surface roughness of a ground surface can be obtained as [6]

$$R_t = \frac{1}{4}\left(\frac{V_w L}{V_s d_s^{0.5}}\right)^2 \tag{7}$$

and

$$R_a = \frac{1}{9\sqrt{3}}\left(\frac{V_w L}{V_s d_s^{0.5}}\right)^2 \tag{8}$$

where

V_w: Workpiece velocity,
V_s: Abrasive wheel velocity,
d_s: Wheel diameter,
L: Distance between successive cutting points.

The "ideal" theoretical surface roughnesses based on Equations (7) and (8) are usually found to be less than the actual values due to other influences such as material side-flow, built-up edge phenomena, vibrations during machining, and the random undefined topography of the wheel surface and abrasive cutting points. However, analyses of "ideal" surface roughness with Equations (7) and (8) provide useful insight into the generation of abrasively machined surfaces and the principal controlling factors. In real AMMG operations, these guidelines are useful in obtaining a fine workpiece surface finish, that is, lower surface roughness:

1. Selecting a small grit size and dense wheel structure,
2. Using higher wheel speeds and lower workpiece speeds,
3. Setting smaller depths of cut and feedrate,
4. Selecting grinding wheels, preferably with larger diameters,
5. Maintaining suitable dressing strategies and operations needed, and
6. Adopting appropriate coolants with favorable cooling, lubricating, and rinsing-off effects.

Surface Integrity

Surface roughness alone does not completely describe a surface profile. Metallurgical or other changes in the altered layer beneath the surface can have a significant effect on the workpiece performance. In AMMG, changes such as hardness, residual stress, microcracks, fatigue strength [19], structural flaws and defects (e.g., voids, dislocations) [20], laps, folds, seams, and the like in the subsurface layer may occur for these reasons:

1. Plastic deformation resulting from the point work of abrasive grits, or
2. Heat generation through cutting, changes of temperature, and its nonuniform distribution in the surface layer.

In comparison with surface roughness, surface integrity is more difficult to control, owing to the variety of factors involved, even although its attainment is a major reason for applying microgrinding and abrasive micromachining techniques. Even with the same roughness, ground and microground surfaces differ. A useful discussion on these differences has been provided by Stauffer [21]. He suggests that the considerable heat generated by grinding gives rise to a disturbed and amorphous layer, which is very unstable. Microgrinding, on the other hand, reaches the base layer of metal. (Stauffer draws an analogy that just as snow on a frozen lake surface cannot support a person's weight, although ice bears the concentrated pressure of ice skates, so the supporting qualities of a ground surface differ from those of a microground surface [21].) The following guidelines have proved useful in resolving these complex issues for the selection of appropriate processing conditions.

1. To minimize thermal damages and microcracks, the abrasive tool has to be frequently dressed.
2. Smaller depths of cut and feedrate, lower wheel speed (V_s), and faster work speed (V_w), are preferable.
3. The right coolants with favorable cooling, lubricating, and rinsing-off effects should be adopted.
4. For brittle materials such as silicon wafers, ferrite crystals, ceramics, and glasses, diamond abrasives ought to be chosen.
5. For hardened tool steels and some aerospace alloys, CBN abrasives are preferable.
6. For normal steel and most cast irons, aluminum oxide abrasives are recommended.
7. For most nonferrous metals, silicon carbide abrasives should be selected.

Many models have been developed for controlling the surface integrity of a ground surface [22]. For metals, they are mostly based on modelling on a macroscale the temperature gradients below the ground surface. It is obviously much more difficult to model the surface integrity at a micrometer, or even smaller, scale for nonmetallic materials such as ceramics, glasses, and silicon crystals, even though some researchers have been making considerable efforts in this area [23,24].

4.4.3 Dimension and Tolerance Control

Two critical features for microgrinding to required dimensional and shape accuracy are low subsurface damage and high figure accuracy [25]. Through microgrinding, it is possible to achieve economically precise, damage-free surfaces in a wide range of shapes for which traditional lapping and polishing are unsuitable. The accurate dimensions and tight tolerances of microground workpieces essentially depend on the high loop stiffness of the grinder and motion control between the abrasive tool and the workpiece, through which the trajectory of the multiple cutting points of the tool can be faithfully copied into the shape and dimensions of the workpiece. Nevertheless, the high loop stiffness of a grinder and motion control are also essential for maintaining critical depth of cut (d_c) in machining brittle materials, and thus ensuring that machining is taking place in the ductile rather than the brittle regime. All brittle materials will undergo plastic flow rather than fracture if the machining depth of cut is sufficiently small (i.e., equal to, or smaller than, the critical depth of cut). This feature determines the significance of ductile regime grinding in the control of both dimensions and surface quality since any fractures will affect both dimensions and mechanical strength of the workpiece as a whole product. The critical depth of cut of a workpiece is related to its material properties such as elastic modulus, hardness, and fracture energy. Microgrinding trials show that the critical depth of cut (at which the transition from brittle to ductile regime machining occurred) for glasses is approximately or less than 0.5 ptin [26]. Brittle materials cannot be machined precisely without surface damage at the brittle mode. To microgrind workpieces with accurate dimen-

Table 1 Recommendations for Achieving Accurate Dimensions and Tight Tolerances in Micromachining

• Work/wheel spindle stiffness:	Larger than 200 N/μm at both axial and journal directions
• Spindle error motions:	Less than 0.05 μm at both axial and journal directions
• Feed resolution:	$d_c/10$ or less
• Slideway straightness:	d_c/travel range
• Work/wheel vibration level:	Less than $d_c/10$
• Wheel trueing accuracy:	Less than d_c,
• Height distribution of cutting points:	Less than d_c,
• The provision of a continuous dressing device and full flood cooling.	

sions and tight tolerances, the conditions listed in Table 1 normally need to be maintained or controlled [8,26,27]. From Table 1, dimensional and tolerance control are noted to rely heavily on control of the accuracy of motion of machine tools, the geometry of the grinding wheel, and the height distribution of the cutting points on the wheel. Control is also dependent on the workpiece material, the effective grit size, and active grit concentration of the wheel, and the undeformed chip thickness.

Dimensions on machined workpieces are measured from surface to surface, and the roughness of each surface is often the main limitation on the dimensional tolerances which can be obtained in production. For example, the roughness of a finished surface can be considered as a measure of the uncertainty in the exact specification of the surface location; that is, the dimensional uncertainty depends upon the combined roughnesses at the measuring points at two end surfaces. For a component, it is therefore generally necessary to have smoother surfaces in order to maintain tighter tolerance control. Other factors affecting tolerance control include machine tool deflection, thermal expansion and distortion of the machine tool and workpiece, and abrasive wear. The dimensional tolerances specified in micromachining are normally 10 to 50 times the arithmetic average roughness (R_a) required [28,29].

The predominant advantage of microgrinding is combining in a single process the fine tolerance achievable in a deterministic grinding process with the superfinished surfaces obtained with nondeterministic polishing or lapping. Microgrinding has the capacity to replace the labor-intensive lapping and polishing, as it is developed, owing to its excellent capacity for tight tolerances and high surface quality control.

4.4.4 Geometrical Form Control

Both dimensional and form accuracy and tolerance (as shown in Table 1) are controlled by similar factors, since the form of a workpiece is generally specified in terms of linear, as well as angular, dimensions. Unlike polishing or lapping, microgrinding is a deterministic process, permitting finely controlled contour accuracy and complex shapes. Most significant form errors in microgrinding are usually caused by abrasive tool wear and the elastic deformation of grains or the wheel, especially in profile or form grinding of cross-sectional shapes with sharp radii or deep grooves. Better form control usually requires a slower-wearing abrasive tool with high contact stiffness such as a vitrified bonded superabrasive wheel even though this can be expected to cause larger machining forces. For instance, there is an increasing use of vitrified bonded CBN wheels in microgrinding of gears and other mechanical components. CBN microground gears are more precise in their geometry, and thus transmit motion or torque more efficiently with less noise [14]. Bronze-bonded diamond wheels are often used in microgrinding nonmetallic brittle materials, primarily owing to their superior wear resistance, which results in highly accurate components of precise geometry, close tolerances, and enhanced component performance, which includes longer fatigue life.

Loose abrasive micromachining can also be used to produce components with a highly accurate shape, although they are usually limited to basic flat, cylindrical, and spherical forms. Selective machining can improve the flatness and surface texture of a workpiece much beyond that obtainable on precision grinding machine tools. For instance, the technique is extensively used in micromachining gauge blocks which have highly accurate surface flatness, parallelism, thickness, and stability. The stability itself requires that the thickness change must be less than 0.02 μm over two years. This condition is achieved by micromachining, close attention also being paid to other processing phases, such as the selection of special materials and heat treatment, throughout the entire manufacturing process.

4.5 INDUSTRIAL APPLICATIONS AND MICROMACHINING EQUIPMENT

The main requirement for precision machining has focused on tighter tolerances, better surface quality, consistent part geometry, and lower total cost, for a variety of uses that include new engineering materials that are difficult to machine. The pressure on manufacturers for tight tolerances and process consistency has drawn close attention to microgrinding and abrasive micromachining and their capabilities. These techniques continue to grow although they remain a small part of machining technology. As shown in Figure 8, however, the techniques are used in machining a variety of engineering materials for electronic, automotive, and optical applications, as well as others. Besides enabling damage-free surface quality, tighter dimensional and tolerance control, and higher geometrical form accuracy, the techniques have additional advantages that distinguish them from conventional grinding, lapping, and polishing:

1. Provision of controlled, predictable machining for difficult, brittle materials,
2. Creation of improved bearing ratios between mating components, either for increased load carrying or improved sealing,

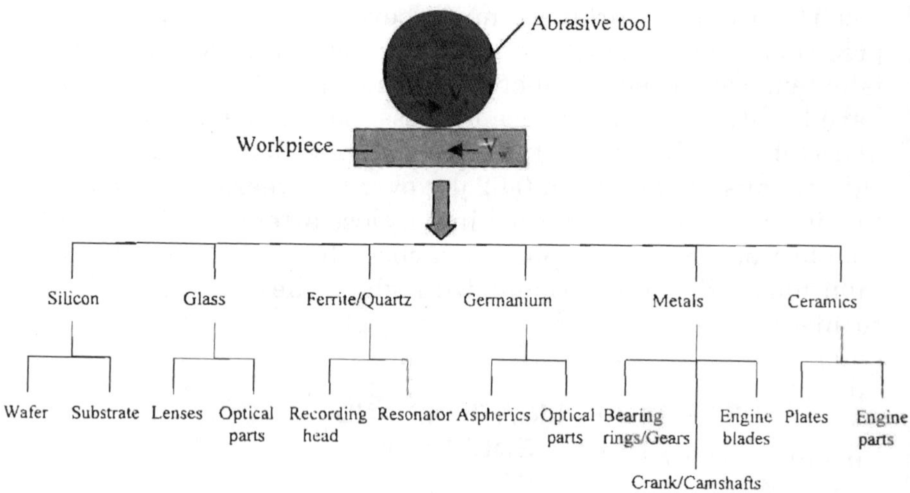

Figure 8 Typical microgrinding and abrasive micromachining applications for a variety of engineering materials.

3. Ability to impart desirable compressive stresses in finished surfaces,
4. Provision of predictable surface finish patterns to meet specific design criteria for wear, sealing, or lubrication, and
5. Broadening the range of machineable materials.

Microgrinding and abrasive micromachining are promising techniques for any potential engineering materials that require machining to high precision, as may be illustrated from the applications described below.

4.5.1 Silicon Wafers

Silicon is the dominant material throughout the semiconductor industry, silicon wafers being used extensively for the microfabrication of integrated circuits (IC), microsensors, and microactuator devices. The wafers must have a high degree of size and flatness accuracy as well as possessing a damage-free surface finish, critical for the manufacture of the high-quality electronic devices listed above. Nevertheless, the larger the diameter of the wafer, the more integrated circuit devices that can be simultaneously produced. The semiconductor industry is continually being moved towards increasingly larger wafer sizes [30]. Figure 9 illustrates a typical fabrication process for IC devices. The machining of silicon wafers is noted to be an essential phase in the fabrication process. Microgrinding is being increasingly used in the direct production of wafers, instead of the process of grinding–lapping–polishing since microgrinding is economically attractive, and offers higher dimensional accuracy (wafer thickness) and flatness control, and can yield a damage-free surface finish of high quality. For instance, microgrinding was successfully used to produce wafers 125 mm in diameter under these conditions [31]:

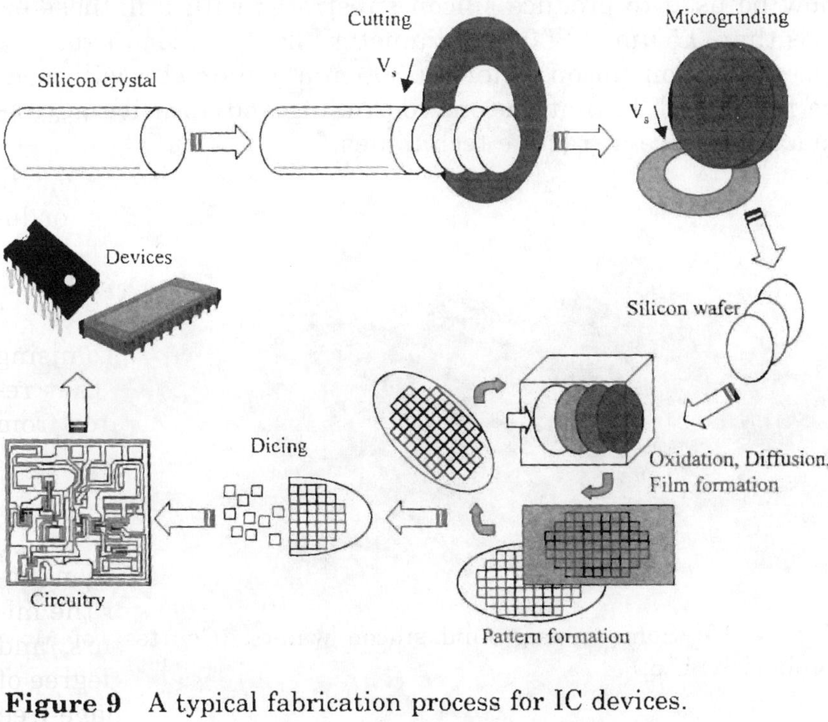

Figure 9 A typical fabrication process for IC devices.

Wafer thickness: 100 to 300 μm
Grinding wheel: diamond cup wheel
Wheel speed: 3000 rpm
Infeed rate: 0.05 mm/min
Spark-out time: 0 s
Coolant: water
Depth of cut: 0.1 μm
Production time: approx. 30 s

The flatness and parallelism of the wafers produced were less than 2 μm, a condition which is difficult to achieve with the conventional production technique of grinding–lapping–polishing. Surface damage, such as scratches and chipping, can be completely avoided with microgrinding. Figure 10 shows microground silicon wafers of various sizes. Microgrinding can now be used to produce silicon substrates with a flatness of less than 0.5 μm in 200-mm diameter [32]. Semiconductor devices based on silicon wafers at 300 mm in diameter will soon be fabricated: to that end microgrinding and abrasive micromachining are attractive techniques.

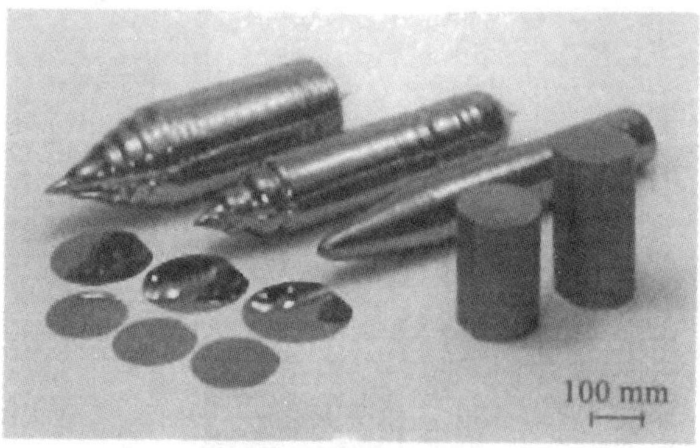

Figure 10 Some microground silicon wafers. (Courtesy of Motorola (UK) Ltd.)

4.5.2 Turbine Engine Blades

Vibratory finishing is a well-known method of surface treatment, in which the abrasive micromachining effect of loose media is maintained in motion by the oscillation of the container which holds both the media and components. The technique has been widely employed for final surface finishing (up to surface roughness $R_a = 0.01$ μm) of products such as gas turbine engine blades, vanes, nozzles, and automobile-door handles, owing to its characteristic finishing consistency with considerably lower manufacturing cost.

Figure 11 shows a CF-6 gas turbine engine blade micromachined in vibratory finishing equipment. The initial blade surface with roughness of 0.2 μm (R_a) exhibits grooves with directional marks left by a previous grinding operation. After 45-min vibratory finishing, the grooves and directional machining marks were completely removed. The final surface roughness was 0.06 μm (R_a). During finishing, the blades move

Figure 11 A turbine engine blade micromachined in a vibratory finishing machine. (Courtesy of GE Caledonian Engines Ltd.)

freely, albeit at different speeds, with the vibratory media. Both the media and blades move in a micro- and macro-orbital motion. As the blades are much heavier than the media, the elements of the latter accelerate faster. The resulting relative movement of the mutually contacting particles of the load produces a finishing action by abrasive microinteraction—mainly abrasive rubbing. The blade surface finish plays a crucial part in the engine performance. Vibratory finishing is widely employed in engine blade production and also for restoration of surface finish during engine maintenance.

Most metals and alloys, ceramics, glass, and plastic parts can be processed by vibratory finishing. Improvement of the surface finish and abrasive microoperations such as deburring, descaling, and radiusing are common objectives. Vibratory finishing can be controlled to yield suitable workpiece surface roughness and stock removal rates, essential for high-precision machining [33]. Figure 12 shows a typical bowl-type vibratory finishing machine of 1000 kg capacity.

Figure 12 A typical bowl-type vibratory finishing machine (1000-kg capacity). (Courtesy of GE Caledonian Engines Ltd.)

Other metal parts, such as camshafts, crankshafts, bearing inner and outer rings, and pinion gears, can be processed by abrasive micromachining on commercially available specialized abrasive finishing equipment [21]. These components usually require improved durability, relying on surfaces of low coefficient of friction, and well-specified geometry, in the attainment of which abrasive micromachining can be an effective technique. For example, abrasive micromachining can result in significant improvement in cross-track geometry and surface quality, the roundness of the bearing inner and outer rings playing a key role for the bearing raceways and consequently in performance.

4.5.3 Glass Lenses

The production of highly aspheric glass lenses can only be performed by ductile grinding, particularly. Moreover, the requirements for the shape accuracy, surface quality, and absence of subsurface damage are much more stringent when aspheric lenses are produced for shorter wavelength applications [34].

Since the early 1990s, the Center for Optics Manufacturing (COM) at the University of Rochester, in collaboration with industrial partners, has developed deterministic microgrinding processes and associated manufacturing facilities that can generate high-quality optical glass surfaces with fine surface figure accuracy [25]. COM has developed Opticam™ machining centers for the fabrication of precision spherical and prismatic optical components through deterministic microgrinding. The first prototype, Opticam SM™ produced by Rank Pneumo, is a multiaxis CNC machining center with a horizontal tool spindle and horizontal workpiece holder. Both the tool and workpiece rotate about their axes, respectively, at about 20,000 and 250 rpm. A spherical concave, or convex, surface is generated in the glass workpiece by positioning the tool spindle axis at an angle to that of the workpiece. Another model, the Opticam SX™ produced by CNC Systems (USA), has been developed based on the Opticam SM™ design. The Opticam SX™ has a vertical workpiece holder; its tool spindle is vertical for planar surfaces, and rotated at an angle to the vertical for manufacture of spherical surfaces. Both models operate in a similar manner, and their tools can be interchanged. On an Opticam™ machine, specular microground surfaces, with less than 1–2-μm subsurface damage and 1-wave peak-to-valley (p–v) figure, can be obtained in five minutes. This notable optics manufacturing process cannot be achieved on conventional processing equipment. Another commercially available ultraprecision grinder is the ASG2500 CNC grinding machine tool manufactured by Rank Taylor Hobson (USA), employed in microgrinding aspheric lenses and other precision optical components.

Texas Instruments has successfully used Opticam™ machines for deterministically microgrinding 10, 47-mm diameter planoconvex hyperboloidal aspheric condenser lenses to a required geometry. A diamond ring tool, bronze-bonded with 15-μm diamond grits, was used to contour the asphere, stepping across the workpiece from its edge to center. The parts had an aspheric deviation of 145 μm from the "best-fit" sphere, with a final figure requirement of approximately 1 μm p–v. These lenses were tested and passed all user requirements, demonstrating the effectiveness of microgrinding [35].

In Europe and Japan, deterministic microgrinding of optical glasses on ultraprecision CNC machine tools has also been developed [8,26,36]. These requirements for microgrinding machines and processes were proposed:

1. A machine tool with high loop stiffness (static and dynamic) between grinding wheel and workpiece to the order of 200 N/μm.
2. The ability to control the motion of the machine tool to ensure that the depth of cut, or undeformed chip thickness, is maintained at the desired value based on the high feed accuracy/resolution to the order of

10 nm. Piezoelectric actuators are normally used to fulfill this capacity [37].

3. The use of grinding wheels, with rigid bonds with the capability to true the cutting edges in situ to a degree of uniformity on the order of 10 nm.
4. The use of continuous dressing techniques to maintain the cutting ability of the grinding wheel and avoid the gross variations in the cutting process which can occur in grinding.
5. The provision of full flood cooling and contained temperature control.

Clearly, the machine tools and processes developed for micromachining silicon and optical glasses can be employed in other advanced engineering materials, such as ceramics, germanium, ferrite, and quartz crystals, owing to the similarity in their machining characteristics and material properties. The use of these materials depends on the achievement of precise, damage-free surfaces in shapes which are difficult to achieve by conventional lapping and polishing. An increasing number of applications for microgrinding and abrasive micromachining for components such as quartz wafers for computer and clock resonators, ferrite parts for electronic data recording equipment, ceramic parts for engines or bearings, germanium aspherics, and infrared windows for optics applications have been cited [11,38,39]. Microgrinding and abrasive micromachining are finding increasingly more industrial applications for machining these particular engineering and electronic materials.

ACKNOWLEDGMENTS

The author particularly thanks Anthony McAdam of GE Caledonian Engines Ltd. and Be Lin of Motorola (UK) Ltd. for their assistance with writing this chapter.

NOTATION AND SYMBOLS

L:	Grit size
θ_a:	Umbling angle
μ:	Nominal coefficient of friction
a_p:	Undeformed chip thickness
ρ:	Radius at the cutting tool or grit tip
ψ:	Effective rake angle of the tool or grit
F_f/F_n:	Friction coeffection
C_R:	Coefficient of proportionality (cm^2/N)
P:	Total load normal to the workpiece surface (N)
ρ_d:	Material density of the workpiece (g/cm^3)
V:	Machining velocity (cm/s)
V_w:	Workpiece velocity
V_s:	Abrasive wheel velocity
d_s:	Wheel diameter
L:	Distance between successive cutting points

REFERENCES

[1] D. Golini, W.J. Rupp, and J. Zimmerman, Microgrinding: New techniques for rapid fabrication of large mirrors. *Proc. International Society for Optical Engineering*, (1113), 204–210 (1989).

[2] K. Phillips, G.M. Crimes, and T.R. Wilshaw, On the mechanism of material removal by free abrasive grinding of glass and fused silica. *Wear*, (41), 327–350 (1977).

[3] D. Golini and S.D. Jacobs, Physics of loose abrasive microgrinding. *Applied Optics*, (30), 2761–2777 (1991).

[4] M.C. Shaw, Precision finishing. *Annals of the CIRP*, (44) 1, 343–348 (1995).

[5] K. Cheng, Q.P. Wang, and M. Zhou, The minimum cutting depth in diamond turning. *Mechanical Design & Manufacture*, (30) 5, 20–26 (1989).

[6] S. Malkin, *Grinding Technology—Theory and Applications of Machining with Abrasives*. Ellis Horwood, New York (1989).

[7] X. Chen and W.B. Rowe, Analysis and simulation of the grinding process—Part II: Mechanics of grinding. *Int. J. Machine Tools and Manufacture*, (36), 883–896 (1996).

[8] M. Miyashita, The way to nanogrinding technology. *Proc. SPIE: Advanced Optical Manufacturing and Testing*, (1333) 7–21 (1990).

[9] M.C. Shaw and R. Komanduri, Attritious wear of silicon carbide. *ASME Transactions: J. Engineering for Industry*, (98), 1125–1136 (1976).

[10] W.J. Rupp, Mechanism of the diamond lapping process. *Applied Optics*, (13), 1264–1269 (1974).

[11] T.G. Bifano and J.B. Hosler, Precision grinding of ultra-thin quartz wafers. *ASME Transactions: J. Engineering for Industry*, (115), 258–262 (1993).

[12] B.C. Crandall (Editor), *Nanotechnology—Molecular Speculations on Global Abundance*. MIT Press, Cambridge, MA (1996).

[13] S. Chandrasekar, M.C. Shaw, and B. Bhushan, Comparison of grinding and lapping of ferrites and metals. *ASME Transactions: J. Engineering for Industry*, (109), 76–82 (1987).

[14] K. Subramanian and R.P. Lindsay, A systems approach for the use of vitrified bonded superabrasive wheels for precision production grinding. *ASME Transactions: J. Engineering for Industry*, (114), 41–52 (1992).

[15] J.A. Borkowski and A.M. Szymanski, *Use of Abrasives and Abrasive Tools*. Ellis Horwood, Chichester (1992).

[16] M.P. Groover, *Fundamentals of Modem Manufacturing—Materials, Processes, and Systems*. Prentice-Hall, NJ (1996).

[17] P.P. Hed and D.E. Edwards, Optical glass fabrication technology: Relationship between surface roughness and subsurface damage. *Applied Optics*, (26), 4677–4680 (1987).

[18] J.C. Lambropoulos, T. Fang, P.D. Funkenbusch, S.D. Jacobs, M.J. Cumbo, and D. Golini, Surface microroughness of optical glasses under deterministic microgrinding. *Applied Optics*, (35), 4448–4462 (1996).

[19] H.K. Tonshoff and E. Hetz, Influence of the abrasive on fatigue in precision grinding. *ASME Transactions: J. Engineering for Industry*, (109), 203–205 (1987).

[20] R. Komanduri, On material removal mechanisms in finishing of advanced ceramics and glasses. *Annals of the CIRP*, (45) 1, 509–514 (1996).

[21] R.N. Stauffer, Taking finish and geometry one step further. *Manufacturing Engineering*, February, 31–33 (1990).

[22] H.K. Tonshoff, J. Peters, and I. Inasaki, Modelling and simulation of grinding processes. *Annals of the CIRP*, (41) 2, 677–688 (1992).

[23] R. Rentsch and I. Inasaki, Molecular dynamics simulation for abrasive processes. *Annals of the CIRP*, (43) 1, 327–330 (1994).

[24] R. Rentsch and I. Inasaki, Investigation of surface integrity by molecular dynamics simulation. *Annals of the CIRP*, (44) 1, 295–298 (1995).

[25] D. Golini and W. Czajkowski, Microgrinding makes ultrasmooth optics fast. *Laser Focus World*, July, 146–150 (1992).

[26] W.I. Wills-Moren, K. Carlisle, P.A. McKeown, and P. Shore, Ductile regime grinding of glass and other brittle materials by the use of ultra-stiff machine tools. *Proc. SPIE Advanced Optical Manufacturing and Testing*, (1333), 126–135 (1990).

[27] T.G. Bifano, T.A. Dow, and R.O. Scattergood, Ductile-regime grinding: A new technology for machining brittle materials. *ASME Transactions: J. Engineering for Industry*, (113) 184–189 (1991).

[28] H. Nakazawa, *Principles of Precision Engineering*. Oxford University Press, Oxford (1994).

[29] T. Masuzawa and H.K. Tonshoff, Three-dimensional micromachining by machine tools. *Annals of the CIRP*, (46) 2, 621–628 (1997).

[30] R.C. Jaeger, *Introduction to Microelectronic Fabrication*. Addison-Wesley, Reading, MA (1990).

[31] S. Matsui and T. Horiuchi, Parallelism improvement of ground silicon wafers. *ASME Transactions: J. Engineering for Industry*, (113), 25–28 (1991).

[32] M. Nakao and Y. Hatamura, Development of an intelligent face grinding machine to fabricate ultraflat surfaces on thin, brittle substrates. *Annals of the CIRP*, (45) 1, 397–400 (1996).

[33] E. Hashimoto, Modelling and optimisation of vibratory finishing process. *Annals of the CIRP*, (45) 1, 303–306 (1996).

[34] M.G. Schinker, Subsurface damage mechanisms at high-speed ductile machining of optical glasses. *Precision Engineering*, (13), 208–218 (1991).

[35] D. Golini, S.D. Jacobs, and W. Kordonsky, Fabrication of glass aspheres using deterministic microgrinding and magneto-rheological finishing. *Proc. SPIE*, (2536), 208–211 (1995).

[36] Y. Takeuchi and T. Sata, Ultraprecision 31) micromachining of glass. *Annals of the CIRP*, (45) 1, 401–404 (1996).

[37] J.D. Kim and S.R. Nam, A piezoelectrically driven micro-positioning system for the ductile-mode grinding of brittle materials. *J. Materials Processing Technology*, (61), 309–319 (1996).

[38] S. Malkin and T.W. Huang, Grinding mechanisms for ceramics. *Annals of the CIRP*, (45)2, 569–580 (1996).

[39] W.S. Blackley and R.O. Scattergood, Chip topography for ductile-regime machining of germanium. *ASME Transactions: J. Engineering for Industry*, (116), 263–266 (1994).

5

Diamond Micromachining

John Corbett
Cranfield University, Bedford, England

5.1 INTRODUCTION

Higher accuracy and performance requirements, coupled with demands to reduce costs, has led to significant developments in advanced CNC diamond turning and grinding machines. A long-term manufacturing trend in which tolerances for many strategic products are decreasing by a factor of 3 every 10 years, on critical dimensions, was highlighted in a USA report [1]. Figure 1 illustrates this trend and the report predicted that between 1980 and 2000, and beyond, geometric tolerances for the most demanding "precision" applications, for example, diamond turning and grinding, would decrease from 0.075 μm to 0.01 μm. The tolerances for the ultraprecision ap-

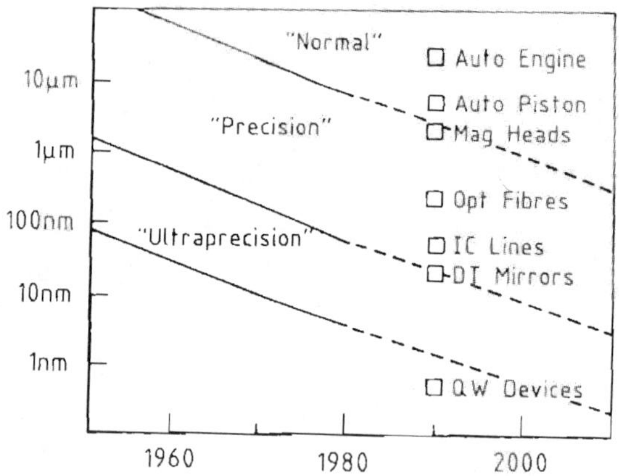

Figure 1 Tolerance trends in manufacturing. (From Ref. 1.)

plications, typically produced by energy beam machining processes, had reached around 0.001 μm (1nm) by the year 2000.

The use of diamond cutting tools has increased in importance as tighter tolerances and greater surface integrities are required for high-value components. Ultraprecision cutting tools need to be hard and sharp and to have enhanced thermal properties in order to maintain their size and shape while cutting. Advantages offered by diamond include:

- Crystalline structure, which enables very sharp cutting edges to be produced,
- High thermal conductivity, the highest of any materials at room temperature,
- Ability to retain high strength at high temperatures,
- High elastic and shear moduli, which reduce deformation during machining.

The earliest documented evidence of diamond machining found to date describes the diamond turning carried out by Jesse Ramsden, FRS in 1779 [2]. Ramsden machined a screw from hardened and tempered steel, with a diamond pointed tool, for use in his linear dividing engine for precision scale making. Diamond is, however, chemically attacked by ferrous materials at high temperatures, and is generally unsuitable for the machining of steels and nickel alloys. This is because of the very high wear rate of the diamond which results in nonviable tool costs.

More recently diamond machining has been used for the machining of nonferrous metals such as aluminum and copper, which are difficult materials on which to obtain a mirror surface by grinding, lapping, or polishing. This is because these metals are relatively soft and the abrasive processes scratch the finished surface and, furthermore, are unable to produce high levels of flatness at the edges of the machined surface [3]. However, diamond grinding has become an important process for the machining of brittle materials, for example, glasses and ceramics. The ability to control precisely the cutting tool position relative to the workpiece is a significant advantage offered by advanced CNC diamond turning and grinding machines. This enables them to produce components that are extremely precise and accurate. On the other hand, the relative position of the tool and workpiece is "force"-controlled with lapping and polishing. This makes it very difficult to obtain precise control of the tool's path for shapes other than simple geometric forms. Diamond machining is therefore proving to be a cost-effective process for the production of complex shaped components that have high accuracy requirements for form and/or surface finish.

Diamond micromachining is of particular interest for the optical and electronic industries. The processes are capable of simultaneously achieving high-profile accuracy, good surface finish, and low subsurface damage in brittle materials needed, for example, for semiconductors, magnetic read–write heads, and optical components.

5.2 MACHINING PRINCIPLES

Single-point diamond turning and ultraprecision diamond grinding are both capable of producing extremely fine cuts and

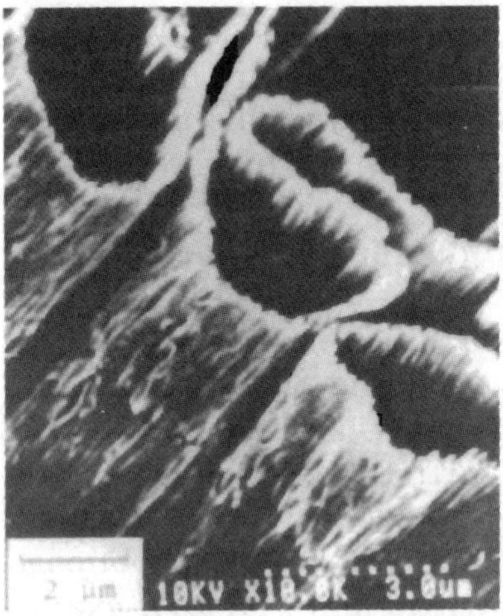

Figure 2 Scanning electron micrograph of electroplated copper, diamond turned at a depth of around 1 nm. (From Ref. 4.)

small chips. Figure 2 shows a scanning electron micrograph of electroplated copper cut by a sharp diamond on an ultraprecision machine tool. The undeformed chip thickness is approximately 1 nm [4]. Because of the very fine chip thickness produced by the microcutting processes, the chip-forming model for turning [5] is different from that for grinding [6], moving from concentrated shear to microextrusion as illustrated in Figure 3.

Important characteristics of materials considered for diamond micromachining are impurities (inclusions) in the material, grain boundaries of polycrystalline materials, and inhomogeneities. These can cause small vibrations of the cutting tool, resulting in a deterioration in surface finish. Another factor affecting the quality of surface finish as well as consistency of form is the high coefficient of expansion coupled with low thermal conductivity of some plastics which are diamond turned. These thermal effects are, to some extent, minimized when cutting with a diamond tool due to its sharp cutting edge, low coefficient of friction, and high thermal conductivity

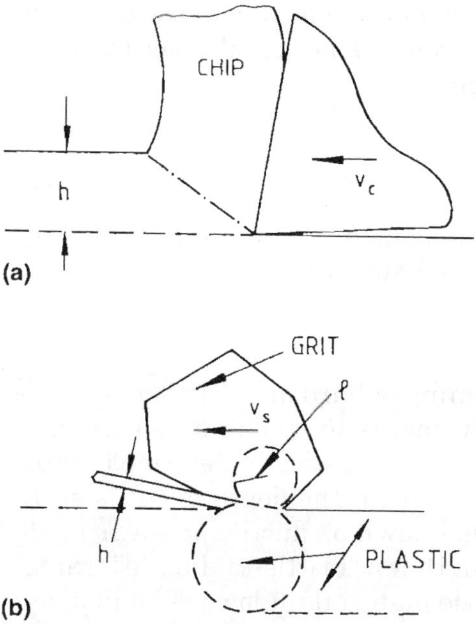

Figure 3 (a) Cutting concentrated shear model (from Ref. 5); (b) fine grinding microextrusion model (from Ref. 6).

which conducts the heat away. The theoretical peak-to-valley surface roughness which can be achieved by diamond turning using a round-nosed cutting tool is limited to [7]:

$$R_t \approx \frac{f^2}{8^*T_r} \tag{1}$$

where

R_t: Theoretical peak to surface roughness (mm),
f: Feedrate per revolution of the workspindle (mm.rev^{-1}),
T_r: Tool nose radius (mm).

However, this equation ignores any of the errors inherent in the machine, and a more accurate equation takes account of the asynchronous error motion of the machine in the direction normal to the component surface. The actual peak-to-valley surface roughness now becomes:

$$R_t \approx \frac{f^2}{8^*T_r} + f\,(Esyn) \tag{2}$$

where *Esyn* is the asynchronous error motion (mm) in the direction normal to the machined surface.

5.2.1 Brittle Materials

The need for the micromachining of hard and brittle materials has led to significant improvements in machine tool technology. It is now possible to produce plastically deformed chips, when machining brittle materials, if the depth of cut is sufficiently small. This process is known as ductile or shear mode machining. It has been shown that a "brittle-to-ductile" transition exists when cutting brittle materials at low load and penetration levels [8]. This "ductile" mode machining is important for the cost-effective production of high-performance optical and advanced ceramic components, with extremely low levels of subsurface damage (microcracking). This enhances their performance and strength significantly and eliminates, or minimizes, the need for postpolishing.

The transition from ductile to brittle fracture has been widely reported and is usually described as the "critical depth of cut." This is generally small (i.e., 0.1 to 0.3 μm), as is the associated feedrate, and this results in relatively slow material removal rates. However it is a cost-effective technique for producing high quality spherical and nonspherical optical surfaces, without the need for polishing.

Figure 4 shows the machining model for turning or fly cutting a brittle material [9], in which the depth of cut and critical chip thickness (*dc*) are illustrated. The location of the critical chip thickness is dependent on feed. For example, it is located towards the upper edge of the shoulder when fine feedrates (f) are used and the microfracture damage zone is

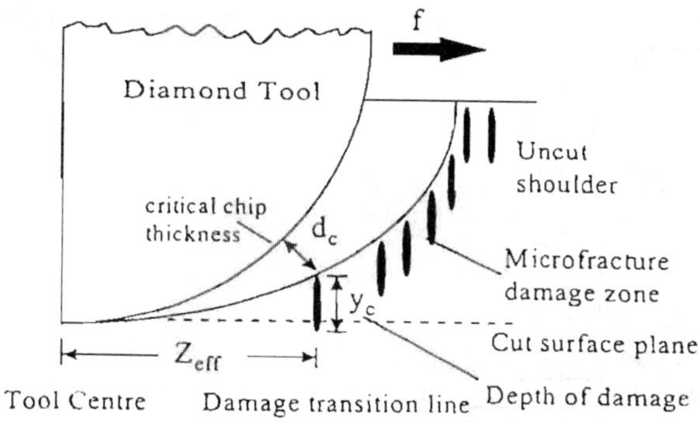

Figure 4 Cutting model for the brittle/ductile regime diamond turning of brittle materials. (From Ref. 9.)

removed during machining. In this case subsurface damage does not extend into the cut surface. However, if the feedrate increases, the critical chip thickness moves down towards the cut surface and this results in the microfracture damage penetrating into the final cut surface. In micromachining it is normally important to ensure that these cracks do not occur by removing the material in a ductile mode.

The mechanism for ductile mode grinding of brittle materials is shown in Figure 5. The process requires careful selection of the machining parameters in order to maximize the material removal rate while maintaining high surface and subsurface integrity. It also demands high-precision, high stiffness machine tools with smooth motions.

5.2.2 Grinding Wheel Conditioning

Extremely fine grit sizes, less than 4 μm (#4000 mesh size), are necessary for predominately ductile grinding. However, fixed abrasive grinding wheels with very small grits can be inefficient through loading, in which the grinding wheel surface becomes filled or clogged with the swarf or chips of the workpiece material. In the past this required frequent dressing of the wheel to produce a new abrasive surface, by removing the

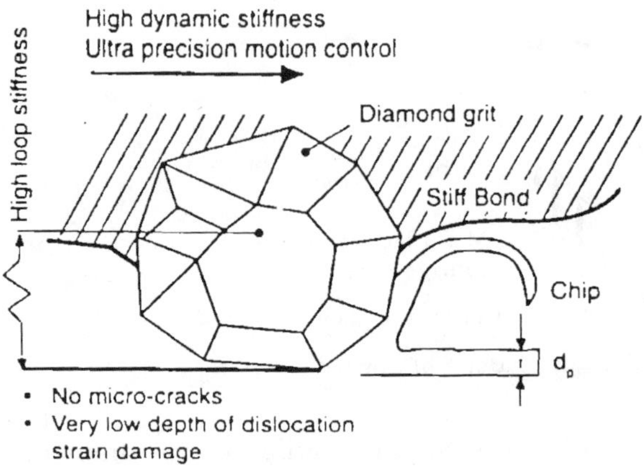

Figure 5 Mechanism of ductile or shear mode grinding of brittle materials.

clogged or loaded layer which reduced the machining efficiency considerably. Loading of the grinding wheel results in large differences in the grinding force level, from one cut to the next, resulting in the process becoming unstable. A new "electrolytic' in-process dressing" (ELID) technique [10] has now been developed to overcome this loading problem. ELID was developed commercially following extensive research by Nakagawa, at Tokyo University, and more recently by Ohmori at the Japanese Physical and Chemical Research Institute, Riken, Tokyo. The technique uses normal water-based coolants as the electrolyte. A pulsed current is generated by directing the coolant through a negative electrode and a positive connection, which is a conducting metal bond grinding wheel. ELID electroetches the metal bond of the grinding wheel continuously, in a highly controlled manner to unclog the wheel and enhance tool life. The continuous nature of the process, which is shown schematically in Figure 6, ensures that a constant fine grain is maintained. Grinding forces are thus kept constant, through the prevention of wheel loading, which results in long wheel life between wheel trueing operations.

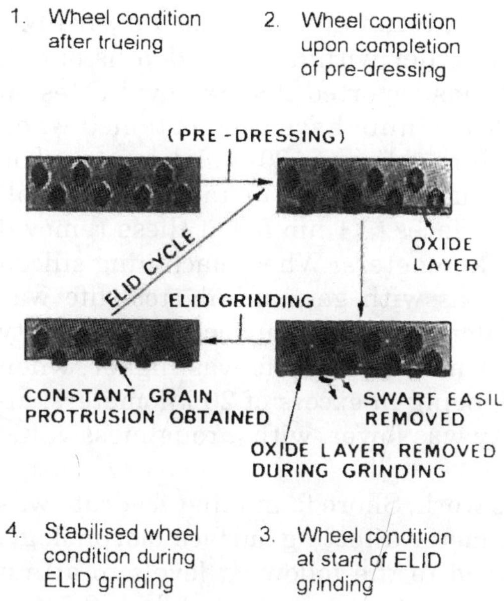

Figure 6 The electrolytic in-process dressing techniques. (From Ref. 10.)

5.3 RATES OF MICROMACHINING FOR RELEVANT MATERIALS

Ductile mode machining is required when machining mirror-like surfaces in hard and brittle materials. However, in order to achieve this condition the actual depths of cut required, to avoid crack generation, can be on the order of 0.1 to 0.01 of those used for cutting "mirror" surfaces in metals.

5.3.1 Diamond Turning

Gerchman and McLain have published the results of early work on the machining of germanium [11] in which they diamond-turned germanium to a surface roughness of 5 to 6 nm R_a. These were spherical surfaces, 50 mm in diameter, for which the removal rate was given in terms of 2.5 μm per revolution of the workpiece together with a 25-μm depths of cut. More recently Shore [12] has reported that removal rates on the order of 2 to 4 mm^3 per minute have been obtained when diamond turning germanium optics of 100-mm diameter. The tool life (expressed as the useful cutting distance of the tool) when producing optical surfaces (<1 nm R_a) at these removal rates was in excess of 12 kilometers. When machining silicon at similar removal rates, as with germanium, tool life was found to be less than 8 kilometers. The surface finish quality was also on the order of 1 nm R_a. Tool life was higher, when machining zinc sulphide, being in excess of 20 kilometers, although the surface quality was lower, with a roughness value of 3.6 nm Ra.

In line with previous work, Shore found that feedrate was the most significant parameter affecting surface morphology. Feedrate had to be reduced to the following levels to attain surfaces free from surface microfracture: zinc sulphide 2.5 μm/rev, silicon 1.5 μm/rev, and germanium 1.2 μm/rev. Shore used 0.5-mm radius tools, with top negative rake angles of $-15°$ for zinc sulphide and $-25°$ for silicon and germanium. The depth of cut was 10 μm and the workholding spindle speed 1000 rpm for each material.

Some typical diamond turning parameters for other materials are given in Table 1 [7].

5.3.2 Diamond Grinding

When diamond microturning a large area of brittle material (e.g., optical devices) the continuous use of a single point tool can result in major problems if it is found necessary to change the cutting tool when partway through a cut. A grinding wheel, however, has innumerable cutting points (grits) which yield a higher machining rate. Diamond microgrinding can therefore be expected to improve the commercial viability for the ductile mode machining of brittle materials.

Table 1 Typical Diamond Turning Parameters [7]

			Roughing		Finishing	
Material	V^a(rpm)	Coolant	D^b (mm)	F^c (mm/min)	D^b (mm)	F^c (mm/min)
Aluminum	800	Light oil	0.05	12 to 25	0.0025	2.5
Copper	800	Light oil	0.05	12 to 25	0.0025	2.5
Electroless nickel	400	Light oil	0.007	5	0.0012	2.5
Plastics (PMMA)[d]	1000	Air/Oil	0.25	15	0.012	3.5

[a] V = work holder spindle speed.
[b] D = depth of cut.
[c] F = feedrate.
[d] PMMA = Polymethylmethacrylate.

Examples of grinding parameters used for the ductile mode grinding of a range of ceramics were [12]:

Grinding wheel speed: 30 to 60 meters per second
Workpiece speed: 0.1 to 1.0 meters per minute
Depth of cut: 0.001 to 0.01 millimeters
Specific removal rate: 0.05 to 0.2 $mm^3/(mm\text{-}s)$
Total power: <1 kW

While grinding is a multipoint process that relies on mechanical actions, it has been shown [13] that chemical effects also play a significant role in material removal rates when using micron-size abrasive grits to grind glasses in a ductile mode. A relatively soft hydrated layer is formed on the glass surface via the chemical reaction between the coolant and the glass.

For the ductile mode grinding of optical glasses and zerodur, Shore [12] obtained material removal rates of 0.75 to 1.55 mm^3 per minute, when normalized for a 100-mm diameter optical component. This value was obtained when producing surface roughnesses of 1 to 3 nm R_a, which are close to what can be achieved by the polishing process. A possible technique to obtain higher removal rates, when ductile grinding, is to utilize very high grinding wheel speeds. These should, theoretically, reduce the undeformed chip thickness, and thus the cutting force per grit, resulting in more ductile flow coupled with less strength degradation [14].

5.4 ACCURACY AND DIMENSIONAL CONTROL

Diamond micromachining is used to produce either:

1. Small workpiece features, by means of tools with cutting features below 100 μm, or
2. Sub-μm or nanometric tolerances and/or surface finishes on macrocomponents.

Very sharp-edged diamond tools have been used in the production of ultrafine optical gratings with an accuracy of 1 nm, and gratings with 1-nm resolution can now be obtained for use on ultraprecision machine tools [15].

The most accurate diamond turning and grinding machines currently available are capable of achieving geometric accuracies of size and profile on the order of 100 nm for dimensions of 250 mm (see Section 5.5). Surfaces of 0.8 nm R_a have also been diamond-machined on several materials, including germanium. Another example of sub-μm tolerances being achieved on macro components is given by Taniguchi [3] who indicates that on the diamond grinding of silicon wafers, as opposed to lapping, a total flatness (TTV) of less than 0.6 μm was obtained on 150-mm diameter wafers.

Diamond micromachining demands extremely smooth movements, particularly between the spindle and the tool. In order to achieve this, hydrostatic oil and air bearings are generally required for the spindles and guideways. Other stringent requirements required from the machine tool are:

1. Extremely high loop stiffness between the tool and workpiece,
2. The ability to apply and maintain very small depths of cut, as low as a few nm in some cases,
3. Low thermal drift, and
4. The ability to operate at uniform feedrates over a wide range.

Other aspects to be considered during the design stage include:

1. The type of coolant and its application, filtration and temperature control,
2. Workholding methods.

5.5 DESCRIPTION OF INDUSTRIAL MICROMACHINING EQUIPMENT

The first commercial ultraprecision diamond turning and grinding machines were developed principally for the manufacture of complex optical elements. The M18 AG three-axis CNC machine, developed by the Moore Machine Tool Company in the USA was the first to be commercially successful. The M18 AG machine incorporated an air bearing work holding spindle, and laser interferometers were used to control the position of the linear axes, which were mounted on precise "opposed vee" type rolling element slideways. More recently Moore Nanotechnology Systems (USA), and Precitech (USA), have developed a wide range of ultraprecision diamond turning and grinding machines. These include the Nanotech® machines from Moore Nanotechnology Systems, and the Nanoform™ series of machines from Precitech. Features of these machines include:

1. Air bearing work holding spindles,
2. Hydrostatic oil bearing or air bearing slideways,

which are more precise (repeatable), and have a smoother motion than rolling element bearing slideways,
3. Laser interferometer or linear scales for controlling the position of the linear axes, and
4. CNC control systems with a 1.0-nm programming resolution.

The machines have a wide range of applications in the fields of consumer electronics, defense, optics, medical devices, and other precision components, as discussed in the following section.

Edge-grinding machines, utilizing ductile grinding, have been developed by the Landis Lund (UK) organization for the finish machining of glass and ceramic disc substrates for the computer magnetic disc drive industry. More recently the company has introduced a machine for the precision edge grinding of silicon wafers up to 300-mm diameter.

The Nanocentre (Figure 7), a three-axis ultraprecision diamond turning and/or grinding machine was developed by Cranfield Precision (UK), and is thought to be the world's most accurate machine tool of its type, with a volumetric accuracy of 100 nm for workpieces up to 250-mm diameter and a machined length of up to 250 mm [16]. The machine is designed for very high loop stiffness between tool and workpiece in both single point diamond cutting and ductile mode grinding. This is achieved through an optimized machine configuration and the use of high stiffness servodrives and hydrostatic bearings throughout.

In order to maximize geometrical accuracy, the Nanocentre, a research machine built under the UK's National Initiative on Nanotechnology (NION), incorporates a "metrology frame" (Figure 8) to optimize the in-process measurement of the linear motions. This automatically compensates for Abbé-offset errors for both pitch and yaw in the *XZ*-plane at the height of the workspindle center line. This system also measures straightness and orthogonality errors, which are also

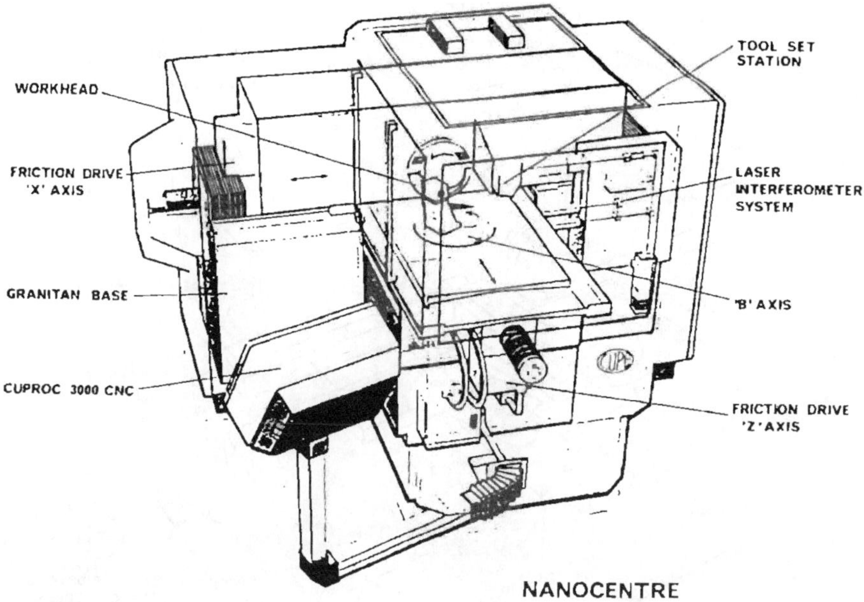

Figure 7 Nanocentre: CNC diamond turning and grinding machine.

software compensated. The highly stiff metrology frame structure utilizes two ultraprecision Zerodur "stick" mirrors for three laser interferometer air paths which are software compensated online, for refractive index changes, by a differential refractometer. The use of these computer software error compensation techniques improved the intrinsic horizontal straightness error motion of the *XZ*-motions from 200 nanometers to 50 nanometers.

Research has continued with the aim of developing the next generation of diamond cutting and grinding machines in order to enter clearly the nanotechnology regime. This has resulted in the review of machine systems and the proposal of new types of structure. For example, the National Physical Laboratory (NPL) has developed the Tetraform structure, which includes a combination of novel concepts for the design

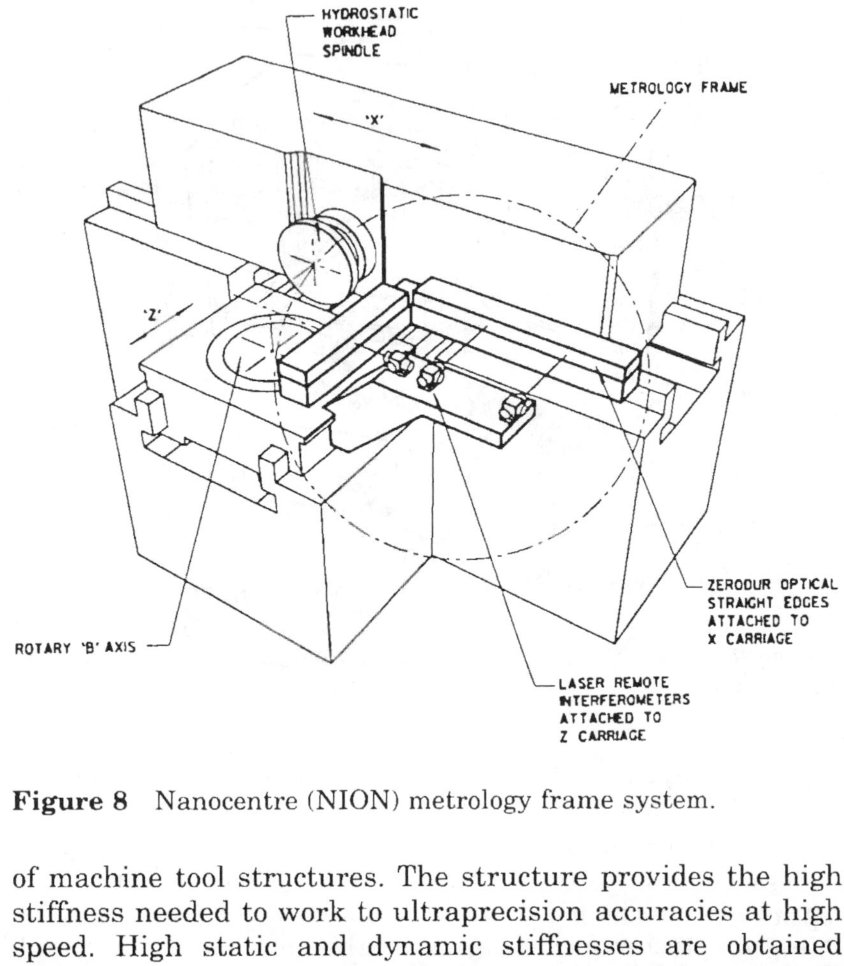

Figure 8 Nanocentre (NION) metrology frame system.

of machine tool structures. The structure provides the high stiffness needed to work to ultraprecision accuracies at high speed. High static and dynamic stiffnesses are obtained through the use of an internally damped space frame with all loads being carried in closed loops via a fully triangulated tetrahedron structure as shown in Figure 9 [17]. Furthermore, the symmetrical form of the structure results in the minimization of thermal effects on the machine performance. Samples of crown glass, fused silica, and single crystal quartz were machined on Tetraform to an extremely high surface quality. The subsurface damage produced in the specimens was on the order of only 1 μm, when taking a 10-μm cut on quartz, which

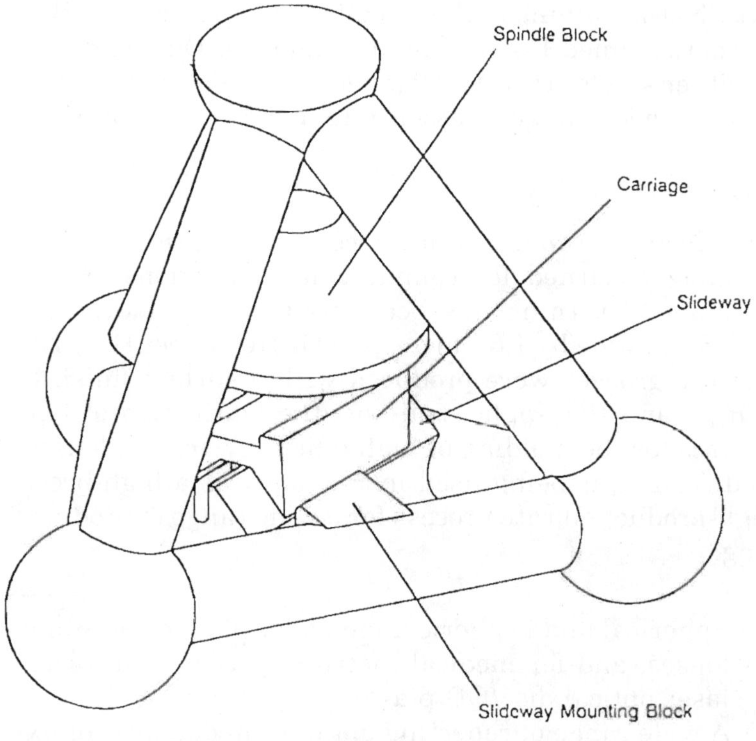

Figure 9 Tetraform ultrastiff diamond grinding machine.

is less than that obtained with the great majority of polished optics. Even with these 10-μm cuts, surface finishes down to 5 nm R_a were produced with very little "roll off" at the edges of the specimens.

5.6 APPLICATIONS

Over the last 30 years the number of applications for diamond turning of a wide range of materials, and the diamond grinding of brittle materials (e.g., glasses and ceramics) has increased significantly following the technological breakthrough in direct CNC machining in the so-called "ductile regime," that is free from retained brittle fracture damage. Diamonds are used in either single crystal or compacted polycrystalline form to machine a wide range of essentially nonferrous materials.

5.6.1 Diamond Turning

Table 2 indicates a range of workpiece materials that are currently diamond turned for commercial applications. Single crystal diamond turning has been used to machine microgrooves 2.5-μm wide by 1.6-μm deep in OFHC copper [18]. The slopes of the grooves were produced with a surface finish to 10 nm R_{max}, and the application was the fabrication of lens master discs for the molding of high efficiency grating lenses. Diamond turning is being used increasingly as a high-precision, high-production rate process for a wide range of products including:

1. Spherical and aspherical molds for plastic opthalmic lenses, and for medical instrumentation and microlaser optical disc/CD players.
2. A wide range of reflecting optics components. For example, aluminum scanner mirrors, space communication and high-power machining laser optics, and aluminum substrates for glancing incidence mirrors for X-ray telescopes.

Table 2 Materials That May Be Diamond Turned

Semiconductors	Metals	Plastics
Cadmium telluride	Aluminium and alloys	Acrylic
Gallium arsenide	Copper and alloys	Fluoroplastics
Germanium	Electroless nickel	Nylon
Lithium niobate	Gold	Polycarbonate
Silicon	Magnesium	Polymethylmethacrylate
Zinc selenide	Silver	Propylene
Zinc sulphide	Zinc	Styrene

3. Infrared hybrid lenses for thermal imaging systems. Typical materials include germanium, zinc sulphide, zinc selenide, and silicon.
4. Aluminum alloy automotive pistons which are machined, in the cold state, to complex profiles with tolerances on the order of 3 to 10 μm.
5. Aluminum alloy substrate drums for photocopying machines.

5.6.2 Diamond Grinding

A range of brittle materials currently being diamond machined, by ductile or shear mode grinding are indicated in Table 3. To date, the greatest demand for this technology has come from the computer peripheral and semiconductor industries. There are, however, numerous new applications where components are required to have lower mass, higher hardness and wear resistance, improved chemical inertness, and higher strength and fatigue life, often while working at higher temperatures than before. The ceramic and intermetallic materials listed in Table 3 can often satisfy these demands.

Following a great deal of worldwide research and development, ceramic and intermetallic materials are ready to be used in gas turbines, pumps, computer peripherals, piston en-

Table 3 Materials That May Be Processed via Ductile Mode Diamond Grinding

Ceramics/Intermetallics		Glasses
Aluminium oxide	Titanium aluminide	BK7[a] or equivalent
Nickel aluminide	Titanium carbide	SF10[b] or equivalent
Silicon carbide	Tungsten carbide	ULE[c] or equivalent
Silicon nitride	Zirconia	Zerodur™[d] or equivalent

[a] BK7 = Borosilicate Crown Glass manufactured by Schott Glass Technologies Inc.
[b] SF10 = Dense Flint Glass manufactured by Schott Glass Technologies Inc.
[c] ULE = Ultra Low Expansion Glass manufactured by Corning Glass Inc.
[d] Zerodur = Ultra Low Expansion Glass Ceramic manufactured by Schott Glass Technologies Inc.

gines, and many other engineering products on a much wider scale. For many of these applications grinding, in the ductile mode, is necessary in order to retain material integrity through the minimization of subsurface damage and microcracking, which reduce the strength and fatigue life of ceramic components.

Diamond micromachining technology for the efficient manufacture of optoelectronics devices is clearly of critical importance in ensuring its progress covering broadband lightwave communication, high-density optical memories, and optical parallel signal processing. Although process development has concentrated on VSLI technologies such as lithography, there is clearly scope for applying ductile mode grinding, slitting, and trenching techniques for the efficient manufacture of monolithic integrated optics components. These techniques are analogous to the ultraprecision grinding of magnetic memory disk file sliders (or flying heads), but now in the ductile regime. The introduction of free abrasives into fixed abrasive processing can be shown to improve surface finish and productivity. Several hundred components can be machined at one set-up to submicron tolerances, producing low-energy-loss contacting optical surfaces (<2 nm R_a and zero surface microcracks) lending themselves to kinematic design for submicron assembly, with large consequent savings in assembly and test labor costs[19].

NOTATION AND SYMBOLS

Esyn	Asynchronous error motion (mm) in the direction normal to the machined surface
f	Feedrate per revolution of the workspindle (mm rev^{-1})
R_a	Surface roughness (μm)
R_{max}	Surface finish (nm)
R_t	Theoretical peak to surface roughness (mm)
T_r	The tool nose radius (mm)

REFERENCES

[1] D.A. Swyte, Challenges to NIST in dimensional metrology: The impact of tightening tolerances in the U.S. discrete manufacturing industry. National Institute of Standards of Technology, Report NISTIR 4757, January, Gaithersberg, MD (1992).

[2] C. Evans, *Precision Engineering: An Evolutionary View*. Cranfield, Cranfield, UK (1989).

[3] N. Taniguchi, *Nanotechnology: Integrated Processing Systems for ultra-Precision and Ultra-Fine Products*. Oxford University Press (1996).

[4] N. Ikawa, R.R. Donaldson, R. Komanduri, W. König, P.A. McKeown, T. Moriwaki, and I.F. Stowers, Ultraprecision metal cutting: The past, the present and the future. *Annals of CIRP*, (40) 2, 587–594 (1991).

[5] M.E. Merchant, Mechanics of the cutting process. *J. Applied Physics*, (216), 267–275, 318–324 (1945).

[6] M.C. Shaw, A new theory of grinding. *Mechanical and Chemical Engineering Transactions. Inst. Engineers (Australia)*, (MC8), 73–78 (1972).

[7] P. Hannah and D. Roaer, Basics of diamond turning. In *Proc. of One Day Workshop*, Cranfield University, 10th October (1988).

[8] T.G. Bifano, T.A. Dow, and R.O. Scattergood, Ductile-regime grinding: A new technology for machining brittle materials. *J. Engineering for Industry. Transactions of ASME*, (113) May, 184–189 (1991).

[9] W.S. Blackley and R.O. Scattergood, Ductile-regime machining model for diamond turning of brittle materials. *Precision Engineering* (12) 2, April, 95–103 (1991).

[10] H. Ohmori and T. Nakagawa, Analysis of mirror surface generation of hard and brittle materials by ELID (electrolytic in-process dressing) grinding with superfine grain metallic bond wheels. *Annals of CIRP*, (44) 1, 287–290, (1995).

[11] M.C. Gerchman and B.E. McLain, Investigation of the effects of diamond machining germanium for optical applications. *SPIE* (929), 94–96 (1988).

[12] P. Shore, Machining of optical surfaces in brittle materials using an ultra-precision machine tool. PhD Thesis, Cranfield University, March (1995).

[13] M.J. Ball, N.A. Murphy, and P. Shore, Electrolytically assisted 'ductile' mode diamond grinding of BK7 and SF10 optical glasses. *SPIE* (1573), 30–38 (1991).

[14] S. Malkin and T.W. Hwang, Grinding mechanisms for ceramics. *Annals of CIRP*, (45) 2, 569–580 (1996).

[15] N. Taniguchi, The state-of-the-art of nanotechnology for processing of ultraprecision and ultrafine products. *Precision Engineering* (16)1, January, 5–24 (1994).

[16] P.A. McKeown, Manufacturing—How small can we go? The challenges and opportunities of the nanometre age. British Association Lecture. Royal Academy of Engineering, London (1996).

[17] K. Lindsey, Tetraform grinding. Commercial applications of precision manufacturing at the sub-micron level. *SPIE* (1573), 129–135 (1991).

[18] K. Goto, K. Mori, G. Hatagoshi, and S. Takahashi, Spherical grating objective lenses for optical disk pick-ups. *Japanese J. Applied Physics*, (26) Suppl. 26–4, 135.

[19] P.A. McKeown, Cranfield University, private communication, March (1997).

6

Ultrasonic Micromachining

D. Kremer
Swiss Federal Institute of Technology, Lausanne, Switzerland

Y. Benkirane
Ecole Nationale Supérieure d'Arts et Métiers, Paris, France

6.1 PRINCIPLES

6.1.1 Introduction to Power Ultrasound

Ultrasonic vibrations are used in many fields of industry for applications such as cleaning, welding, and machining, including assistance to cutting or grinding, nondestructive control, mixing, and separation.

Applications involving ultrasonic waves can be placed in two groups:

The transmitted power is small (mW to W); it is mainly used for nondestructive applications, in which ultrasonic waves must not damage either the item or a live body part that is to be treated;

The transmitted power is an important factor (W to kW), and the waves are used to cause modifications to the part.

Waves can have a very strong effect. Assuming the law of displacement is $u = a \sin(\omega t)$ (with a, half amplitude of vibrations; ω, pulsation; t, time), we can find velocity and acceleration to be respectively:

$$V = a\ \omega \text{ cosine } (\omega t) \quad \text{and} \quad \Gamma = -a\ \omega^2 \sin(\omega t)$$

Maximum velocity and acceleration are: $V_{max} = a\ \omega$ and $\Gamma_{max} = a\ \omega^2$. For a 20-μm half amplitude of vibration at a 20 kHz frequency, which is common for ultrasonic machining, welding, or cleaning, we obtain $V_{max} = 2.5\ ms^{-1}$ and $\Gamma_{max} = 312\ 000\ ms^{-2}$. These very high values are the bases for the effects that these vibrations can have on parts to which they are exposed: for example, shocks, friction, cavitation, and erosion.

6.1.2 Principle of Ultrasonic Machining (USM)

The ultrasonic machining process relies on the projection of very hard abrasive particles on the part to be machined, by use of a tool (sonotrode) vibrating at an ultrasonic frequency (20 kHz or more). These particles are conveyed by means of a fluid, water in most cases (Figure 1).

Two different mechanisms can be observed [1–5]:

The mechanical action of particles on the surface of the workpiece, which is effective for the machining of hard and brittle materials such as glass, ceramics, composites, quartz, precious stones, and semiconductors; and

Cavitation erosion caused by rapid changes in pressure inside the fluid carrying the particles, which has a significant effect when machining fragile and porous materials such as graphite and porous ceramics.

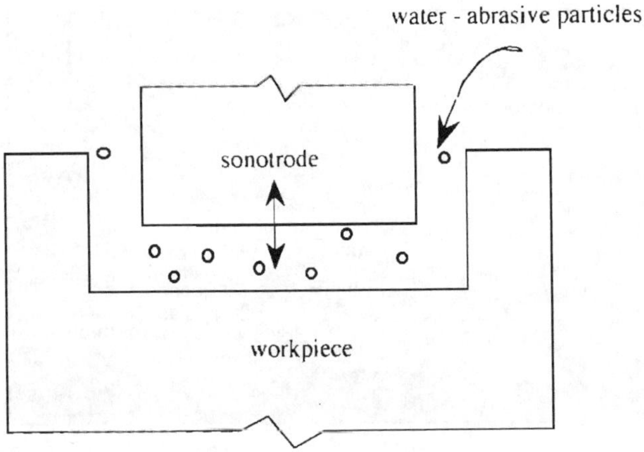

Figure 1 Principle of ultrasonic machining.

Particles affect both part and sonotrode: material removal takes place at the workpiece; wear occurs on both the sonotrode and particles. The process is therefore characterized by removal rate on the workpiece, sonotrode wear, and abrasive wear.

Ultrasonic machining can deal efficiently with brittle materials. The sonotrodes should be made from materials that can resist wear: they should be either ductile (aluminum alloys, steel, titanium alloys, nickel alloys) or extremely hard (diamond). The abrasive particles have to be harder than the workpiece material: aluminum oxide, silicon carbide, boron carbide, and diamond are used.

6.2 MATERIALS FOR WORKPIECE, SONOTRODE, AND ABRASIVE GRAINS

6.2.1 Workpiece Materials

The USM process is able to machine any material, but is more efficient on brittle materials. Materials respond differently to

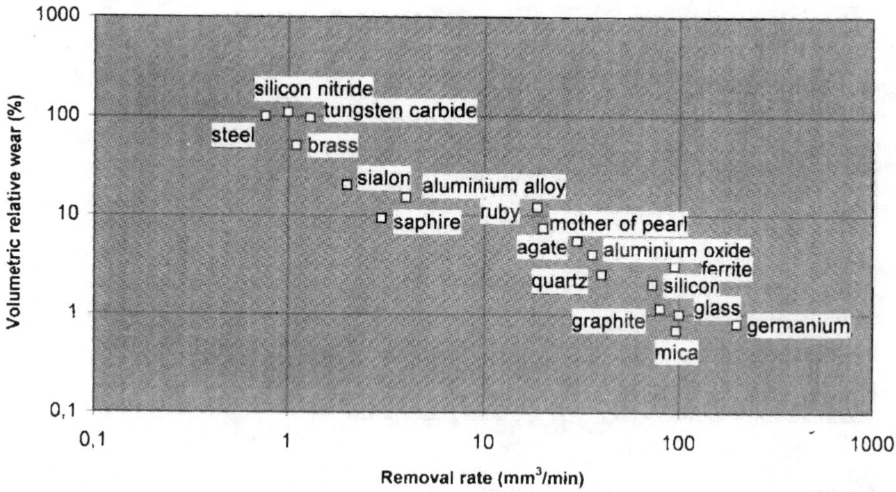

Figure 2 Ultrasonic machinability. (From Refs. 6 and 7.)

the process, as shown in Figure 2 [6,7]. By inspection of this figure, which relates removal rate to sonotrode wear, we can divide materials into several categories:

1. Materials that are machined very easily, due to their high fragility and porosity, for example, glass, germanium, and silicon. High removal rates can be obtained, up to hundreds of mm^3/min. The corresponding wear of a steel tool is very small ($<1\%$), and steel sonotrodes can be used. USM should be used for machining these materials.
2. Materials that are more difficult to machine, since they are less fragile or porous, for instance, ceramic oxides or carbides and precious stones. Removal rate can be as much as dozens of mm^3/min; tool wear varies from 2% when machining silicon oxide up to 10% for ruby (steel tool). It is preferable to use other materials for sonotrodes (titanium or nickel alloys, tungsten carbide, or diamond). USM could be used for ma-

chining such materials, and could compete well with other processes.

3. Materials that are well resistant to the process, since they are either very hard (hard ceramics) or very ductile (metal alloys). Poor removal rates are obtained (a few mm^3/min), and high tool wear is observed (30 to 100% for a steel tool). These materials should be machined with USM only when no other process is effective (e.g., silicon nitride or diamond).

6.2.2 Abrasive Materials

The actual cutting tools are the abrasive particles. Their characteristics and dimensions have to be adapted to the material to be machined and to the specific application intended.

The hardness of the abrasive particles has to be higher than that of the workpiece material. For example, silicon carbide can machine glass, graphite, silicon, aluminium oxide, or precious stones; boron carbide has to be used for harder materials such as silicon carbide and silicon nitride. Diamond is the only abrasive able to machine even harder materials like diamond. The hardness of different materials is shown in Figure 3. For ease of use, and owing to cost, boron carbide is often chosen for machining every material except diamond, and diamond is used only when boron carbide is insufficiently hard.

Wear of particles is a crucial factor, since removal rate depends on particle diameter [6]. Grains can be worn rapidly, as shown in Figure 4 which gives mean diameter distribution for particles, before and after machining glass [8].

6.2.3 Sonotrode Materials

Two main considerations arise in the selection of the sonotrode material: fabrication and cost. Sonotrode wear during ultrasonic machining also has to be taken into account. Usually, the sonotrode (or its working end) has to be produced for every specific application. In order to facilitate this condition, materials selected are often taken from those that can be machined

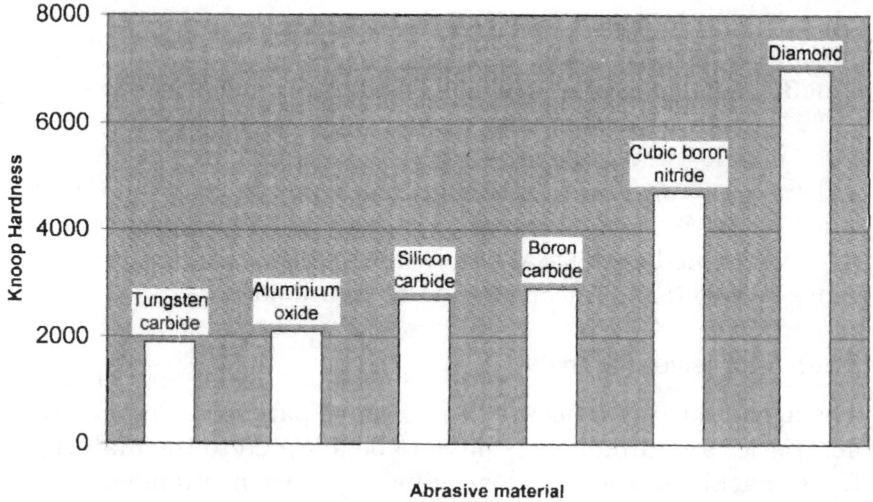

Figure 3 Abrasive hardness.

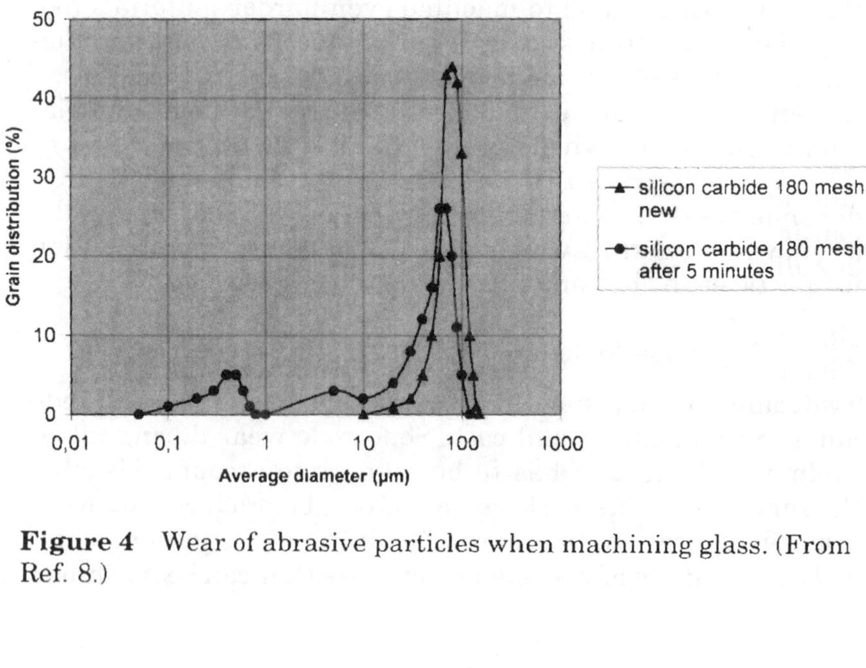

Figure 4 Wear of abrasive particles when machining glass. (From Ref. 8.)

by conventional cutting such as steel, titanium alloy, and nickel alloy. They will encounter some wear during ultrasonic machining.

In some cases, when the working end of the tool has a very simple shape, diamond can be used as tool material: it will resist wear very well and thus have a long life. It can be machined by either diamond grinding or EDM.

Wear is a significant feature of ultrasonic machining, since it directly affects machining accuracy. In the machining of ceramics, the relative amount of wear requires more than one sonotrode to be used if it is made from metal alloys (for roughing and finishing operations).

Metal alloys are used for machining small series of parts by the sinking method (complex shapes, holes, slots); diamond is used for producing very accurate shapes, by means of either the sinking or the contouring technique, for small or key series of parts.

6.3 EFFECTS OF PROCESS VARIABLES ON REMOVAL RATE

6.3.1 Abrasive Parameters

The effect of abrasive particles also depends on their shape and dimension. It is difficult to adopt shape as a parameter, and main diameter is often taken as the particle dimension parameter. The influence of sharpness of edges on grains is not discussed here.

In general, it has been found that increasing the main diameter d of the impinging particles leads to a rise in the amount of eroded material. Sometimes there is a major effect: material removal rate can be proportional to d^4 [9]. Often in ultrasonic machining, a behavior is observed that is composed of first a very significant increase for small diameters, then no further increase, and eventually a decrease, as shown in Figure 5 [10]. This behavior can be explained by the close relationship between this parameter and amplitude of vibrations,

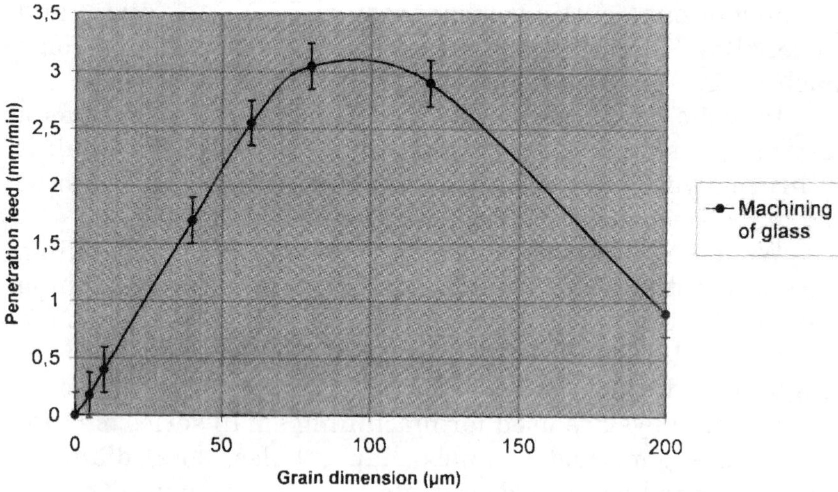

Figure 5 Removal rate variation with grain dimension. (From Ref. 10.)

and for conditions where amplitude is either not sufficiently high to project grains with adequate efficiency, or too great.

In micromachining, the grain main diameter has to be very small, which means that its size has to be adapted to the surface frontal area to be machined. A few studies have been made that give some data on the relationship between frontal area and grain size: in order for grains to move and not be blocked, they should be at least 10 times smaller than the smallest passage [11]. In ultrasonic microdrilling, particles of 2 µm or smaller are used for 0.2-mm diameter tools [12]. They lead to a relatively small removal rate, due to the small working surface, although a reasonable penetration feed (around 0.1 mm/min) is obtained. It is possible to work with even finer grains. In a recent study [13], 0.2-µm particles have been used to drill holes as small as 9 µm in glass or silicon, by use of a 5-µm diameter tool.

Grain concentration has an effect on removal rate: raising the grain concentration leads to an increase in removal rate,

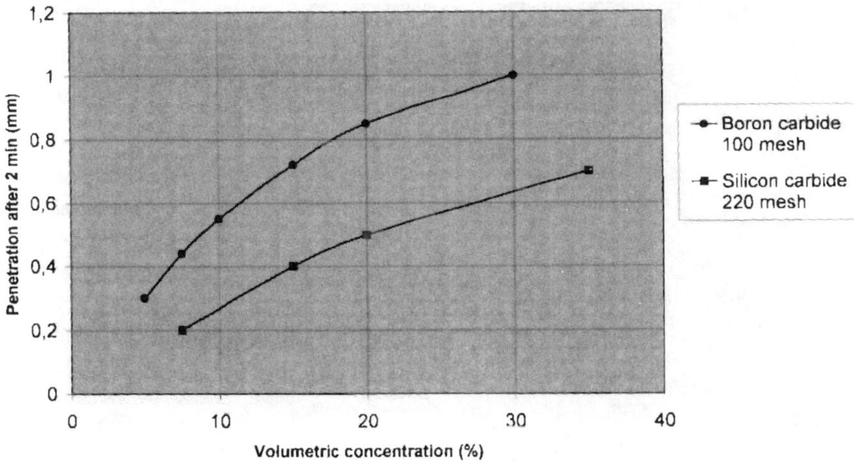

Figure 6 Removal rate as a function of concentration for machining of glass. (From Refs. 4 and 7.)

as shown in Figure 6 [4,7]. However, too high a concentration has an adverse effect: the movement of particles and debris from the part inside the working gap becomes impeded. In general, volumetric concentration is kept between 20 and 30%.

The actual concentration in the working gap can be quite different from that in the tank. For example, when drilling deep holes, inadequate renewal of the flushing fluid can lead to an increased concentration which makes machining less efficient. A decrease in penetration feed is observed with the increase in depth, as shown in Figure 7 [4]. A blockage can even be encountered for very deep holes.

6.3.2 Ultrasonic Process Variables

Amplitude and frequency of vibrations are significant variables.

Frequency greatly influences the efficiency of the process. Removal rate increases with frequency, as shown in Figure 8

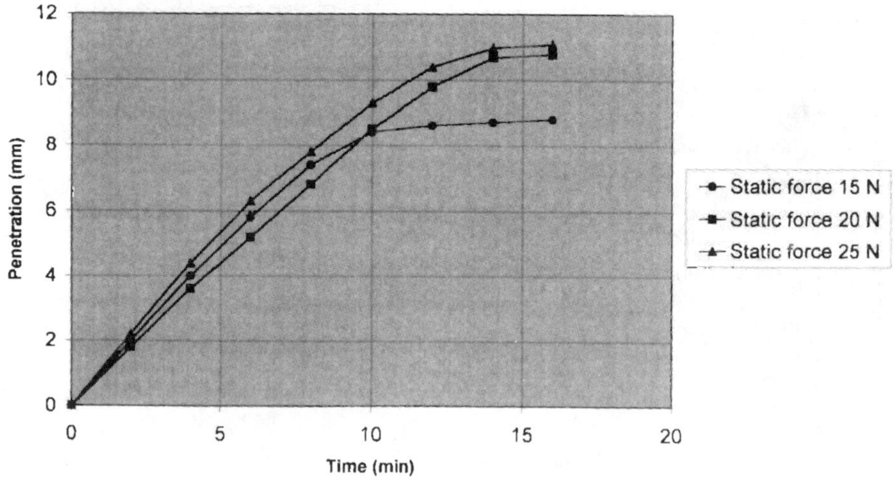

Figure 7 Dependence of penetration on time for machining of graphite. (From Ref. 4.)

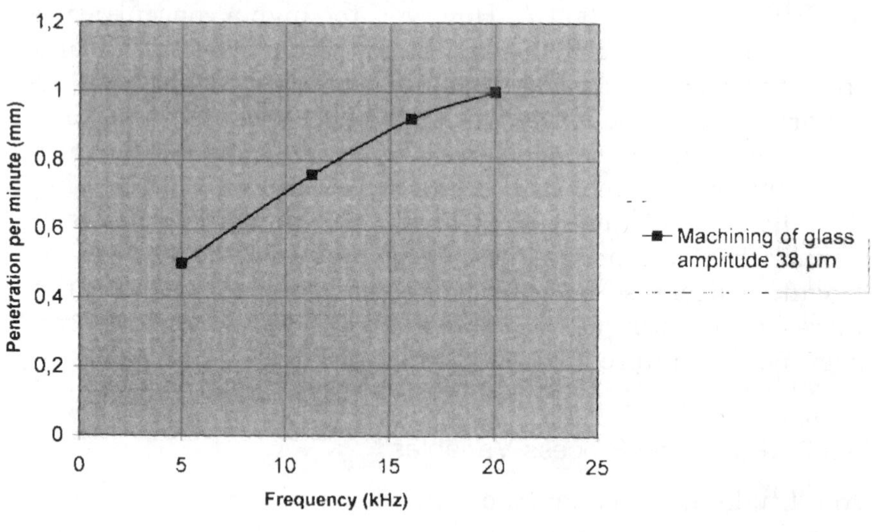

Figure 8 Relationship between penetration and frequency. (From Ref. 7.)

[7]. But it is noted that increasing frequency leads to higher demands on the acoustic system, giving rise to greater heating and thus smaller fatigue life of components; increasing frequency requires smaller amplitude, which does not benefit removal rate. At 20 kHz, an amplitude of 20 to 50 μm can be obtained; at 40 kHz, it is limited to 5 to 20 μm for the same mechanical resistance.

The second effect of increasing frequency is a decrease in the maximum dimensions of the acoustic system that can be used: at 20 kHz, the sonotrode length is around 125 mm (half wavelength in steel); at 40 kHz, it is 62.5 mm. The lateral dimensions are also dependent on frequency. A 20-kHz frequency allows working up to 1 dm^2 cross-section, which is unattainable at 40 kHz. Nonetheless this limitation is not a main concern for micromachining, where the dimensions of the machined shapes are small. For micromachining, higher frequencies could thus be used [14].

Amplitude is a key variable for the process. Too little an amplitude gives the grains a smaller energy, which can be insufficient to cause material removal; if the amplitude is too great, a larger working gap arises, the grains then have to run through a higher distance, and thus the process operates at lower efficiency.

The relationship observed between removal rate and velocity of particles is $Q = V^n$, where n depends on material, angle of impact, grain shape, and grain dimension ($n = 2$ to 3 for ductile materials, 5 to 6 for brittle materials) [9]. As velocity is directly proportional to amplitude, increasing amplitude should lead to an increase in removal rate. This condition arises with other abrasive processes where velocity can be controlled, such as high-pressure water jet cutting. However, it is not observed in ultrasonic machining, as particles are moved between sonotrode and part in a confined environment, and do interact.

Furthermore, it is known that amplitude of vibrations and particle dimensions are strongly interdependent. It is generally observed that a maximum removal rate is obtained when amplitude is equal to grain diameter, as shown in Figure 9 [7].

6.3.3 Machine Conditions

The machine conditions have to be carefully selected in order for the working gap to provide for effective action by the abrasive particles. These conditions in turn depend on the other variables and on the material to be machined. In sinking, the gap between sonotrode and part is adjusted by static force; in contouring, there is no adjustment of gap, which depends on the choice of the variables used. In sinking, there is an optimum value of static force, as shown in Figure 9 [7]. A smaller static force gives rise to a larger gap, particles are not adequately efficient, a higher static force is associated with a smaller gap, and particles have a restricted motion. In contouring, the main variables are depth of cut and lateral feed speed. An increase in these two variables causes more material to be removed per unit of time and thus decreases the

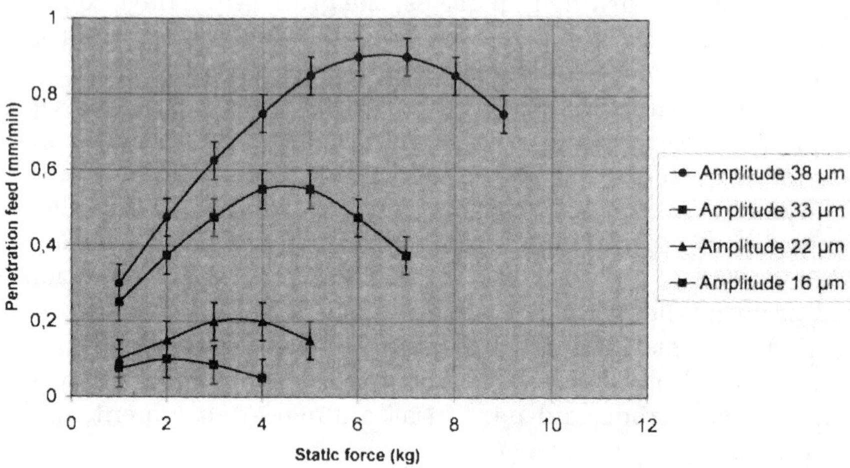

Figure 9 Removal rate as a function of static force. (From Ref. 7.)

working gap. In this case also, an optimum has to be found for selecting all variables that are interdependent.

6.4 ACCURACY AND TOLERANCE IN USMM

6.4.1 Relationship Between Quality and Abrasive Characteristics

The most significant condition is grain diameter which has a major effect on quality. The smaller the grain, the lighter is the impact, and the better the quality. A relationship can be found between grain diameter and crater dimensions: crater diameter is about $d/3$; crater depth is approximately $d/10$ [9]. For the achievement of higher quality, very fine grains should be used, and of course values of other variables have to be appropriately adapted: smaller amplitude, smaller static force (sinking), smaller depth of cut, and larger lateral feed speed (contouring).

In general, needed accuracy entails both roughing and finishing, since quality can seldom be obtained in a single operation. Roughing is performed with large grains (20 to 120 μm) to give sufficient removal rate; finishing is achieved through grains fine enough (0.2 to 10 μm) to obtain the desired quality. Drilling of very small holes is performed in a single operation.

Tool wear is of major consideration for accuracy, since it affects both tool geometry and dimension. The use of metal alloys leads to noticeable wear, for example, from less than 1% up to 50% when machining, respectively, graphite and glass, and silicon nitride. Carbide provides greater resistance, but still gives some wear. The solution to tool wear rests with using diamond, which has been proved to give a wear so small that it could be considered nonexistent. Microdrilling with diamond as abrasive gives a 1% maximum wear ratio, whereas carbide and steel can lead to a 75% ratio when machining hard ceramics [13].

6.4.2 Accuracy

Accuracy strongly depends on the machining mode (Fig. 10). In sinking, it is the result of sonotrode initial accuracy, sonotrode wear, abrasive dimensions and working parameters. The lateral gap between sonotrode and part is found between one and two times more than the abrasive main diameter. The frontal gap is a little larger, due to amplitude of vibrations. Fluctuations of gap are smaller for smaller grains. In general, when drilling, the use of roughing (40 μm) and finishing (5 μm) can provide +/−10-μm accuracy. When finer grains, about 1 μm or less, are used, accuracy can be better than +/−5 μm. This includes the sonotrode accuracy. It is difficult to provide estimates of accuracy for very small holes (10 to 200 μm), because of difficulties with tool accuracy. In contouring, accuracy can even be better, since tool imperfections can be compensated by 3-D movements.

In ultrasonic micromachining, since very fine grains are used, a +/−5 μm (or better) accuracy can be obtained when

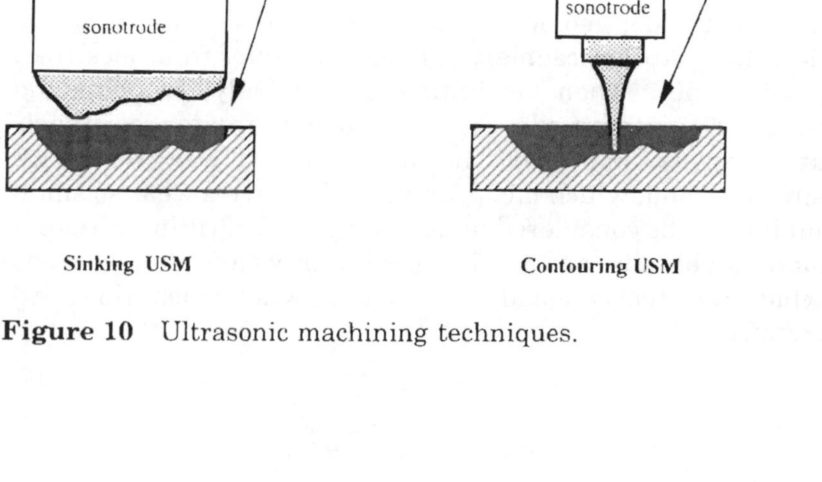

Figure 10 Ultrasonic machining techniques.

tools are conventionally made. A higher accuracy can be achieved by using specially manufactured tools (e.g., the tool form is produced on the USM machine, by use of wire EDM).

6.4.3 Surface Roughness

As it does not impart significant heat, ultrasonic machining produces virtually stress-free surfaces. The quality of the surface produced exhibits fewer stress effects and less surface damage than encountered with other processes, either mechanical (diamond grinding) or thermal (CO_2 and YAG lasers) material removal techniques.

As discussed above, surface roughness depends on particle size. However, it also depends on the material of the part being treated. Very brittle materials will propagate microcracks more readily than less brittle materials, resulting in large craters and poor surface roughness: graphite and glass, for example, yield a roughness of approximately 1.5 to 2 μm *Ra* when 40-μm grains are used; silicon carbide, boron carbide, and silicon nitride produce a roughness of about 0.4 μm *Ra* [4].

Reduction in roughness always requires the use of finer grains. In a recent study on small hole drilling, a roughness of 0.3 μm *Ra* has been obtained on glass when 0.2-μm particles have been used [13].

Ultrasonic machining can also be used for polishing surfaces [15,16]. In this case, the energy given to the active particles has to be decreased so they only affect the surface peaks without removing much material from the part. Sinking and contouring can be used. Superior results are obtained with the contouring method in which an injection of particles into the gap through the sonotrode is used; this procedure imparts a complementary lateral motion to these particles during their passage along the workpiece surface. The method has been applied to fragile, as well as ductile, materials. An improvement in surface quality has been observed when polishing aluminum alloys or steels, from 2 μm *Ra* to 0.2 μm *Ra*, enabling application of the ultrasonic flow polishing process to molds and dies.

6.4.4 Contour Machining with Diamond Tools

There are two main methods that can be used, sinking and contouring:

Sinking is a reproduction process using a form sonotrode which reproduces its shape in negative into the workpiece: machining motion is done on one axis only, parallel to the longitudinal axis of the sonotrode, and is obtained by means of either static force or pressure;

Contouring is a generating process using a simple shape sonotrode working end; machining motion is 3-D NC axes [5,17,18].

In sinking, the gap between sonotrode and part is adjusted by static force, which is the main machine parameter. There is an optimum value of static force. A smaller static force gives a larger gap, and particles are not efficient enough; a higher static force gives a smaller gap, and particles have a restricted motion.

In contouring, there is no adjustment of gap, which depends on parameter choice. The machine parameters will be depth of cut and lateral feed speed. Increasing these two parameters gives more material to remove per unit of time and thus decreases the working gap. Here also, an optimum has to be found for selecting all parameters, which are interdependent.

Increasing lateral feed speed leads to an increase in material removal rate, as shown in Figure 11 [5,17].

The use of diamond (polycrystalline or monocrystalline) allows working with almost no wear, diamond being harder than boron carbide, the most often used abrasive. Three-

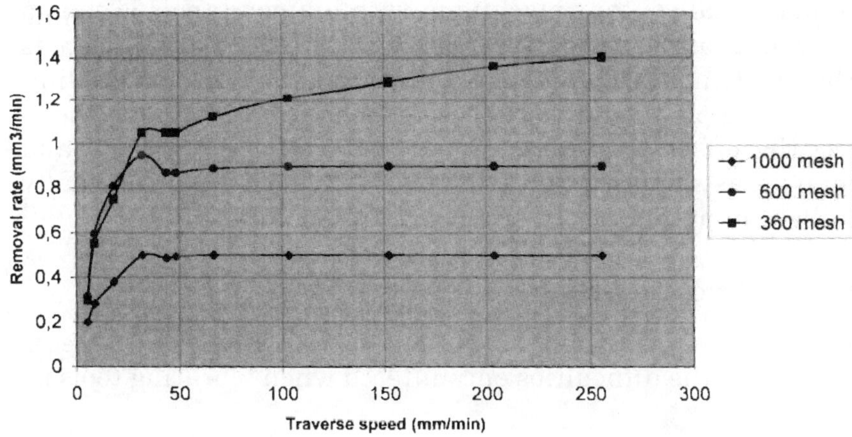

Figure 11 Removal rate as a function of traverse speed in contour machining of glass. (From Refs. 5 and 17.)

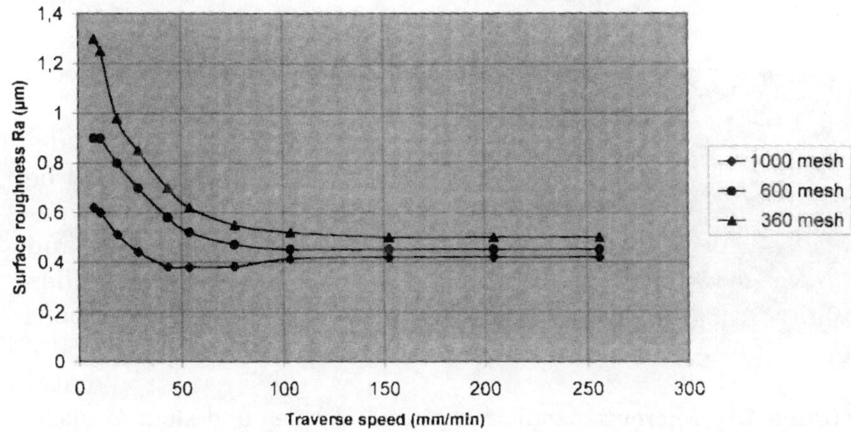

Figure 12 Surface roughness as a function of traverse speed in contour machining of glass. (From Refs. 5, 17.)

dimensional contour machining with diamond tools is thus a very accurate method of machining. It allows producing parts that could not be obtained by conventional ultrasonic machining. A complete study has been made of this technique, showing the effect of parameters on material removal rate and quality, as shown in Figure 12 [5,17]. Shapes that are usually milled in metals can be produced in ceramics.

6.4.5 Microhole Drilling

It is not possible to drill holes smaller than 20 μm in diameter because of the difficulties encountered when vibrating tools as

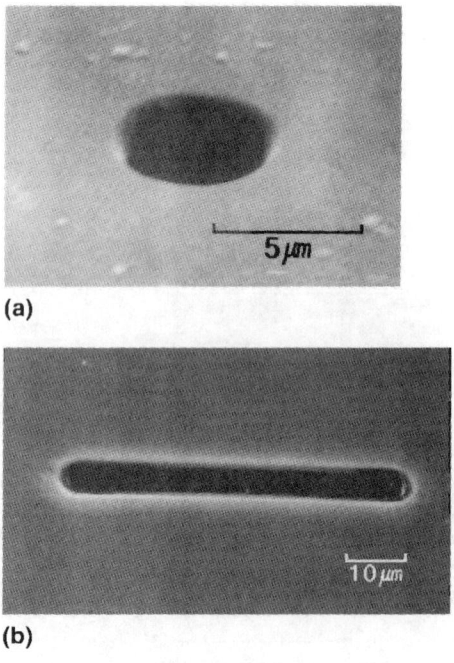

Figure 13 Microultrasonic machining of holes and slots in glass: (a) microhole machined in quartz glass (tool: ϕ 4 μm WC-alloy; depth: 10 μm; (b) trench machined in glass (tool: WC-alloy; depth: 18 μm; abrasive: silica). (From Ref. 13.)

small as 15 μm in diameter [12]. In a recent study, it has been proposed to vibrate the part and rotate the tool [13]. This allows handling the tool with a high-quality spindle and a very precise attachment, since they do not need to be vibrated. The rotation eccentricity is said to be smaller than 0.5 μm.

This study has shown that ultrasonic machining can produce holes as small as 5 μm in diameter, in glass and silicon. For example, Figure 13(a) shows a hole drilled in quartz glass with a 4-μm diameter WC-alloy tool, to a 10-μm depth, and Figure 13(b) shows a trench machined in glass with a WC-alloy tool, to a 18-μm depth. Very fine grains have to be used: 0.2-μm diameter diamond abrasive particles. In this case, amplitude has to be decreased in order to be adapted to particle diameter. Drilling 20-μm holes with 0.2-μm grains requires amplitude to be smaller than or equal to 1 μm. Machining rate is dependent on amplitude (Fig. 14). Machining penetration feed ranges from 1 up to 25 μm/min (hole diameter 9 to 40 μm). Micromachining requires very small static forces to be applied, from 10 mN down to less than 0.1 mN (Fig. 15).

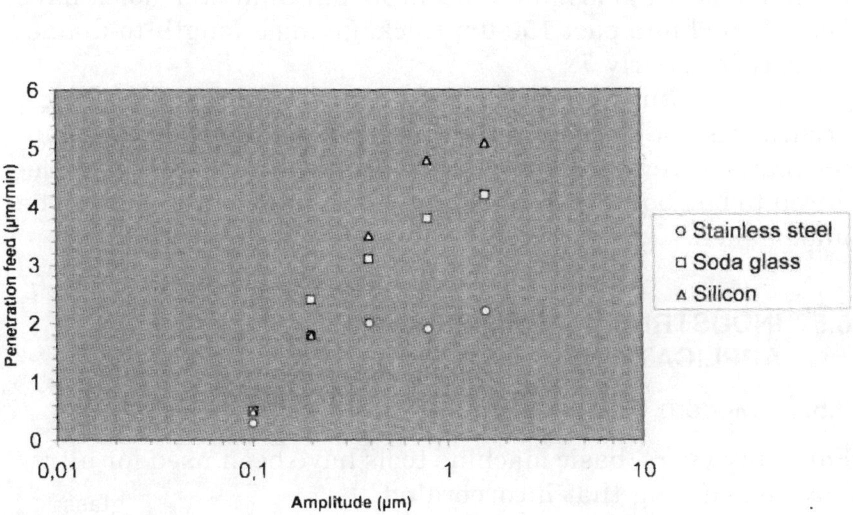

Figure 14 Removal rate as a function of amplitude: Tool diameter 10.5 μm; static force 0.2–0.4 mN. (From Ref. 13.)

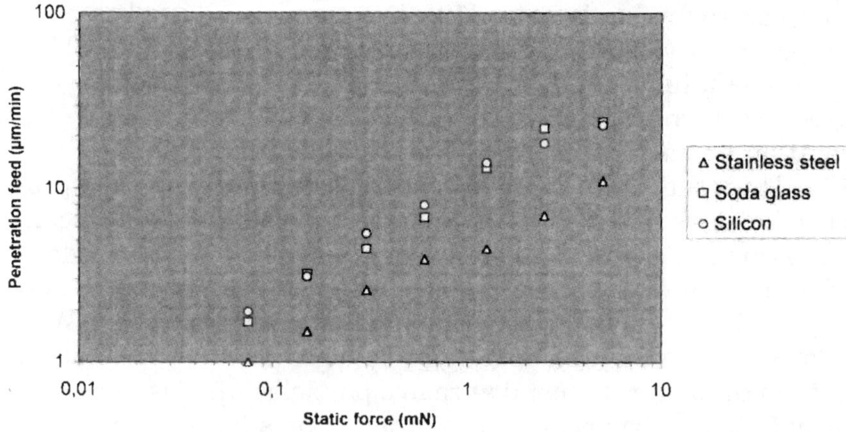

Figure 15 Variation of removal rate with static force: tool diameter 10.5 μm; amplitude 1 μm. (From Ref. 13.)

Holes as small as 5 μm in diameter and 37-μm depth have been drilled in glass and silicon; 21-μm diameter holes have been drilled in a part 150-μm thick, giving a length-to-diameter ratio of nearly 7.

In this study on drilling of 10- to 40-μm diameter holes, accuracy is not discussed. However, photos shown in the chapter indicate very good circularity, from which accuracy is deduced to be about 1 μm or better. Additional work needs to be undertaken.

6.5 INDUSTRIAL MACHINES AND APPLICATIONS

6.5.1 Modern Machines

For many years, basic machine tools have been used for ultrasonic machining that incorporated:

Standard generators, with a fixed frequency, which did not readily permit identification of resonance,

One-axis displacements, obtained by either counterweight or air pressure.

The last 10 years have seen more advanced machines proposed by machine tool builders, employing resonance-following generators and 3-D-axis NC movements with force sensors and adaptive control.

6.5.2 Resonance Following

In ultrasonic machining, to obtain sufficient amplitude at the end of the sonotrode requires exciting the entire acoustic head at one of its natural vibration modes, usually a longitudinal mode. At resonance, efficiency is maximum and no significant difficulty in operation is observed (no excessive heating). When the system operates away from its resonance frequency, efficiency is low and heating occurs. With the standard generators, operators had to make adjustments of sonotrodes so that the entire head could vibrate in a narrow range of frequencies (+/−200 Hz), which was a time-consuming task, and they had to adjust the operating frequency by means of a potentiometer. This method does not yield an optimum, and sonotrodes are not correctly vibrated.

Modern machines use microprocessor-based generators that are able to search the correct resonance frequency and maintain it during the entire machining operation. If frequency changes (because of either heating or wear), the generator automatically makes the necessary adjustment [19]. This gives specific advantages:

The preciseness of vibrating the acoustic head at its resonance, in real-time, which brings high efficiency and very small heating,

The possibility of increasing the frequency range, which gives more flexibility when designing and producing sonotrodes: +/−1000 Hz, for example,

The assurance of maximum security, the microprocessor-based technology being able to detect any anomaly or malfunction and instantly stop the operation.

The use of such a generator has greatly improved the operation of machines.

6.5.3 Adaptive Control Movements

The second major improvement in machines concerns motion. The early machines worked with solely a one-axis motion, and feed movement was obtained by counterweights. This system permitted only sinking operations, with a reproduction of the sonotrode shape in the workpiece.

Modern machines incorporate three-axis motions with numerical control. Feed movement is obtained by adaptive control through use of force sensors. Material is removed by a succession of cuts. The advantages are:

The prospect of working with tools of simple shape and small dimensions; they can then be made from very hard materials, such as diamond, tool wear being controlled;

A very high accuracy, dependent on machine quality and not on tool reproduction, tool wear, and gap fluctuation; and

The capacity to work with very small static forces, the working end of the tool acting locally: very fragile parts can be produced and fine details can be obtained with very small tools.

However, since the tool is working locally, a consequence of this condition is a relatively small removal rate (a few mm^3/min).

6.5.4 Applications

Ultrasonic machining has always been used for producing small details in very hard materials such as glass, precious stones, ceramics, and composites. Owing to the relative slowness of the process, it has been used only when items could not be manufactured by the other techniques. Thus most applications lie in the aerospace and electronics industries.

The process has therefore been focused on machining small details (small diameter holes, slots) and sometimes more complex shapes, obtained with the sinking method. Many applications concern the production of several details in a single operation, for example:

Simultaneous drilling of 750 holes (diameter 1.2 mm +/−0.05 mm) in glass wafers (thickness 3 mm);

Simultaneous machining of washers (silica glass): external diameter 6 mm, internal diameter 2.5 mm, thickness 0.12 mm, accuracy +/−0.02 mm (series are produced in one operation);

Simultaneous machining of slots (0.6-mm wide) in optical glass, accuracy +/−0.25 mm; and

Simultaneous machining of rectangular slots (0.88 × 0.75 mm) in a graphite plate: thickness 1 mm, dimension of ribs between slots 0.2 mm.

Other applications involve the production of one detail for each operation:

Drilling of monocrystalline and polycrystalline diamond dies ranging in diameter from 0.1 to 8 mm, by use of diamond powder;

Preparation of specimens for transmission electron microscopy, from materials such as refractory metals, ceramics, and composites: disks are produced from wafers as thin as 50 μm;

Machining of an aluminum oxide component for measurement purposes: 6 cylinders (diameter 0.6 mm, height

Figure 16 Aluminum oxide component. (Courtesy LPMO-CTM, Besançon, France.)

Figure 17 Hollow cube in silicon oxide. (Courtesy ONERA, France.)

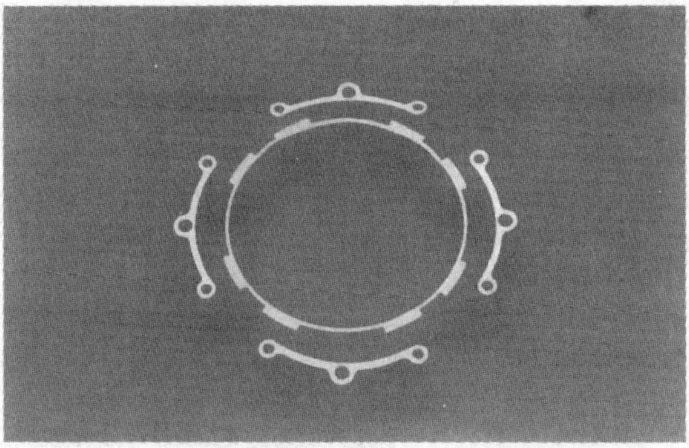

Figure 18 Aluminum oxide parts. (Courtesy ONERA, France.)

2 mm) attached to a membrane of thickness 85 μm (Fig. 16).

The use of the 3-D contour machining method with diamond tools allows the production of parts that cannot be obtained by conventional ultrasonic machining. This method has been used to produce parts from quartz and aluminium oxide used for measuring acceleration forces [18]:

Hollow cube in silicon oxide: used as a component of an electrostatic accelerometer (Fig. 17);

Aluminium oxide parts: components of a capacitance sensor for roundness measurements (Fig. 18), diameter 67.5 mm, thickness 0.4 mm, accuracy better than 10 μm, machining time 2 hours.

6.6 ULTRASONIC ASSISTANCE TO MACHINING PROCESSES

6.6.1 Background

The main practical relevance of using ultrasonic vibrations is to obtain a decrease in machining forces. This condition is obtained mainly by the reduction in friction coefficient observed when one of the two surfaces in contact is vibrated. Other advantages are observed: ultrasonic vibrations facilitate evacuation of chips and provide cleaning of the tool. This procedure is used in either grinding or cutting, for producing parts from metal alloys or ceramics, and also for medical applications (dentistry, surgery).

6.6.2 Rotary Ultrasonic Machining (RUM)

This technique involves the vibration of a small grinding tool, and is used for machining brittle materials (glass, ceramics, stone). A large reduction in machining forces is obtained (two to four times), which permits either an increase in machining speed or a decrease in machining force, a useful condition for machining with very fragile tools.

Removal rate can be two to four times higher depending on the material of the component [20]. It is possible to drill 1-mm diameter holes to a 75-mm depth in glass and to machine narrow slots in ceramics. Ultrasonic vibrations give a reduction in drill wear, by evacuating chips blocked between abrasive grains, and also worn grains. Productivity is drastically increased by ultrasonic assistance, as shown in Figure 19 [20].

6.6.3 Ultrasonic Cutting Assistance

In cutting, the main required effect of ultrasonic assistance is the reduction in friction among tool, chips, and part. The first effect is a large reduction in cutting forces, in turning as well as in drilling, which has been observed for metal and ceramic materials. The second benefit is a significant improvement in surface quality, the machined surface showing less tearing when ultrasound conditions are applied. This effect is particularly relevant in the treatment of materials that are difficult to machine which have an inherent tendency to adhere to the tool edge (the so-called "built-up edge" phenomenon): aluminum copper, stainless steel, nickel and titanium alloys. A recent study [21] has been conducted on the machining of stain-

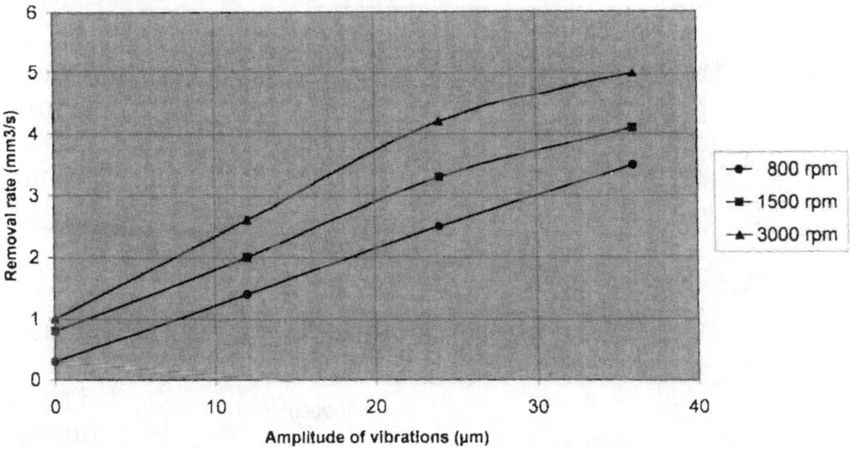

Figure 19 Effect of ultrasonic assistance on productivity. Grinding of ceramics. (From Ref. 18.)

less steel with diamond tools, which is known to be a very difficult operation owing to the chemical activity between tool and part. Encouraging results have been obtained: surface roughness can be as small as 0.1 μm R_{max} (Fig. 20).

6.6.4 Ultrasonic Assistance for Medical Applications

By applying ultrasonic vibrations to a surgical tool, the same advantages of smaller forces and better evacuation of chips can be obtained. Several applications in medicine involve ultrasonic cutting tools.

In dentistry, ultrasonic vibrations impart to the cutting edge of the tool small movements that are less likely to harm and are more acceptable to a patient in comparison with the conventional rotary tools previously used. The tip of the tool is also cooled with water or special fluids. The high tool oscillation frequency and very small amplitude (in the μm range) allow plaque to be removed without damage to the tooth.

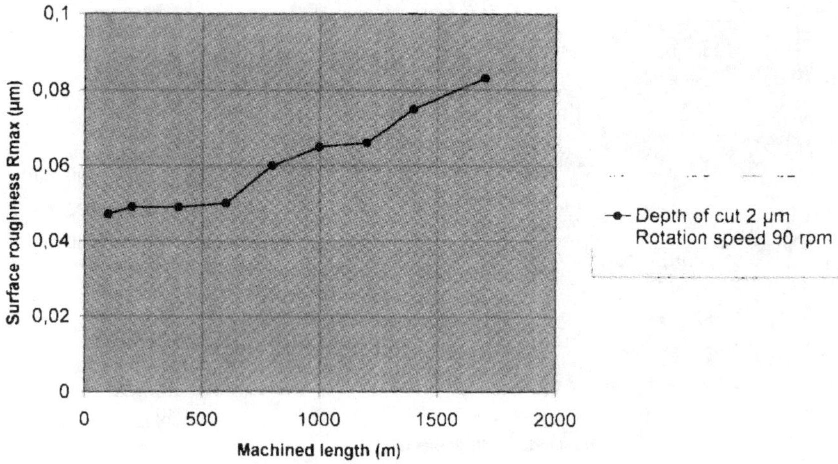

Figure 20 Effect of ultrasonic assistance on surface quality. (From Ref. 19.)

In surgery there is growing interest in the high-frequency, small amplitude action of the cutting edge of scalpels for cutting tissue (e.g., in eye surgery), and also for coagulating blood.

New applications of ultrasonic systems are being proposed such as removal of cement during hip replacement prosthesis and removal of brain tumors. Research is proceeding on an ultrasonic apparatus to clear blocked passages in arteries. The tool is a titanium wire (diameter 0.8 to 0.4 mm, length 1.2 m) which is inserted into the body along the femoral artery to the heart or to the leg. Its distal working end is a 1.4- to 2.5-mm diameter titanium ball: when vibrated, this ball can penetrate the occlusion and "recalibrate" the artery to a satisfactory diameter whereby required operations can then be undertaken, avoiding open thorax surgery. Results on tests made on coronary and leg arteries have shown the possibility of clearing 70% of blockage [22,23].

Another example is the use of a scalpel which is vibrated at ultrasonic frequency (55,000 Hertz, 100-µm amplitude): the rapid movement causes separation of the tissue ahead of the blade, and aids dissection [24]. There is minimal tissue charring and thermal injury.

Furthermore, ultrasonic assistance is used with phacoemulsification hand pieces and in systems for the surgical removal of cataracts. This method is based on a cataract extraction that uses ultrasound waves to break the cataract-obstructed lens of the eye into small fragments that can then be removed through a hollow needle. It requires only a 3- to 4-mm incision compared to incisions of up to 12 mm for other techniques [25].

6.7 SYNTHESIS

As noted earlier, in most applications ultrasonic machining has been mainly used for producing small details on parts. Its physical principles make the process useful for machining brittle and hard materials, and disadvantageous for ductile materials. It has thus been used for graphite, glass, quartz, ceramics, precious stones, semiconductors, and diamond.

The need to work at very high frequencies (at least 20 kHz) restricts the dimensions of the working end of the tool to 25 to 50 mm in general. This restriction means that ultrasonic machining can be applied to micromachining. To this end, since the result to be obtained on the part depends on tool dimensions and working variables, it is necessary to be able to use tools of size from as small as a few micrometers to one millimeter, and adjust the process variables: grains from 0.2 μm to 20 μm in diameter; amplitudes from 0.1 μm to 20 μm; and forces from 0.1 mN to 1 N.

Under these conditions, holes from 1 mm to as small as 5 μm have been obtained, with a depth-to-diameter ratio up to 7. Corresponding accuracy can be +/− 10 μm for larger holes of 1 mm in diameter, and even greater for smaller holes of 9 μm in diameter.

Applications of ultrasonic micromachining can be found in electronics, aerospace, biomedicine, and surgery.

NOTATION AND SYMBOLS

A	Amplitude of vibration (μm)
V	Velocity of vibration (m/s)
Γ	Acceleration of vibration (m/s^2)
f	Frequency of vibrations (Hz)
F	Static force (N)
Φ	Main grain diameter (μm)
C	Grain concentration (%)
Ra	Rugosity (μm)
R_{max}	Rugosity (μm)
d	Crater main diameter (μm)
Q	Removal rate (mm^3/min)

REFERENCES

[1] G.S. Kainth, A. Nandy, and K. Singh, On the mechanics of material removal in ultrasonic machining. *Int. J. Mach. Tool Des. Res.*, (19), 33–41 (1979).

[2] E.V. Nair and A. Ghosh, A fundamental approach to the study of the mechanics of ultrasonic machining. *Int. J. Prod. Res.*, (23) 4, 731–753 (1985).

[3] V. Soundararajan and V. Radakrishnan, An experimental investigation on the basic mechanisms involved in ultrasonic machining. *Int. J. Mach. Tool Des. Res.*, (26) 3, 307--321 (1986).

[4] D. Kremer, A. Moisan, S.M. Saleh, and S.R. Ghabrial, The state of the art of ultrasonic machining, *Annals of CIRP*, (30) 1, 107–110 (1981).

[5] H. Kamoun, Application des ultrasons aux procédés de fabrication. Modélisation et expérimentation. PhD Thesis, ENSAM Paris, March (1994).

[6] M. Adithan and V. C. Venkatesh, Effect of system parameters on tool wear in ultrasonic drilling. *IE(I) J. ME* (57) July 33–35, (1976).

[7] METCUT. *Machining Data Handbook. Machinability Data Center*, Vol. 2, 3rd edition. METCUT Research Associates, 1043–1063 (1980).

[8] A. Malouli, Modélisation de l'abrasion assistée par ultrasons. Etude de la fragmentation des particules et de son influence sur l'enlèvement de matière. PhD Thesis, ENSAM Paris, December (1999).

[9] A.G. Evans and T.R. Wilshaw, Quasi-static solid particle damage in brittle solids. Observations, analysis and implications. *Acta Metallurgica*, (24), 939–956 (1976).

[10] V.F. Kazantsev and L. D. Rozenberg, The mechanism of ultrasonic cutting. *Ultrasonics*, October December, 166–174 (1965).

[11] L. Rhoades, Abrasive flow machining. *Manufacturing Engineering*, November, 75–78, (1988).

[12] K. Egashira, T. Masuzawa, and M. Fujino, Micro ultrasonic machining method by precise tool rotation and workpiece vibration. In *Proc. of the IEEE Electro Mechanical Systems Workshop*, San Diego, February (1996).

[13] K. Egashira and T. Masuzawa, Micro ultrasonic machining by the application of workpiece vibration. *Annals of CIRP*, (48) 1, 131–134 (1999).

[14] K. Suzuki, S. Mishiro, Y. Sishido, T. Kitajima, S. Hirai, and T. Uematsu, Development of 400 kHz micro ultrasonic grinding device and its grinding performance. In *Proc. of the IEEE Electro Mechanical Systems Workshop*, San Diego, February (1996).

[15] Y. Benkirane, Etude de faisabilité d'un procédé de polissage par pâte abrasive assisté par ultrasons. PhD Thesis, ENSAM Paris, May (1998).

[16] Y. Benkirane and D. Kremer, Ultrasonic flow polishing process. *Int. J. Forming Processes*, (1) 3 (1999).

[17] Y. Benkirane, D. Kremer, and A. Moisan, Ultrasonic machining: An analytical and experimental study on contour machining based on neural network. *Annals of CIRP*, (49) 1, 135–138 (1999).

[18] G. Campergue, R. Gouhier, D. Horriere, and A. Thiriot, Machine d'usinage par abrasion ultrasonore. Patent No. 87-05142 (1987).

[19] B. Thirion and P. Poupaert, Procédé et dispositif d'alimentation électrique d'un transducteur de vibrations tant sonores qu'ultrasonores. Patent No. 2,586,883 (1985).

[20] Z. Pei, Rotary ultrasonic machining of ceramics: Characterization end extensions. PhD Thesis, University of Illinois at Urbana Champaign (1995).

[21] T. Moriwaki and E. Shamoto, Ultraprecision diamond cutting of hardened steel by applying elliptical vibration cutting. *Annals of CIRP*, (48) 1, 441–444 (1999).

[22] G. Drobinski and D. Kremer, Ultrasonic percussion device. Patent No. 5,649,935 (1997).

[23] G. Drobinski, D. Brisset, F. Philippe, D. Kremer, C. Laurian, G. Montalescot, and M. D. Thomas, Effects of ultrasonic energy on total peripheral artery occlusions: Initial angiographic and angioscopic results. *J. Interventional Cardiology*, (6) 2, 157–163 (1993).

[24] A. Coulson, The use of the harmonic scalpel in minimally invasive coronary artery bypass surgery. www.inreach.com (1999).

[25] Zevex, Sensors and surgical devices. http://www.zevex.com.

7

Microelectrodischarge Machining

David M. Allen
Cranfield University, Bedford, England

7.1 INTRODUCTION

Microelectrodischarge machining (also known as micro-EDM, μ-EDM, and electrodischarge micromachining) has been developed in the past 30 years from the nonconventional manufacturing technique of electrodischarge machining (EDM) commonly known as spark erosion (see Section 7.2).

While EDM has been used as a production tool for over 50 years, true μ-EDM only commenced in 1967 [1] when Kurafuji and Masuzawa succeeded in machining 6-μm diameter circular holes through GTi 10 cemented carbide 50-μm thick, thus demonstrating the rapid production of high aspect-ratio holes [2]. Since that time, there has been a concerted effort to improve the micromachining rates of various materials, without loss of accuracy, and to improve the excellent surface finish and dimensional control already associated with the EDM technique (see Sections 7.3 and 7.4).

As might be expected, commercial μ-EDM equipment has been produced by companies in Switzerland and Japan, acknowledged centers of excellence in microtechnology and precision engineering (see Section 7.5). μ-EDM is now being used to machine a wide variety of miniature and microparts from electrically conductive materials such as metals, alloys, sintered metals, cemented carbides, ceramics, and silicon (see Sections 7.3 and 7.6). μ-EDM may also be used to produce molds and dies that can themselves be utilized to manufacture other microparts from both conductive and nonconductive materials such as plastics (see Section 7.6).

7.2 PRINCIPLES OF EDM AND μ-EDM

7.2.1 EDM Principles and Methods of Machining

In essence, the principle of EDM is simple.

1. Two electrodes are separated from each other by a dielectric fluid.
2. A voltage difference is applied between the two electrodes, a cathode (negatively charged) and an anode (positively charged).
3. If the two electrodes are moved close enough together and the voltage is high enough, the dielectric fluid will break down and conduct an electrical current, causing an electrical discharge (a spark) between them.
4. The sparks will produce an extremely high temperature (of the order of 10,000 K) at localized spots on the electrodes such that the electrode materials (especially the anode) will be vaporized, leaving craters

behind on the surfaces as evidence of material removal.

5. By careful choice of the dielectric, voltage generator, and electrodes this material removal process can be used as a manufacturing tool in the following ways.

7.2.2 EDM Die-Sinking

By fabricating a cathode (tool) in a conductive material such as copper or graphite and feeding the tool under servocontrol towards a planar (workpiece) anode immersed in kerosene, the shape of the tool can be impressed into the workpiece. Thus a (male) tool can be used to form a (female) workpiece of inverse shape, of great use in the production of a mold which can be used to replicate many "male" parts economically. The concept is illustrated in Figure 1.

7.2.3 Wire EDM (WEDM)

By using a continuous wire as the cathode guiding it under NC, and keeping it flooded with a dielectric such as deionized

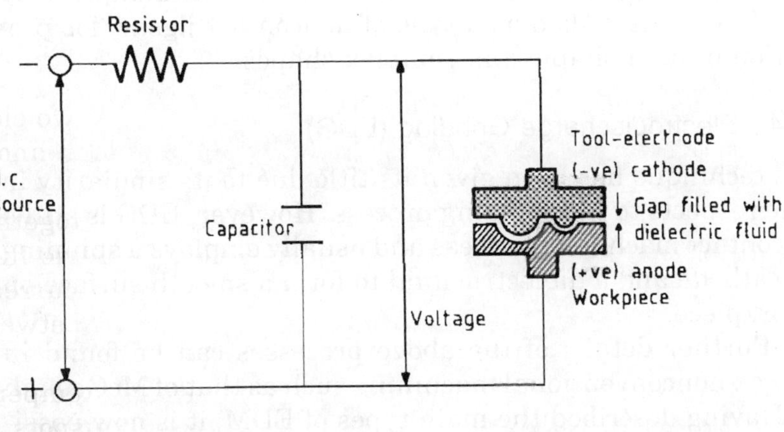

Figure 1 EDM by die-sinking connected to a relaxation (RC) circuit.

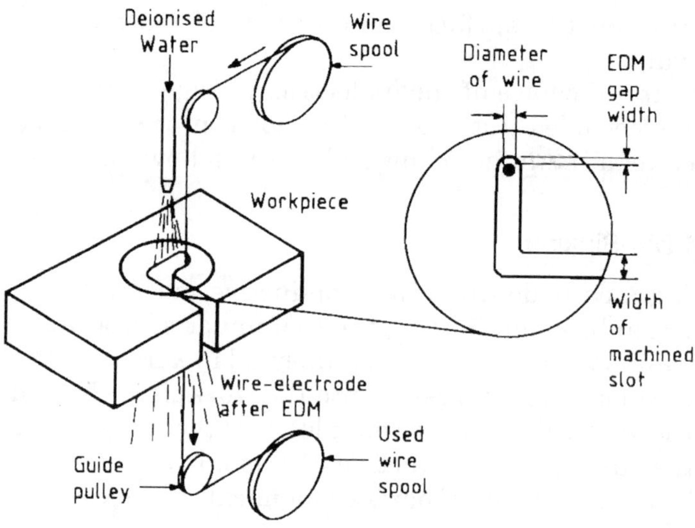

Figure 2 Wire EDM.

water, shapes may be machined into conductive materials, such as flat sheets and plates as illustrated in Figure 2.

Typically wires comprise copper, brass, or tungsten with diameters ranging from 50 to 250 μm. The technique may therefore be described as a type of noncontact jigsaw for production of both simple and complex shapes.

7.2.4 Electrodischarge Grinding (EDG)

This technique has been given its title due to its similarity in some respects to the grinding process. However, EDG is still a noncontact machining process and usually employs a spinning flat cathode and a dielectric fluid to form a smooth surface on a workpiece.

Further details of the above processes can be found in texts on nonconventional machining such as that of McGeough [3]. Having described the main types of EDM, it is now possible to examine the micro-EDM techniques that have developed from them.

7.2.5 The Origins of μ-EDM

As microelectronics are essential to the national economy of Japan, it is not surprising that microengineering has attracted great attention in its universities and research institutes. μ-EDM as we know it today can be traced back to 1967 when Kurafuji and Masuzawa used EDM to machine high aspect-ratio holes through cemented carbides [2]. An English langeuage account of these experiments appeared in 1968 and described the equipment used and the importance of controlling the energy of the discharges to less than 10^{-8} J [4]. The equipment comprised a copper wire electrode of 160-μm diameter, mounted on an optical microscope frame in close proximity to a workpiece of STi 20 cemented carbide alloy immersed in a bath of kerosene dielectric fluid. The electrode was hand-fed by the microscope focus control.

An RC circuit (see Fig. 1) was used to generate the discharges of a form shown in Figure 3, with capacitances (C) ranging from 2 pF to 5 μF. It was shown that the electrode wear ratio (cathode wear/anode wear in terms of length) decreased to less than 10% if C was smaller than 100 to 1000 pF. If the polarity was reversed such that the tool-electrode formed the anode, then the wear ratio was as large as 250 to 1700%.

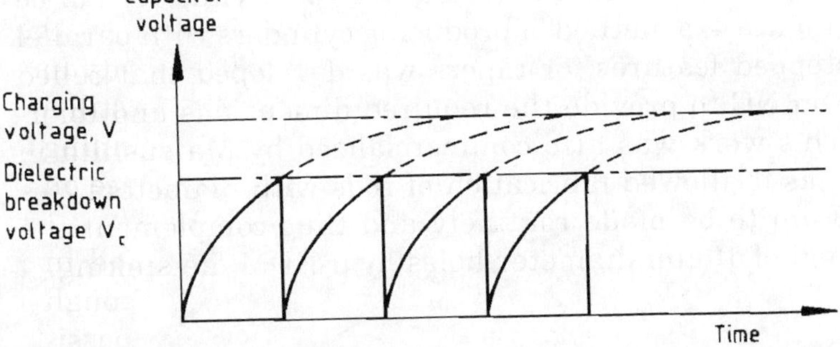

Figure 3 Variation of voltage with time using an RC circuit.

The discharge energy (E) can be calculated from the formula $E = CV^2/2$, where C = capacitance and V = voltage.

These initial investigations were to pave the way for the first μ-EDM by die-sinking. Using a copper-plated tungsten wire, microholes were machined into 50-μm thick GTi 10 cemented carbide alloy. They had an entrance diameter of 9 μm and an exit diameter of 6μm. Similarly, rectangular microholes 40 μm × 45 μm were also machined through the same material in only 10 to 15 s. using a 400-μm long rectangular copper strip. Steel, brass, and other materials could also be machined. This research was used as the basis for a commercial μ-EDM machine developed at the Matsushita Research Institute, Tokyo in 1983 [1]. These machines are also marketed under the Panasonic and Charmilles brand names.

It is important for the production of holes by μ-EDM that the tool-electrode is accurately made. Thus work was started on a method of producing such tools, replacing EDG by WEDG (wire electrodischarge grinding).

Prior to 1984, EDG had been used to fabricate thin rods by spinning a rod (the anode) on its axis using a block electrode as a cathode in an EDM cell. However, block wear, debris build-up, and dimensional control of the gap between rod and block proved problematic. Masuzawa et al. [5] replaced the block with a continuous wire of constant diameter, thus ensuring a fresh cathode surface for each discharge and also introduced z-axis movement of the wire electrode. Thus, as can be seen in Figure 4, a method of producing cylinders with parallel sides, stepped features, or tapers was developed that relied heavily on NC to provide the required dimensions and tolerances. This work was also commercialized by Matsushita in 1988 [1] as it allowed fabrication of rods with diameters less than 10 μm to be made routinely and thus complement the production of 10-μm diameter holes by μ-EDM die-sinking.

7.2.6 Other Forms of Miniaturized EDM

By reducing the size of the diameter of electrode wires in WEDM down to say 30 μm, small devices (see Section 7.6.3)

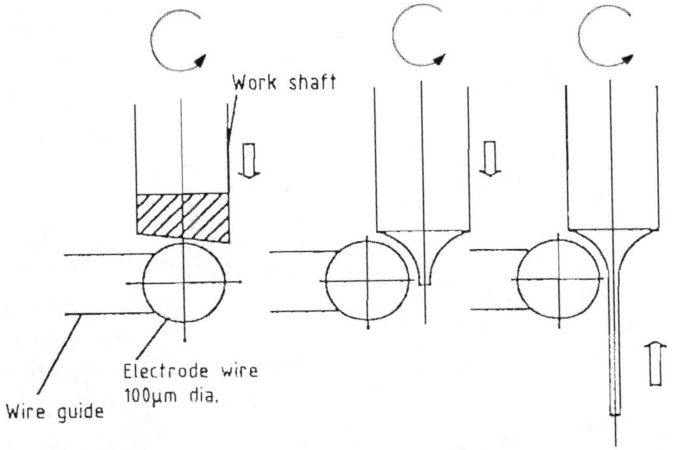

Figure 4 Principle of WEDG. (Courtesy of T. Masuzawa.)

can be manufactured provided that internal radii exceed half the diameter of the wire (in this case, 15 μm). Such machines have been developed by Agie of Switzerland. The technique is now becoming known as μ-WEDM.

By reducing the size of a cathode in an EDG system, miniature features such as precision slots can be machined. Applications of μ-EDG are discussed in Section 7.6.4.

7.3 RATES OF MICROMACHINING FOR RELEVANT MATERIALS

The rate of machining is dependent on the discharge energy, which for the attainment of fine surface finish is kept low ($<10^{-7}$ J per pulse) in μ-EDM. Table 1 shows rates of machining for sulphur-killed (SK) steels, stainless steels, and silicon calculated from published information.

Various techniques have been utilized to try to increase the material removal rate such as

Switching a larger capacitance into the RC circuit, thus producing a large discharge energy [7],

Table 1 Rates of Micromachining by EDM

μ-EDM	Feature	Feature size (μm)	Dielectric	Material	Thickness (μm)	Machining time (s)	Metal removal rate (mm^3/min)	Ref.
Die-sinking	Circular hole	$\phi = 100$	D. I. water	SK steel	1000	120–260	$1.8–3.9 \times 10^{-3}$	6
Die-sinking	Circular hole	$\phi = 55$	D. I. water	SK steel	500	21–25	$2.9–3.4 \times 10^{-3}$	6
Die-sinking	Circular hole	$\phi = 100$	D. I. water	AISI 304 stainless steel	1000	600	0.79×10^{-3}	7
Die-sinking	Circular hole	$\phi = 38$	Kerosene	17-7PH stainless steel	80	182	0.03×10^{-3}	8
Die-sinking	Circular hole	$\phi = 85$		AISI 304 stainless steel	1000	40	8.5×10^{-3}	9
μ-WEDM	Slot	Width = 150	D. I. water	50 mΩ.cm silicon	350	Wire cutting at 0.175mm/s	550×10^{-3}	10

Dielectric flushing,
Jump action and vibrofeeding of the tool,
Premachining holes to allow debris to flow away from the electrode [8], and
Use of controlled pulse generators in WEDM as discussed in Section 7.5.2.

7.4 ACCURACY AND DIMENSIONAL CONTROL

The accuracy and dimensional control of parts made by μ-EDM varies with the types of machines used.

In general terms, the data in Tables 2 and 3 have been abstracted from manufacturers' literature for parts made by μ-EDM.

7.5 DESCRIPTION OF INDUSTRIAL MICROMACHINING EQUIPMENT

7.5.1 Micro-EDM by Die-Sinking and WEDG

The Charmilles Micro-EDM EL-10 and HO-10 machines have been developed from Masuzawa's research as described in Section 7.2.4. The two machines are often used for sequential fabrication. The EL-10 is a WEDG machine and can be used to

Table 2 Accuracy of Holes Made from μ-EDM Die Sinking and μ-WEDG

Machine	Charmilles HO-10/EL-10 (μm)	Panasonic MG-ED 72 W (μm)
Maximum hole diameter	300	300
Minimum hole diameter	10	5
Machined surface R_{max}	0.1	0.1
Positioning resolution	—	0.1
Repeatability of positioning	Not Specified	Not Specified

Table 3 Accuracy of Parts Made by μ-WEDM

Machine	Agiecut 250SF + F HSS	Agiecut 150 HSS
Wire material	Tungsten	Tungsten
Wire diameter	30 μm	30 μm
Part	Injection die	Spray nozzle
Part material	Inox	Sintered carbide 0.2 μm
Surface finish (R_a)	0.15 μm	0.2 μm
Max. dimensional variation (μm)	±1	—
Machining time	47 min	2 min 36 s

make microelectrodes, as well as microshafts, pin gauges, and micropunches. The HO-10 is a μ-EDM die-sinking machine and can be used to make microholes, microcavities, and microslots.

The EL-10 comprises the μ-EDM machine connected to a programmable NC system. Machining of wires or rods is achieved by WEDG. Cylindrical, conical, stepped, or semicylindrical forms can be produced with diameters between 10 to 300 μm. Typically tungsten wires can be machined to produce microelectrodes suitable for transferring directly into the HO-10 machine to manufacture high aspect-ratio holes (see Section 7.6). The machining of a tungsten rod or wire is achieved as shown in Figure 4. The rod is spun at 3000 rpm, the end levelled by moving the continuous wire in the x-axis, and the correct diameter of the rod achieved by several passes of the rod in the z-axis while moving the continuous wire in the x-axis. In this operation the rod is the anode in the cell.

When the tungsten electrode is transferred into the HO-10 machine, it is connected as the cathode, with the material to be machined as the anode. While rotating, the electrode is lowered in the z-axis to effect machining of a hole. The HO-10 comprises the μ-EDM machine connected to a programmable NC system and an optional external PC if simple contouring (e.g., slots) is required.

The NC system controls the z-axis motion of the microelectrode and the x- and y-axes motions for positioning of the holes. The specifications of the two machines are as shown in Table 4. The machines are equipped with low impedance microdischarge generators giving very fine surface finishes of R_{max} = 0.1 μm and electrode positioning is controlled by d.c. stepping motors.

Although the HO-10 and EL-10 are fitted with X20 inspection microscopes, a X100 positioning microscope is fitted only to the HO-10.

7.5.2 Micro-WEDM

One of the essentials for production of precise microparts is a microdischarge generator. Agie has developed the Agiepuls HSS generator for μ-WEDM. This pulse generator draws only 4.5 kW and has a metal removal rate of over 300 mm^2/min for steel. The pulse generator differs from an RC circuit comprising only a d.c. source which is fed via a resistor and an electronic switch to the machining gap.

The voltage/time characteristic shown in Figure 5 is very different from that of the RC circuit (Fig. 3). With a duty cycle

Table 4 Specification of HO-10 and EL-10

Machine		HO-10	EL-10
Size (l × w × h)	mm	750 × 500 × 420	500 × 500 × 420
Weight	kg	200	70
Travel (x-axis)	mm	200	5
(y-axis)	mm	50	—
(z-axis)	mm	3	3
Speed of electrode rotation	rpm	3000	3000
Work tank (l × w × h)	mm	300 × 140 × 45	
Stepping resolution			
(x-axis)	μm	1	0.1
(y-axis)	μm	1	—
(z-axis)	μm	1	1

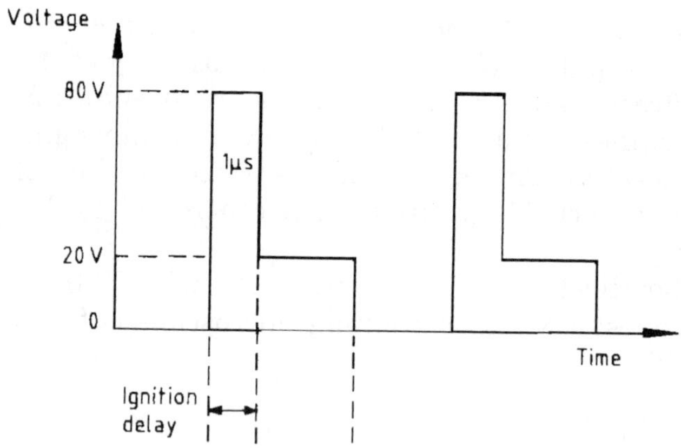

Figure 5 Variation of voltage with time using a controlled pulse generator.

(ratio of "pulse on" to "pulse off") that can be as high as 2000, it can be seen that metal removal rates can far exceed those attained using an RC circuit where considerable machining time is lost during capacitor charging.

As water is used as the dielectric, an anodic workpiece can be subject to electrolytic attack. The HSS generator is designed so that a negative voltage is applied to the positively charged workpiece during the intervals between the pulses lasting at most 3 μs. This gives the workpiece a high degree of cathodic protection and it is the continuous wire electrode that suffers any electrochemical attack.

A true micropart such as the injection die for precision gears mentioned in Section 7.6 was fabricated with an Agiecut 250SF + FHSS system utilizing a 30-μm diameter tungsten wire as the continuous cathode. The same wire was also used to cut paint spray nozzle orifices in sintered carbide using an Agiecut 150 HSS system. It should be remembered that the process of μ-WEDM requires wire-threading for each aperture cut.

7.6 APPLICATIONS

The applications of μ-EDM are many and various. They are described below according to the method used for fabrication.

7.6.1 Application of WEDG

Microelectrodes for μ-EDM Die-Sinking

Typically a WEDG machine is used to make cylindrical electrodes with diameters down to 5 μm. If these electrodes are to be used for μ-EDM die-sinking precision holes, then the μ-electrodes need to be straight and have a high length-to-diameter (*L*/*D*) ratio so that many holes can be fabricated from the same electrode before it wears away. The *L*/*D* ratio obtained varies according to the electrode material and the diameter required. Laboratory investigations by Masuzawa et al. [11] indicate that tungsten carbide performs better than tungsten, which outperforms AISI 304 stainless steel. If straightness values <0.2 μm are required, *L*/*D* ratios of 10 can be obtained in all three materials for $\phi = 10$, 20, and 30 μm; *L*/*D* ratios of 30 can only be obtained in tungsten and tungsten carbide; and *L*/*D* ratios of 50 can only be obtained in all three diameters using tungsten carbide.

Electrode geometries other than cylindrical can also be made by WEDG. Segments can be removed from the cylindrical geometry on the stationary electrode such that half-moon, barrel, and square geometries can be fabricated. Half-moon and barrel electrodes were reputed to machine cylindrical holes faster than cylindrical electrodes, as dielectric turbulence in the machining gap was enhanced, leading to faster removal of debris and more efficient machining [6] but further investigations showed little advantage to be gained if any [7].

Electrodes having a square cross-section (50 × 50 μm) are now being used to fabricate sharp-cornered cavities and slots with widths greater than the electrode width. (see Section 7.6.2) [12].

Microshafts and Pins

These parts are especially important in the assembly of miniature devices. The problem of alignment has been solved by combining μ-EDM processes with modular machining assembly (MMA) units. For instance, a pin can be machined by WEDG and then its polarity reversed so that it can machine a hole in the MMA. The pin, containing a suitable tapered feature, can then be inserted with the help of ultrasonic vibration into the hole that it has machined and fixed in position. The pin is then separated from its supporting fixture by twisting and may then be further machined by μ-EDM as a second process or be used itself as an electrode to machine a second component dependent on the polarity given to the pin.

A description of these techniques is given in the fabrication of a micropipe assembled into a macrocylinder and the fabrication of cavities into the ends of cylinders [13,14].

Micropipes

A novel process has been developed for the manufacture of micropipes, combining WEDG and electroforming, as shown in Figure 6. The core of such a pipe is prepared by WEDG in exactly the same way as one would prepare a microelectrode. The surface of the core is then treated with a release coating and electroplated with a suitable metal such as nickel or copper. Further WEDG after metal deposition produces the cylindrical outer diameter of the pipe. The core is now extracted from the pipe, taking care not to fracture it. Best results are obtained from using austenitic stainless steel as the core and polishing it to a mirrorlike finish by wire electrochemical grinding (WECG) after WEDG [15]. (WECG relies on the use of deionized water with a specific resistance of 0.6 MΩcm as an electrolyte to replace a dielectric fluid such as kerosene as normally used for WEDG.)

As the core of the pipe is produced by WEDG, the core need not necessarily be cylindrical. The core can be tapered

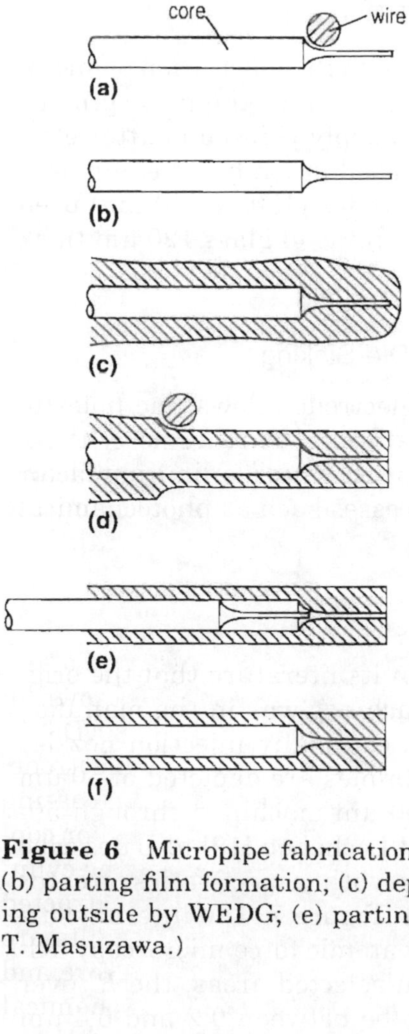

Figure 6 Micropipe fabrication: (a) Core preparation by WEDG; (b) parting film formation; (c) deposition (electroforming); (d) forming outside by WEDG; (e) parting; (f) finished nozzle. (Courtesy of T. Masuzawa.)

provided that there are no reentrant features that will prevent extraction of the core from the electroform. A single constriction may be machined into the core but this means the core must be broken at the constriction and extracted in two pieces from opposite ends of the electroform.

Other Products Made by WEDG

Various references have been found to the fabrication of micro-punches with a diameter of 70 μm used for the mass production of inkjet printer heads [5], sharply pointed emitter electrodes with a diameter of 50 μm used in ion beam equipment [5], and needles and square electrodes (150 × 150 μm) used for the electrochemical micromachining of glass 120-μm thick [16].

7.6.2 Applications of μ-EDM by Die-Sinking

Using a thin rod or wire as an electrode allows fine holes to be machined into conductive materials. Furthermore, these precision holes can attain high aspect-ratios, a limiting factor of some rival manufacturing processes such as photochemical machining (photoetching) [17].

Inkjet Nozzles

Charmilles Technologies states in its literature that the original application of its μ-EDM machines was in the manufacture of parts for inkjet printers, especially injection nozzles for bubble jet color printers. Examples are depicted of 50-μm diameter holes with a pitch of 60 μm machined through 50-μm stainless steel [18]. Allen and Lecheheb [19] also describe inkjet nozzle fabrication results with aspect-ratios greater than two. Such a hole is shown in Figure 7. The internal surface finish has been examined by atomic force microscopy. Although R_a can be 54 to 75 nm in selected areas, the R_a over a 24 × 24-μm area was found to be between 0.2 and 0.4 μm [20].

High Aspect-Ratio Holes

Charmilles Technologies also depicts 15-μm diameter holes in 100-μm stainless steel, giving an aspect-ratio of 6.67, with a smooth internal surface finish (R_{max} = 0.1 μm) [18].

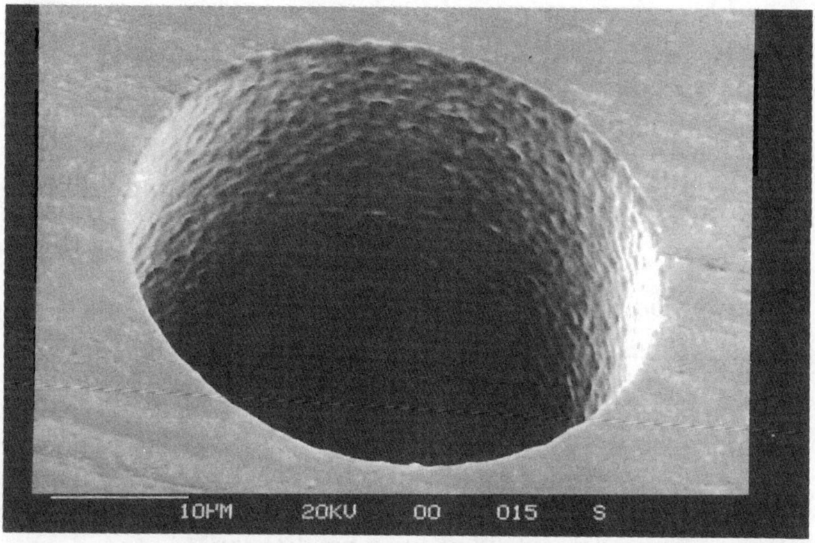

Figure 7 Inkjet nozzle.

High Aspect-Ratio Slots

Movement of a cylindrical electrode along one axis under PC control, enables the fabrication of a slot. Charmilles [18] depicts two slots 40-μm wide and 300-μm long and each 100-μm deep intersecting each other at right angles. These slots have been machined into the surface of a pen ballpoint.

Square-Cornered Cavities

As mentioned in Section 7.6.1, a square cross-section tungsten carbide electrode (50 × 50 μm) has been utilized to machine a 300 × 300-μm cavity with a flat bottom and vertical sides into AISI 304 stainless steel. To maintain the required geometry in the cavity the electrode must be used in such a way that the wear on the electrode is uniform over its surfaces. The scanning paths used to machine the steel in the x- and y-

axes are therefore strictly controlled and designed to ensure uniform electrode wear and minimize surface roughness of the machined surface.

By careful programming of the feed conditions in all three axes, cavities may also be machined where the sidewalls are no longer vertical but inclined to the surface at any angle required [12].

Other Products Made by μ-EDM Die-Sinking

The following products containing holes have also been fabricated by μ-EDM die-sinking.

Gasoline injector spray nozzles,
Dies for extrusion,
Spinnerettes for artificial fiber production,
Liquid and gas microfilters, and
Medical implants.

Holes may be machined through tubular as well as flat materials. Charmilles [18] has machined a 20-μm hole through a stainless steel hypodermic needle, and 200-μm diameter holes and slits (made by WEDM) have been machined successfully through nickel–titanium shape memory alloy tube walls 340-μm thick [21].

Micro-EDM die-sinking has also been carried out using complex microelectrodes fabricated by LIGA. For instance, Ehrfeld et al. [22] describe the formation of a gear wheel impression (mold) in stainless steel using a copper gear wheel electrode manufactured by the LIGA process, and a series of dies for hot embossing microreactor elements [23]. This same process has also been used in the fabrication of devices in a thermoplastic polymer that allows the precise alignment of optical fibers. The polymer ribbon ferrules are mass-produced from a mold containing precision grooves and inserts made by μ-EDM and LIGA [22].

7.6.3 Applications of μ-WEDM

Dies

WEDM has been used extensively in the past to make dies. Similarly μ-WEDM can be used for the manufacture of precision blanking, drawing, and injection dies. Details of an injection die are given in Table 3.

Microgear Wheels

The development of miniature electromagnetic motors has led to a demand for miniature gearboxes less than 2 mm in diameter with microgear wheels. Micro-WEDM using 20 to 30-μm wire has been employed to cut the complex tooth profiles required in the gearwheels. Ehrfeld [22] shows a ring gearwheel for a planetary gearbox with an outer diameter of 1.9 mm.

Special Orifices

Details of an orifice for a paint spray nozzle are given in Table 3. Micro-WEDM can be used to make angled cuts through conductive materials as the wire can be manipulated independently both above and below the metal being cut. Thus an orifice resulting in a fan-shaped spray can be produced.

Miniature Neurosurgical Instruments

Microforceps have been fabricated from 0.63- and 0.39-mm diameter nickel–titanium shape memory wires by bisecting the wire in a sawtooth pattern using μ-WEDM with a 30-μm diameter tungsten wire as the cathode. Microscissors were also made from the 0.63-mm diameter wire by a similar technique [24].

7.6.4 Applications of μ-EDG

Micro-EDG appears to be the least common form of μ-EDM. However, it has been reported that 60-mm long channels, 900-μm deep and 60-μm wide with closed ends have been machined

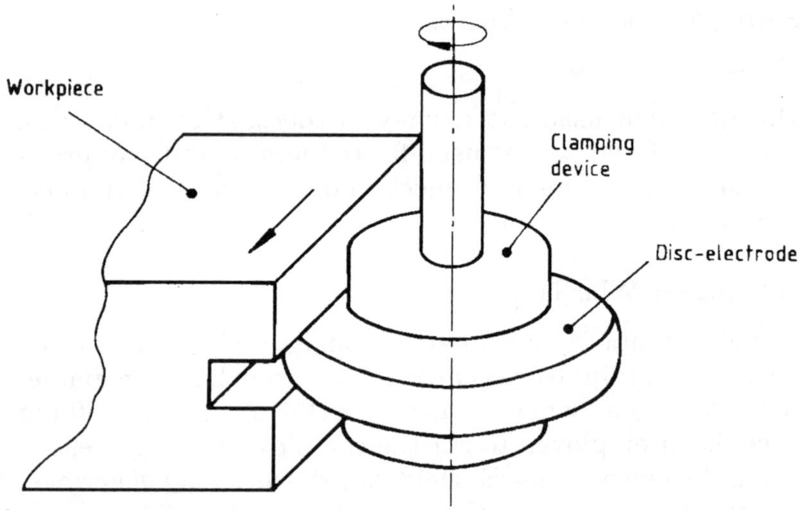

Figure 8 Micro-EDG.

into both sides of a stainless steel plate to form part of a microreactor [22]. Such microreactor structures are also used in mixing chambers, heat exchangers, and pumping systems, and in such materials as titanium diboride [23].

The electrode used to form the grooves comprises a round flat disc which is rotated as the workpiece is moved along its circumference in the opposite direction to the peripheral movement (see Fig. 8). A cylindrical electrode has been used to machine microgrippers for high-precision assembly of RF circuits by pick-and-place units [23].

ACKNOWLEDGMENT

Prof. Allen would like to thank Takahisa Masuzawa for his help in writing this chapter. The technique of μ-EDM is inextricably linked to his name as its inventor and a collection of his early papers on the topic was much appreciated.

REFERENCES

[1] T. Masuzawa, Personal communication, July 9th (1997).

[2] H. Kurafuji and T. Masuzawa, Micro-EDM of cemented carbide alloys. *J. Japan Society of Electrical-Machining Engineers* (2) 3, 1–16 (in Japanese with English Abstract) (1968).

[3] J. A. McGeough, Electro-discharge machining. In *Advanced Methods of Machining*, Chapter 6, Chapman and Hall, London, 128–152 (1988).

[4] H. Kurafuji and T. Masuzawa, Micro-energy EDM of cemented carbide alloys. *Bulletin of the Japan Society of Precision Engineering* (3) 1, 11–12 (1968).

[5] T. Masuzawa, M. Fujino, K. Kobayashi, and T. Suzuki, Wire electro-discharge grinding for micro-machining. *Annals of the CIRP* (34) 1, 431–434 (1985).

[6] T. Masuzawa, J. Tsukamoto, and M. Fujino, Drilling of deep microholes by EDM. *Annals of the CIRP* (38) 1, 195–198 (1989).

[7] T. Masuzawa, C.-L. Kuo, and M. Fujino, Drilling of deep microholes by EDM using additional capacity. *Bulletin of the Japan Society of Precision Engineering* (24) 4, 275–276 (1990).

[8] D. M. Allen and S. X. Huang, The reduction of tool wear and machining time for the micro-electrodischarge machining of micro-holes by using copper vapour laser machining as a roughing process. *Int. J. Electrical Machining* (2), 9–11 (1997).

[9] K. Takeda and F. Unno, Electrodischarge machining microholes with high ratio of length-to-diameter. *Electrical-Machining Technology* (18) 59, (in Japanese) (1994).

[10] D. Reynaerts, Personal communication, September 30th (1997).

[11] T. Masaki, K. Kawata, T. Sato, T. Mizutani, K. Yonemoti, A. Shibuya, and T. Masuzawa, Micro electro-discharge machining. In *Proc of the International Symposium for Electro-Machining (ISEM-9)*, Nagoya, 26–29 (1989).

[12] Z. Yu, T. Masuzawa, and M. Fujino, 3D micro-EDM with simple shape electrode; Part 1: Machining of cavities with sharp corners and electrode wear compensation. *Int. J. Electrical Machining* (3), 7–12 (1998).

[13] H. H. Langen, T. Masuzawa, and M. Fujino, Modular method for microparts machining and assembly with self-alignment. *Annals of the CIRP* (44) 1, 173–176 (1995).

[14] H. H. Langen, T. Masuzawa, and M. Fujino, Reverse micro-EDM and its applicability to microassembly. *Int. J. Electrical Machining* (1), 53–57 (1996).

[15] T. Masuzawa, C.-L. Kuo, and M. Fujino, A combined electrical machining process for micronozzle fabrication. *Annals of the CIRP* (43) 1, 189–192 (1994).

[16] H. Langen, I. Ceausoglu, M. v.d. Meer, E. Lehmann, H. Bleuler, and P. Renaud, Electrochemically micromachining of glass using micro-EDMed microtools. In *Proc. of the 9th Int. Precision Engineering Seminar, 2,* 672–675 (1997).

[17] D. M. Allen, Micro-electrodischarge machining. *PCMI J.* (66) Fall, 7–8 (1996).

[18] Charmilles Technologies, Micro EDM. Undated publication.

[19] D. M. Allen and A. Lecheheb, Micro electro-discharge machining of ink jet nozzles: Optimum selection of material and machining parameters. *J. Materials Processing Technology* (58), 53–66 (1996).

[20] S. X. Huang and D. M. Allen, AFM quantitative data provides new understanding of microelectro-discharge machined surfaces. In *Proc. of the 9th Int. Precision Engineering Seminar,* 1, 285–287 (1997).

[21] D. M. Allen, T. Leong, S. H. Lim, and M. Kohl, Photofabrication of the third dimension of NiTi shape memory alloy microactuators. *Proceedings of Microlithography and Metrology in Micromachining III, SPIE,* (3225), 126–132 (1997).

[22] W. Ehrfeld, H. Lehr, F. Michel, A. Wolf, H.-P. Gruber, and A. Schmid, Micro electrodischarge machining as a technology in

micromachining. *Proceedings of Micromachining and Microfabrication II, SPIE*, (2879), 332–337 (1996).

[23] A. Wolf, W. Ehrfeld, H. Lehr, F. Michel, M. Nienhaus, and H. P. Gruber, Combining LIGA and electro discharge machining for the generation of complex micro structures in hard materials. In *Proc. of the 9th Int. Precision Engineering Seminar*, *2*, 657–660 (1997).

[24] A. E. Guber, O. Baldinus, N. Giordano, M. Loser, R. Moryson, and P. Wieneke, μ-EDM technique for the fabrication of miniaturized neurosurgical instruments. In *Proc. of the 9th Int. Precision Engineering Seminar*, *2*, 665–668 (1997).

8

Laser Micromachining

Johan Meijer
University of Twente, Enschede, The Netherlands

8.1 INTRODUCTION

Laser micromachining is based on the interaction of laser light with solid matter. As a result of a complex process, small amounts of material can be removed from the surface of the solid. Two different phenomena may be identified: pyrolithic (thermal) and photolithic processes. In both cases short to ultrashort laser pulses are applied in order to remove small amounts of material in a controlled way.

Pyrolithic processes are based on a rapid thermal cycle, heating, melting, and (partly) evaporation of the heated volume. In the case of photolithic processes the photon energy is sufficient for direct breaking of the chemical bonds in a wide variety of materials. It is applied mostly on polymers by use of ultraviolet lasers in wavelengths of 157 to 351 nm. Because the photon energy is converted directly in breaking chemical bonds there is almost no thermal interaction with the product itself. The reaction products escape as gas or small particles.

An overview of application areas is given in Figure 1. In this diagram the absorbed power density I is related to the time that the laser beam is in contact with the material. A presentation of laser processes is given in Table 1.

Laser micromachining includes a wide range of processes where material is removed accurately but the term is also used to describe processes such as microjoining and microadjust-

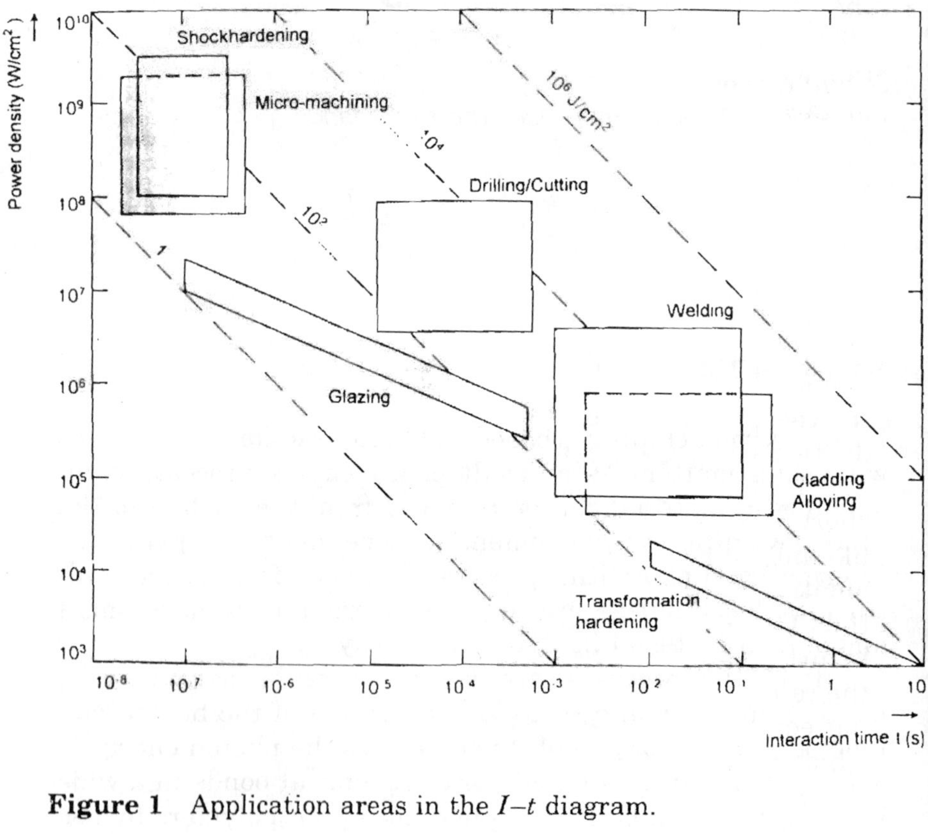

Figure 1 Application areas in the I–t diagram.

Table 1 Laser Micromachining Processes

Ablation	Wire stripping	Soldering
Cutting	Marking	Trimming
Drilling	Welding	Hardening
Decoration	Structuring	Texturing

ment by laser beam. Most applications are found in the electronics industry in high-volume production. The earliest industrial applications occurred in the 1960s in the cutting of trim grooves on conventional resistors and drilling small holes in diamonds. In the 1970s laser spot welding was applied to the production of lamps and parts of television monitors. The 1980s saw the beginning of laser micromilling and laser ablation with excimer lasers, while in the 1990s laser microadjustment was developed for use in industry. With the development of new lasers such as ultrashort pulsed lasers and passively Q-switched microlasers, new applications continue to arise. The development of smaller features is still proceeding. From the beginning of laser technology in the 20th century a reduction in size by a factor of two every seven years has been observed. The cost of production equipment is growing much faster: the complicated optics of a step-and-repeat camera for semiconductor production is now over a million dollars. Nevertheless the cost of products is being reduced, owing to the higher production volume.

Lasers used for micromachining are characterized by short pulse lengths from the millisecond range for applications like microwelding to the pico- and even femtosecond area for ablation of metals. Wavelengths vary from $\lambda = 10.6$ μm for the CO_2 laser to 157 nm for a fluorine excimer laser. The beam is further characterized by the (half) divergence angle Θ and the radius w of the beam waist. For the ideal beam the following equation applies: $\Theta \cdot w = \lambda/\pi$. This quantity is invariant which means that with "ideal" optics this relation is valid over the whole beam trajectory. A significant expression for the beam quality is the M^2 number, which is the ratio between the $\Theta \cdot w$ product for the real beam, compared to an "ideal" Gaussian beam. For a real beam the product becomes $\Theta \cdot w = M^2 \cdot \lambda/\pi$ where $M^2 \geq 1$. With focusing optics (Fig. 2) the "divergence" angle after passing the lens becomes $\Theta_f = D/2f$ resulting in a minimum spot diameter δ:

$$\delta = \frac{4}{\pi} \cdot M^2 \lambda \frac{f}{D}. \tag{1}$$

For micromachining a small spot is usually required. This can be obtained with a high beam quality $M^2 \cong 1$, with a short wavelength and a short focal length lens. Some examples are given in Table 2.

Special low power Nd:YAG lasers can be provided in nearly monomode which improves the beam quality considerably thereby enabling the smallest spot dimensions of about 6 μm to be obtained. Further improvement is possible by frequency doubling (530 and 265 nm). The copper vapor laser can also be frequency doubled (λ = 255 nm). Other properties, important for the machining process, are given in Table 3.

The introduction of passively Q-switched micro lasers, described by Guillot [1], has opened up new avenues for micromachining. A continuous wave diode laser of about 1 W is used to pump a laser material with a saturable absorber on the output window. When this solid-state Q-switch reaches the

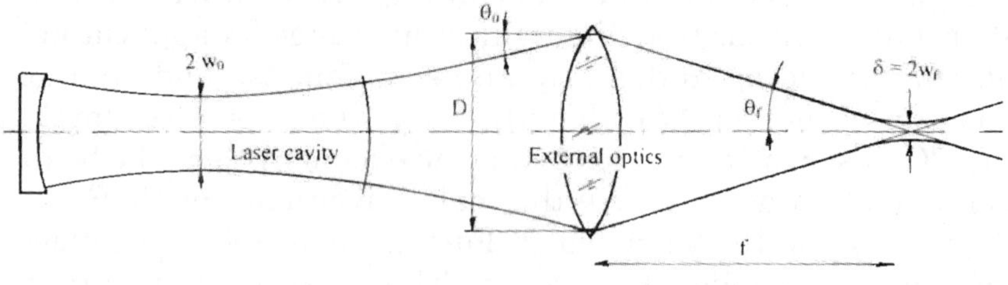

Figure 2 The product of beam waist (w) and divergence angle is invariant.

Table 2 Examples of Beam Properties

Laser	Wavelength λ (μm)	Power P (W)	$w \cdot \theta$ (mm·mrad)	Beam quality M^2 (−)	Spot diameter with f/4 lens δ (μm)
HeNe	0.63	0.002	0.2	0.98	3
Nd:YAG[a]	1.06	100	6	10	50
Nd:YAG[b]		1000	25	80	500
Q-Switched		10	2	3	15
Nd:YAG	1.06	100	6	10	50
		1000	25	80	500
CO_2	10.6	1000	10	1.5	80
Copper vapor	0.51	20	0.5	3	8
Ti:Sapphire	0.78	1			
Excimer	0.193–0.351	100	20	200	—

[a] Fine drilling or cutting mode.
[b] Normal industrial laser.

Table 3 Pulse Duration and Repetition Rate for Micromachining

Laser	Pulse duration	Frequency (kHz)
CO_2	200 μs	5
Nd:YAG	100 μs	100
Q-Switched Nd:YAG	100 ns	1
Excimer	20 ns	0.5
Copper vapor	30 ns	4–20
Ti:Sapphire	200 fs	1

threshold it becomes transparent within a nanosecond and a short pulse (0.3 to 1.5 ns) is delivered. Repetition rates are between 2 and 50 kHz.

8.2 PRINCIPLES OF LASER MATERIAL REMOVAL

The mechanism of laser beam interaction and material removal is shown in Figure 3(a). Laser energy is focused on the material surface and partly absorbed. The absorptivity depends on the material, the surface structure, the power density, and the wavelength. With a CO_2 laser about 20% is absorbed with laser micromachining while with shorter wavelengths (Nd:YAG and excimer lasers) 40 to 80% is absorbed. The remaining part is reflected. Absorption occurs in a very thin surface layer, where the optical energy is converted into heat. The optical penetration depth is defined as the depth for which the power density is reduced to $1/e$ of the initial density. For steel this depth is on the order of 15 nm for CO_2 radiation or 5 nm for Nd:YAG radiation. The absorbed energy diffuses into the bulk material by conduction. For short pulses the heat flow is approximately one-dimensional. The temperature at

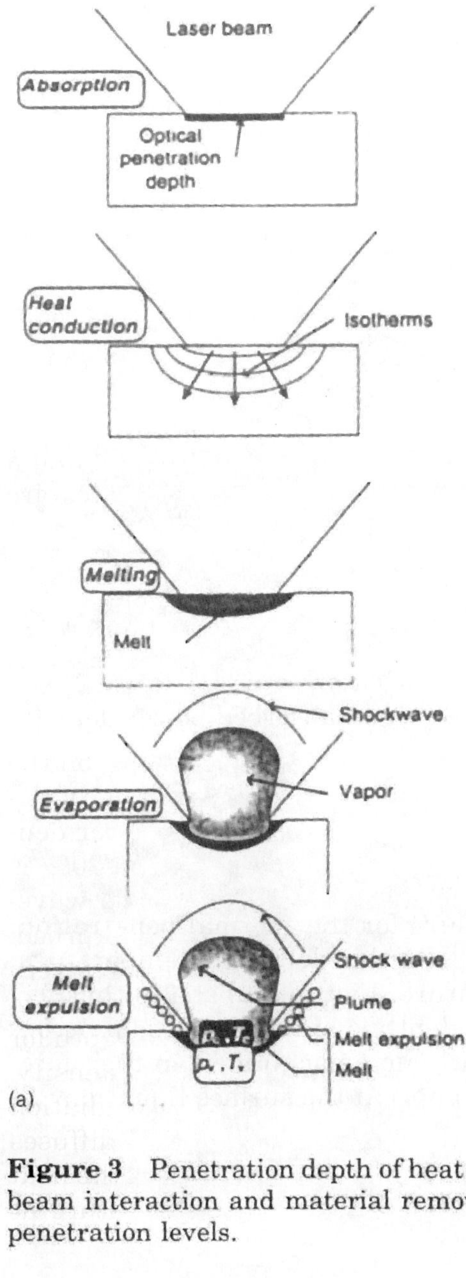

Figure 3 Penetration depth of heat for short laser pulses: (a) laser beam interaction and material removal; (b) micron and submicron penetration levels.

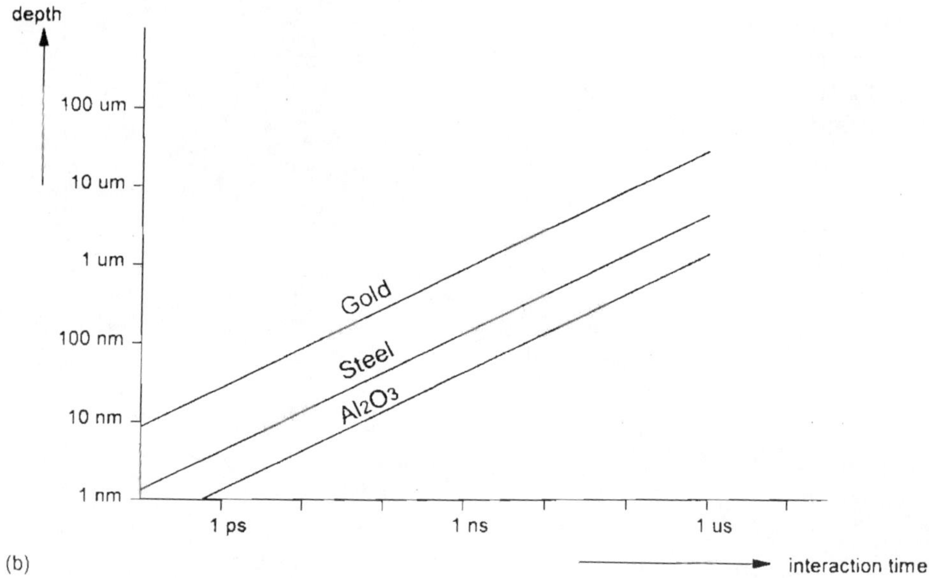

Figure 3 Continued.

the center of the spot follows with an absorbed power density I_a from:

$$T_{z,t} = \frac{I_a \delta}{\lambda}\, ierfc\left(\frac{z}{\delta}\right) \tag{2}$$

in which $\delta = \sqrt{4at}$ is a measure for the thermal penetration depth during the pulse time t. At this depth the temperature is 9% of the surface temperature. Figure 3(b) shows that for short pulses the penetration depth is at micron or submicron level. The heat penetration in micromachining can therefore be considered as one-dimensional. At the surface Equation (2) reduces to

$$T = \frac{I_a}{\lambda}\sqrt{\frac{4at}{\pi}}. \tag{3}$$

With, for example, $I_a = 10^9$ W/cm^2 on steel the melting point is reached in 300 ns. However, the time to melt is reduced by a factor of 100 to only 3 ns if the power density is increased tenfold. The high vaporization rate (vapor speeds have been reported in the range of 3 to 10 km/s) causes a shock wave and a high vapor pressure at the liquid surface considerably increases the boiling temperature. Finally the material is removed as a vapor by the expulsion of melt, as result of the high pressure and by an explosivelike boiling of the superheated liquid after the end of the laser pulse. In metals a rim of resolidified material caused by laser micromachining is clearly evident (Figs. 4 and 5). In plastics, however, the process is quite different; here the material is removed by breaking the chemical bonds of the macromolecules, and is dispersed as gas or small particles. No melt is found (Fig. 6).

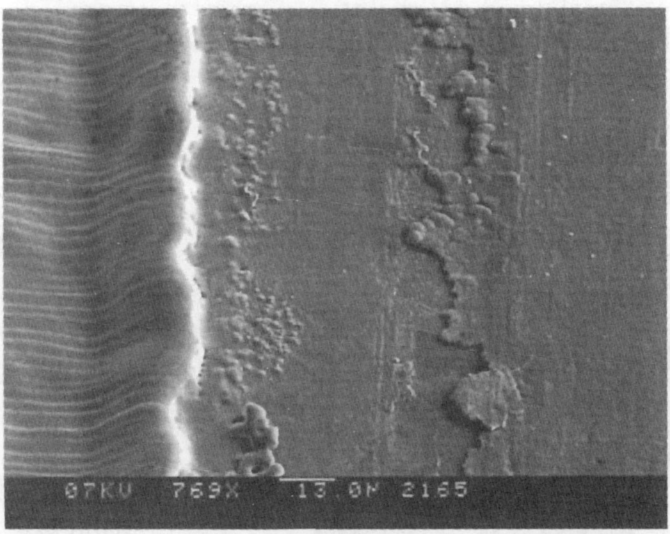

Figure 4 Rim of resolidified material after micromachining of stainless steel. Also some redeposition is evident (right): excimer laser, KrF 248 nm, 20 ns, 12 J/cm^2, 90 Hz. The ablation depth is 15 μm; rim height is 10 μm.

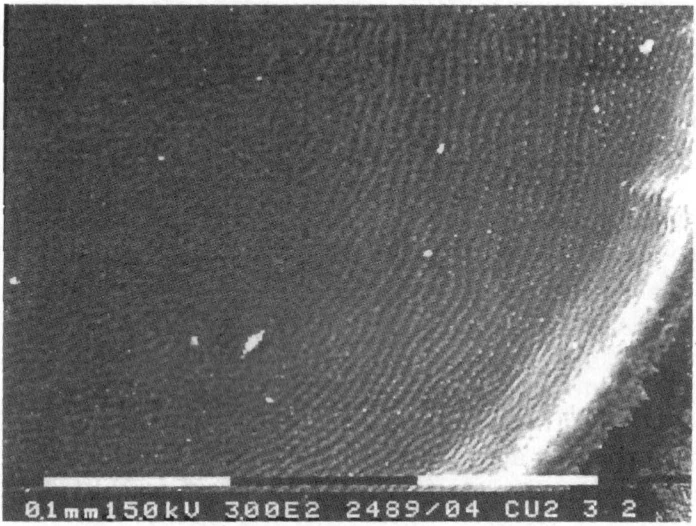

Figure 5 Rim of resolidified material after micromachining of copper: 5000 pulses, 3.5 J/cm^2, 248 nm, 20 ns, 50 Hz. Ablation depth 50 μm, rim width 20 μm, height 40 μm.

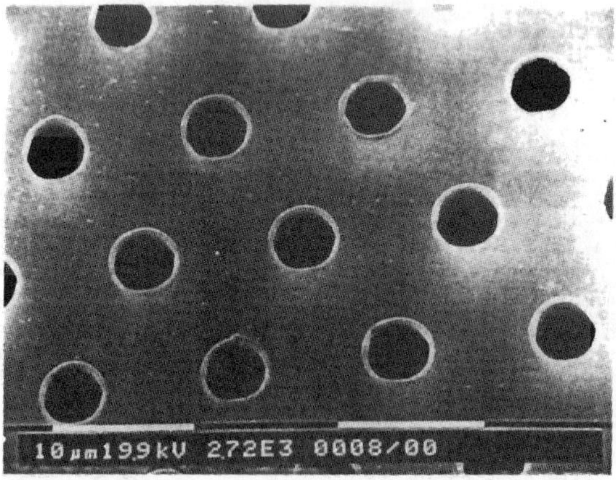

Figure 6 Fine hole drilling in polyamide: diameter 5 μm, depth 25 μm. No significant liquid phase could be observed. (Courtesy of Photonics Spectra.)

8.2.1 Plume Development

A plume takes some time to develop. If a detectable plume is assumed to form at a time t_p when the surface has reached the temperature T_v, then it follows from Equation (3) that

$$t_p = \frac{\pi E}{4 I_a^2} \tag{4}$$

in which E is the erosion resistance of the material ($E = \lambda\rho c T_v^2$), a material constant. Plume initiation times, measured by Wisselink [2], for the machining of different steels, are given in Table 4.

The measurement of t_p can be used to monitor the micromachining process. This time is a measure of the most significant process variable, the absorbed power density. The incident power density is mainly determined by the focus-surface position (Fig. 7). The optimal power density I_{mach} is identified experimentally resulting in a corresponding plume initiation time which in turn can be used in a closed loop control to maintain the distance z in the optimal processing area. Other engineers such as Van Krieken et al. [3] have developed a process based on the existence of a machining threshold. Material is removed with a constant focal position, and the removal stops when the depth z_{thresh} is obtained. This procedure is known as a self-regulating process.

8.2.2 Propagation of the Melt Front and Ablation by Evaporation

Even during short (nanosecond) laser pulses there is a heat flow into the material, which in general generates a melt pool.

Table 4 Plume Initiation Time

Laser	t_p
CO_2 laser 6 kW, 150 μs	40–100 μs
Nd:YAG, 500 W, 100 μs	2–25 μs
Nd:YAG, Q-switched, 100 ns	40–80 ns

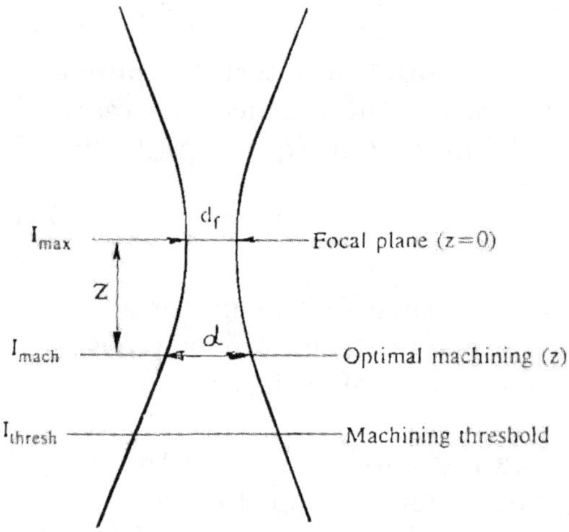

Figure 7 Machining areas. In the focal area the process is mainly drilling. Smooth micromachining is obtained at a distance z_{opt}. The machining rate decreases at lower intensities.

Only in the case of ultrashort (femtosecond) pulses is no melt generated. In the calculation of temperature profiles in the material several conditions have to be taken into account: first, the surface is moving due to evaporation of material, second, the heat of fusion creates a moving boundary, (the Stefan problem), and third, the liquid material loses its metallic properties above the critical temperature. Then the liquid metal exhibits dielectric behavior and becomes transparent.

The temperature distribution in the material is described by the Fourier equation:

$$\rho\, c_p \frac{\partial T(z,t)}{\partial_t} - \rho\, c_p \upsilon \frac{\partial T(z,t)}{\partial_z} = \lambda \frac{\partial^2 T(z,t)}{\partial z^2} + I_a \alpha e^{-\alpha z} \qquad (5)$$

in which the first term represents the temperature rise and the second describes convection, taking into account the moving surface (with velocity υ). The third term accounts for the diffusion of heat into the bulk material and the final expression gives the heat input from the laser; I_a is the absorbed laser power on the surface while the absorption in depth is given by $\alpha e^{-\alpha z}$. Under normal conditions the heat is absorbed in a very thin surface layer of about 10 nm ($\alpha \cong 10^8$ m^{-1}). At the critical temperature, however, $\alpha \to 0$ which implies that there cannot be any laser absorption in the superheated layer; all the absorption is concentrated in a small region where the temperature is just below the critical temperature (Fig. 8). Material is removed at the top of the dielectric layer by evaporation. At the sidewalls material is forced away by the plasma pressure on the liquid while at the end of the pulse, when the pressure suddenly drops, material is removed by the boiling of superheated liquid. Both effects result in redeposition around the processing area.

The origin (z = 0) is taken at the surface of the dielectric liquid which moves downward with a velocity v. The latter quantity can be calculated from Langmuir's method of vaporization from a free surface, by means of a solution of the Clausius–Clapeyron equation of the equilibrium pressure, discussed by von Allmen [4].

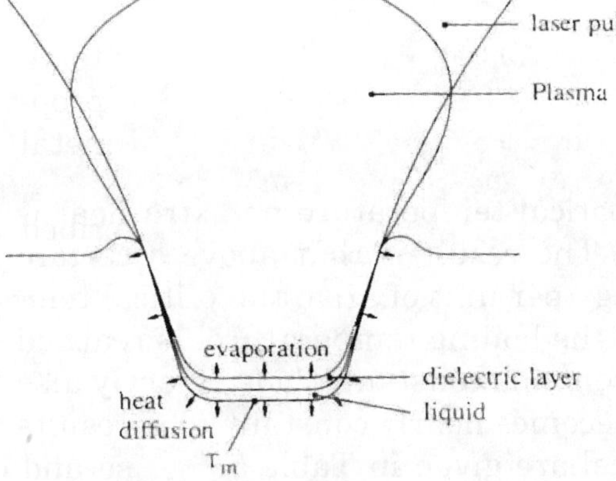

Figure 8 Ablation process (schematic). The heat input is on the boundary between the dielectric layer and the melt.

$$\upsilon(T) = \frac{1 - c}{\rho}\sqrt{\frac{m}{2\pi RT}}\,p_n \exp\left[\frac{L_\upsilon m}{RT_n}\left(1 - \frac{T_n}{T}\right)\right]. \tag{6}$$

In this equation the quantity c is a recondensation factor which is usually taken as 0.18 [see von Allmen [4] and Tokarev et al. [5]. Equations (5) and (6) can be solved analytically by calculating T for a given velocity and using this temperature to determine the corresponding υ by iteration. An analytical solution to Equation (5) is given by Meijer et al. [6,7] although it is not possible to implement analytically the absorption α as a function of the temperature and the heat of melting at the moving melt front. This situation results in surface temperatures above the critical temperature, which is not realistic.

Better results are obtained with a numerical solution in which the enthalpy H is used as the explicit variable instead of the temperature T. Then the increase in H at the melt front can be properly taken into account. The drop in absorptivity at the critical temperature is implemented by a smooth reduction of α from a constant value (about 100 μm^{-1}) to zero. The specific heat for evaporation L_υ decreases with temperature according to

$$L_\upsilon = L_{\upsilon 0}\sqrt{1 - \left(\frac{T}{T_c}\right)^2}. \tag{7}$$

This means that at the critical temperature no extra heat is required for evaporation. The result is that above a certain fluence, which is the energy per unit of area, the critical temperature (about 1.4 times the boiling temperature) is reached in a short time and then remains constant. Consequently also the evaporation velocity becomes nearly constant. The results for micromachining of steel are given in Table 5.

From the table ablation still occurs at the end of the pulse owing to the presence of a relatively thick superheated layer, which continues to evaporate while the surface is already cool-

Table 5 Numerical Results for Laser Micromachining of Steel[a]

Fluence	1 J/cm^2			2.5 J/cm^2			5 J/cm^2		
Time (ns)	5	10	20	5	10	20	5	10	20
Surface temperature (K)	2000	2500	3200	4000	4300	4300	4300	4300	4300
Ablation velocity (m/s)	0	0	0.8	5	12	12	12	12	12
Melt layer (μm)	0.03	0.15	0.3	0.2	0.3	0.65	0.35	0.8	1.4
Dielectric layer (μm)	0	0	0	0	0.1	0.4	0.3	0.6	1.3
Ablation depth at end of pulse (μm)		0			0.18			0.24	
Total ablation depth (μm)		0			0.22			0.40	

[a] Pulse Length 20 ns.

ing down but still remaining above the normal boiling temperature. From the model it was found that with a given (total) fluence the ablation depth has an optimum (maximum) for a given pulse length. For 2.5 J/cm^2 this optimum is obtained with 15 ns pulses, for 5 J/cm^2 with 50 ns, and for 10 J/cm^2 with 150 ns pulses. The ablation depth given in Table 5 is based on material removal only by evaporation. It could be expected that melt is pushed away by the vapor pressure and that boiling of the superheated melt after the laser pulse will also remove droplets of material. On the other hand the phenomenon of redeposition ensures that the melt is not removed, finally reducing the ablation rate. Nevertheless the experimental results are close to the quantities given in Table 5. In special applications such as directly focused high-quality excimer laser beams considerably higher ablation rates have been achieved, up to 20 μm per pulse with energies and pulse lengths greater than, respectively, 100 J/cm^2 and 200 ns.

8.2.3 Ablation of Metals with Ultrashort Laser Pulses

The laser–material interaction consists of a set of physical steps each characterized by its typical time constant. The laser energy is first transferred to the electrons, especially in the case of metals. The electrons will transfer the energy to the lattice and finally, within the lattice the heat is distributed further by atomic lattice collisions.

The first step, the absorption of a photon by an electron, requires about 10^{-15} s (1 femtosecond). The relaxation time of a high-energy electron, that is, the time to transfer the energy to the lattice, is about 10^{-12} s (1 picosecond). The time to diffuse the heat in the lattice by thermal conduction, over a distance equal to the optical penetration depth, is also on the order of 1 picosecond.

Three different processes may be identified, based on these characteristic times.

Femtosecond Ablation

In this case there is no energy transfer to the lattice during the pulse; all energy is stored in a thin surface layer. If this energy is more than the specific heat of evaporation there will be vigorous evaporation after the pulse. The ablation depth per pulse is given by Chichkov et al. [8] as

$$z_a \approx \alpha^{-1} \ln\left(\frac{F_a}{F_{th}}\right) \tag{8}$$

in which F_a is the absorbed and F_{th} the threshold fluence. Here the threshold fluence is the energy required to evaporate the irradiated volume of material ($F_{th} \approx \rho L \alpha^{-1}$) in which L represents the heat needed for evaporation. With $\alpha^{-1} \approx 10$ nm a threshold fluence of 0.1 J/cm^2 is obtained for ultrashort ablation of metals. From Equation (8) it is clear that at the threshold fluence there is no ablation. The fluence should be about three times the threshold to remove the irradiated layer with thickness α^{-1}. The ablation process has to be considered as a direct solid–vapor transition of a thin layer. The energy is transferred from electrons to the lattice (after the laser pulse) in a picosecond. This transfer converts the layer in a dense vapor or plasma, which expands rapidly. No time is available for heat transfer to the lattice during this series of processes. The outcome is a very precise and pure laser ablation of metals; this result has been demonstrated experimentally but has not yet been well established in production.

Picosecond Ablation

With picosecond pulses the pulse length is on the same order as the time to transfer the energy from electrons to the lattice. The lattice temperature at the end of the pulse is approximately equal to the femtosecond ablation. In this case there is also considerable evaporation and the ablation depth per pulse as given by Equation (8) can be applied. Although the heat conduction into the lattice may be neglected there will be a considerable heat flow by the free electrons during the pulse. This results in the formation of a melted zone inside the material. At the surface there is a direct solid–vapor or solid–plasma transition but deeper in the material a liquid phase is present. This condition reduces the precision of the ablation of metals compared to femtosecond pulses.

Computer simulation by molecular dynamics, discussed earlier in Chapter 3 and also by Ohmura and Fukumoto [9], has been applied in a detailed study of the evaporation process. At power densities of 10^9 to 10^{10} W/cm^2 metal is vaporized by particles, which are composed of clusters of atoms. The study is useful only for short pico- and femtosecond pulses as thermal conductivity is not considered. The absorption of laser energy in the lattice generates shock waves resulting in many small voids, which then combine into larger groupings. Finally the surface region "bounds out" and clusters of material are erupted. Slip planes arise in the solid phase due to the tensile effect of the erupted material. The speed of the ejected particles has been calculated to be as much as 6 km/s. Fragments of particles scatter forming finer particles at higher power densities. In particular, when explosion of the surface vapors occurs, a large compression field arises. The propagation velocity of the thermal shock wave is equal to the elastic wave velocity and independent of laser power density.

Nanosecond Ablation

In terms of the thermal processes during laser–material interaction the nanosecond pulses have to be considered as long pulses. The absorbed laser energy first heats the work specimen to its melting point and then to the vaporization temperature. During the interaction the main energy loss is by heat conduction into the solid. The threshold fluence for long pulses can be estimated in the same way as for ultrashort pulses. The layer thickness α^{-1}, the absorption depth in Equation (8), now has to be replaced by the thermal diffusion depth $\sqrt{at}$, which is about 0.5 μm for 20-ns pulses on metal. This condition results in a typical threshold fluence of 4 J/cm^2 for strong evaporation. Generally droplets or crater walls are observed around the machined area in the nanosecond domain.

8.2.4 Ablation of Organic Polymers

The mechanism of material removal for plastics is based on photochemical reactions with photons. The typical bonding energy for many macromolecules is in the 3 to 15 eV range, which corresponds approximately with the photon energy in the ultraviolet. The process consists of three steps: the UV photons are absorbed in the top layer typically of 0.2-μm thickness, the long chain molecules in this layer are broken into parts, and finally they are removed from the processing area in the form of vapor and small particles.

The photon energy obtained from different lasers is listed in Table 6. Only photons with higher energies can release the chemical bonds. Machining of Teflon, for instance, requires at least the photon energy as produced by the fluorine laser. Higher photon energies, however, require more expensive coatings on the applied lenses or mirrors while for short wavelengths from about 200 nm the absorption in air becomes sig-

Table 6 Photon Energies from Different Laser Sources and Required Dissociation Energies for Several Chemical Bonds

Laser	Wavelength λ (nm)	Photon energy E (eV)	Chemical bond	Bond energy E (eV)
CO_2	10600	0.12		
Nd:YAG	1064	1.16		
XeF	351	3.53	Si–Si, Cl–Cl	1.8–3
XeCl	308	4.03	C–N, C–C	3–3.5
Nd:YAG 4th harm.	266	4.65		
KrF	248	5.00	C–H, O–H	4.5–4.9
KrCl	222	5.50		
ArF	193	6.42		
F_2	157	7.43	C=C	7

Table 7 Ablation Parameters for Drilling 100-μm Holes

Type of material	Wavelength (nm)	Fluence (J/cm^2)	Ablation rate per pulse (μm)
Polycarbonate	248	4	0.4
Polyester	248	4	0.8
Polyethylene	248	3.7	1.0
	193	6	0.4
Silicone rubber	308	10	1.5
	248	30	0.2
Kapton foil	308	10	1.2
Plexiglass	193	1	0.3
Hostaform	248	2.8	0.6

nificant. Photons with more than 5.1 eV have sufficient energy to break apart oxygen molecules in their path. This aspect is illustrated by a 193-nm ArF beam passing through air and producing the characteristic ozone smell.

The threshold fluence for a wide variety of plastics is about 120 mJ/cm^2. At low fluence the walls become tapered from about 2° at 500 mJ/cm^2 to 20° at 150 mJ/cm^2. Useful fluences are given in Table 7.

Ultrashort Chirped Pulse Amplification (CPA)

In terms of basic material removal principles ultrashort laser pulses are more suitable for micromachining. A currently available type is the Ti:Sapphire laser. The optical layout of a chirped pulse amplification laser system consists of oscillator, grating stretcher, regenerative amplifier, and grating pulse compressor. The optical energy is pumped in the Ti:Sapphire crystal by argon or frequency-doubled Nd:YAG lasers. The alignment tolerances are stringent, and therefore specialized laser technical knowledge and competence are needed to operate and maintain such lasers. For this reason, CPA ultrafast lasers at present are used exclusively in scientific laboratories; see Liu et al. [10].

The rapid developments in diode-pumped all solid-state laser systems will bring ultrafast laser technology into industry. It has been demonstrated that low-cost, compact, turnkey, ultrafast laser systems can be designed for industrial applications. Laser materials suitable for direct diode pumping include Nd:glass and Yb:YAG. The Yb-doped laser materials such as Yb:YAG and Yb:glass are ideal candidates for high- and average-power ultrafast lasers. They can be efficiently diode pumped in the 900- to 980-nm band, where inexpensive diodes are available. Other components such as compact-to-align pulse stretchers and compressors also need to be developed.

8.3 INDUSTRIAL MACHINING EQUIPMENT

Currently used industrial lasers are mainly CO_2, Nd:YAG, excimer, argon ion, and copper vapor types. An overview of the pulse duration and peak power for the different lasers is given in Figure 9. General-purpose machining equipment consists of a stationary laser beam with a product holder on a horizontal *xy*-stage and a lens capable of moving in the vertical direction. The solid state Nd:YAG laser is the main vehicle for micromachining applications (Fig. 10). The energy is pumped by flash lamps into the Nd:YAG rod. The laser beam with about 6-mm diameter can be focused by lenses directly on the surface to spots of diameter 50 μm for fine drilling or cutting, to about 0.5 mm for spot welding.

For Nd:YAG laser welding a large number of glass fibers are used to transport the laser energy to the workpiece. Up to 12 fibers can be connected to one laser allowing 12 semisimultaneous welds over a single period. Alternatively fibers can be coupled to different welding stations with the same laser used on a time-sharing basis. Such machining equipment is highly

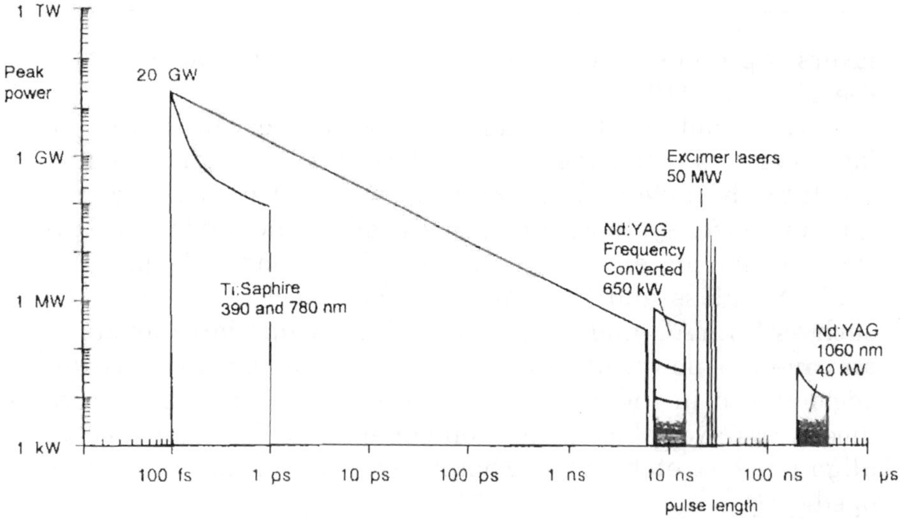

Figure 9 Overview of peak power, pulse length, and wavelength range for Nd:YAG, excimer, and titanium:sapphire lasers. (Courtesy of LZH.)

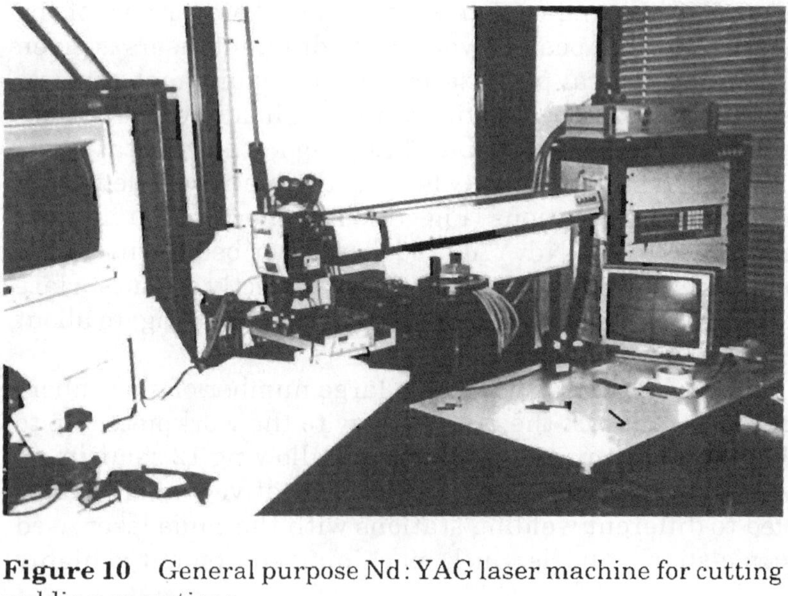

Figure 10 General purpose Nd:YAG laser machine for cutting and welding operations.

automated. The parts are supplied by vibration feeders, clamped on a rotating product holder, and then removed in short cycle times.

8.4 APPLICATIONS

8.4.1 Laser Microdrilling

Laser drilling of small holes is a widespread application. Holes are drilled in hard materials such as metals, ceramics, or diamond, in softer materials for microelectronics or medical purposes, and also in plastics to perforate foils or for ventilation. Two different techniques are used: direct focusing of the beam to the desired (small) diameter of the hole and alternatively by imaging a mask.

Direct focusing is the easier procedure as only a lens is needed. The focal length should be chosen so that the focal diameter (Eq. (1)) corresponds to the required hole diameter. This condition requires a well-defined and stable laser beam, because the beam quality M^2 directly affects the focal and hole diameter. The laser fluency should be in accordance with its required amount, about one J/cm^2 for plastics to tens of Joules per cm^2 for metals. Laser energy may be applied effectively by using an array of microlenses to obtain simultaneously tens or hundreds of holes. Several lasers are used for drilling by direct focusing. CO_2 lasers are used for high numbers of holes in thin materials (production rates over 10,000 per second). Nd:YAG lasers are employed for drilling precision holes in hard materials such as metals and diamond, while excimer lasers are used mainly for composites and ceramics.

A special application of micromachining on large-scale products is drilling holes for boundary layer suction on airplanes, pioneered by British Aerospace. Holes of 20- to 80-μm diameter are drilled in aluminium by a 1-kW excimer laser with 200-ns long pulses of high quality, $M^2 = 1.5$. About 100 holes are drilled simultaneously by means of an array of microlenses. Up to 20 pulses are required to penetrate a 1-mm thick aluminum plate. The diameter of the holes is controlled by the fluence, 20-μm holes requiring 250 J/cm^2 and 80-μm holes being drilled at 80 J/cm^2.

The other excimer laser technique, the mask projecting method (Fig. 11), images the laser beam through a mask on the substrate. This procedure enables the manufacture of shapes from simple round, square, or rectangular holes to more complex configurations. The mask is produced at a scale of about 3 times the product geometry for plastics, to over 10

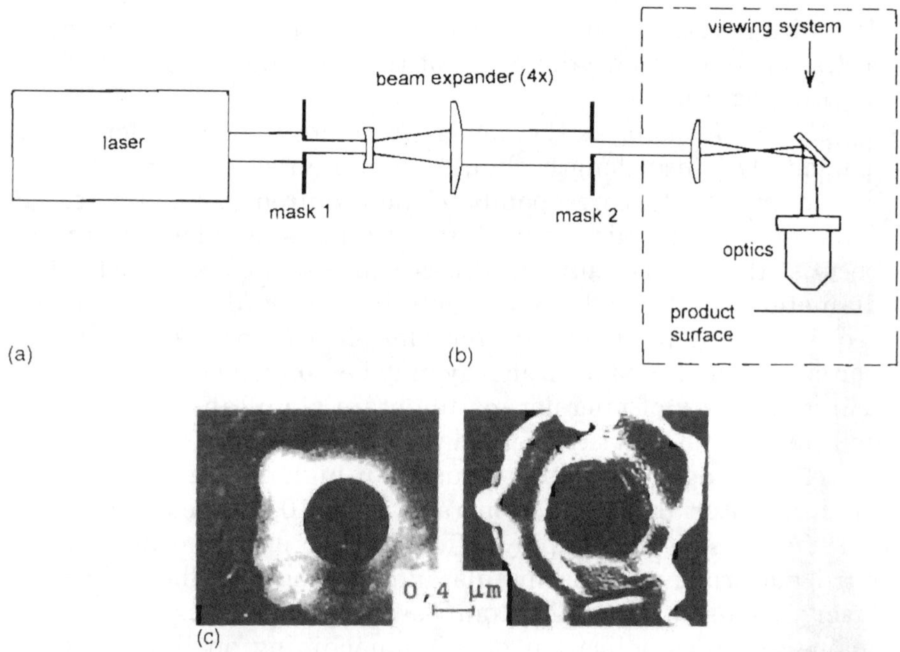

Figure 11 (a) Ablation by mask projection technique. A condenser lens might be applied to concentrate the beam in the aperture of the projection lens; (b) mask set-up for drilling small holes. Mask 1 selects the homogeneous part of the laser beam; mask 2 is imaged to the product; (c) hole, diameter 0.5 μm, drilled in molybdenum. The expelled liquid is clearly seen on the front side (right).

times for metals. Masks for holes are drilled or lasercut from sheet metal (Fig. 11), whereas masks for complex holes are produced by metal films on quartz. The demagnification rate follows primarily from the optimum laser fluence on the substrate. In particular, with high demagnification rates only small areas can be processed at one time. Larger products are processed by moving the product simultaneously with the mask, an electronic "gearbox" being used between the axes. By this technique masks of about 25 × 25 cm^2 are applied in practice (Fig. 12).

8.4.2 Laser Machining of Diamond

Diamond is difficult to machine as the material is optically transparent over a wide range of wavelengths. At high-power densities, however, the diamond is transformed into graphite, which absorbs the laser power and is subsequently removed

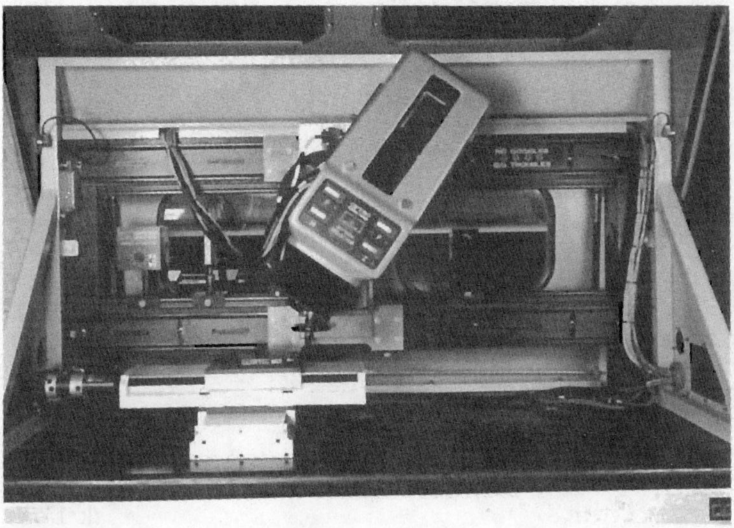

Figure 12 Computer-controlled excimer laser ablation station with flying mask.

by ablation. Diamond machining is currently undertaken by microsecond pulse Nd:YAG and nanosecond pulse excimer lasers, as discussed by Windholz and Molian [11] and Shirk and Molian [12].

Examples of applications are drilling holes in wire drawing dies (Fig. 13) and cutting knife blades for eye surgery by

Figure 13 Diamond wire drawing die "drilled" by a *Q*-switched Nd:YAG laser. Wire opening 50 μm. (Courtesy of Diamond Tools Group, Netherlands.)

Q-switched Nd:YAG lasers. Such lasers deliver high-quality 150-ns pulses of about 1 mJ at a frequency of 1 to 8 kHz and are focused to 15-μm spots. Thin layers of graphite or amorphous carbon are found on the surface after laser machining which requires an extra polishing operation to remove the graphite. A new technique is the use of ultrashort femtosecond lasers. According to Shirk and Molian [13], no evidence of graphite is found as the thermal diffusion depth is only 50 nm. Some processing data are given in Table 8.

8.4.3 Laser Microwelding

Although the technique differs from micromachining a brief discussion of laser microwelding is appropriate as it is probably the most common application. It is employed in the mass-production of electronic products such as television sets, quartz lamps, and other consumer products. An example is given in Figure 14. Wires with a diameter of 0.1 mm and even the connector feet of surface-mounted devices are welded directly on the printed wiring (Fig. 15). The welding time is 2 ms per weld whereas with a moving laser beam 50 welds/s are

Table 8 Laser Ablation of Diamond

Laser	Q-switched Nd:YAG	Excimer	Femtosecond laser
Wavelength	1.06 μm	248 nm	248 nm
Pulse length	150 ns	20 ns	500 fs
Thermal diffusion depth	30 μm	10 μm	50 nm
Fluence (J/cm^2)		Ablation per pulse	
	(μm)	(nm)	(nm)
0.8		5	5
2		10	15
4		20	35
6		30	50
10	0.5	45	
20	1.0	60	
50	2.5		
150	4.0		

Figure 14 Laser-welded grids of a television monitor.

Figure 15 Detail of a laser-welded connector pin of a surface-mounted device at a 30-μm thick copper track on a print plate.

obtained. Almost all microwelds are made by pulsed Nd:YAG lasers. The microstructure of the weld mostly shows a fine dendrite structure. Warm cracking is induced by shrinkage, which is not compensated for by freedom in the geometry. Elements such as phosphor and sulphur, particularly in combination with nickel, increase the cracking sensitivity.

8.4.4 Laser Microadjustment

The basis of laser adjustment is the generation of thermal–mechanical stresses in metal structures. An example is the fine adjustment of an audio head enabling precision adjustment in a short time, as discussed by Van der Meer [14]. The accuracy after (laser) welding of the head on the support is about 6 μm. The misalignment is measured and corrected in real-time by laser pulses on the positions given in Figure 16. A final accuracy of 0.3 μm is reached within three seconds by this method.

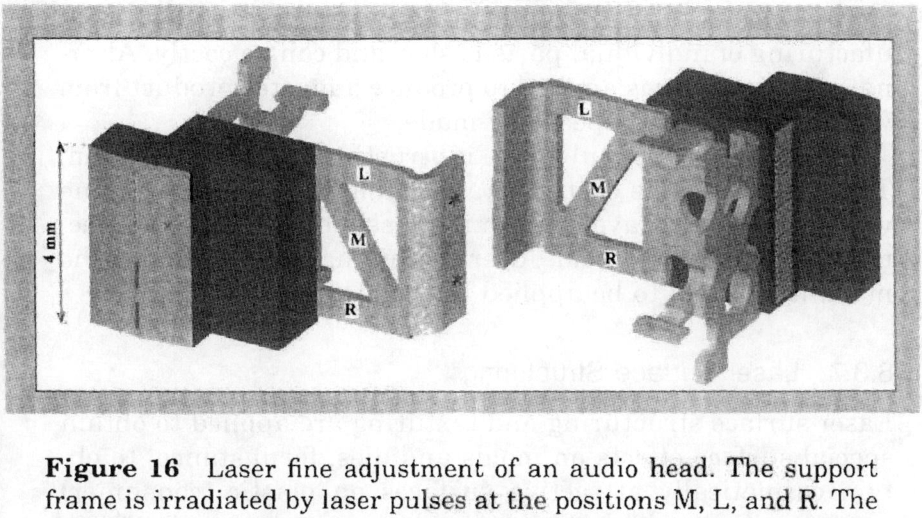

Figure 16 Laser fine adjustment of an audio head. The support frame is irradiated by laser pulses at the positions M, L, and R. The response is a downward movement, a cw rotation, or a ccw rotation, respectively. Pulses on L + R give an upward movement.

8.4.5 Laser Cleaning

Laser cleaning is very fine excimer laser machining applied to remove particle contamination and single-layer residuals from microelectronics and other particle-sensitive surfaces. Particles below 1-μm size become strongly adherent to engineering surfaces owing to Van der Waals and static forces. They are easily ablated by 248-nm excimer laser pulses of 350 mJ/cm^2, as discussed by Elliot [15]. A stream of reactive gas-like oxygen over the surface is used to remove the ablation products. A few pulses are sufficient to remove aluminum particles from a contaminated mask. Processes such as photoresist stripping may require less energy.

8.4.6 Laser Microstructuring

Microstructuring by excimer lasers is used effectively for producing fine surface structures on parts. The shape is produced by modulating the laser intensity across the sample, as is done frequently in cornea shaping for myopia correction, or alternatively by stepping a small beam across the surface thereby controlling the fluence and the number of pulses applied. Manufacturing of individual parts is slow and can be costly. Alternatively the laser is applied to produce a master product from which replicas are repeatedly made.

First a polymer substrate is structured by laser ablation. Then the surface is coated by a vacuum deposition technique with a thin metal layer, which is subsequently grown by electroplating to form a mold. After machining the rear side of the mold, it is ready to be applied to produce replicas (Fig. 17).

8.3.7 Laser Surface Structuring

Laser surface structuring and texturing are applied to obtain special surface effects on molds and dies, for instance, to obtain cosmetically attractive surfaces on plastic (consumer) products. *Q*-switched Nd:YAG and excimer lasers are used. Products can also directly be textured, marked, or colored by

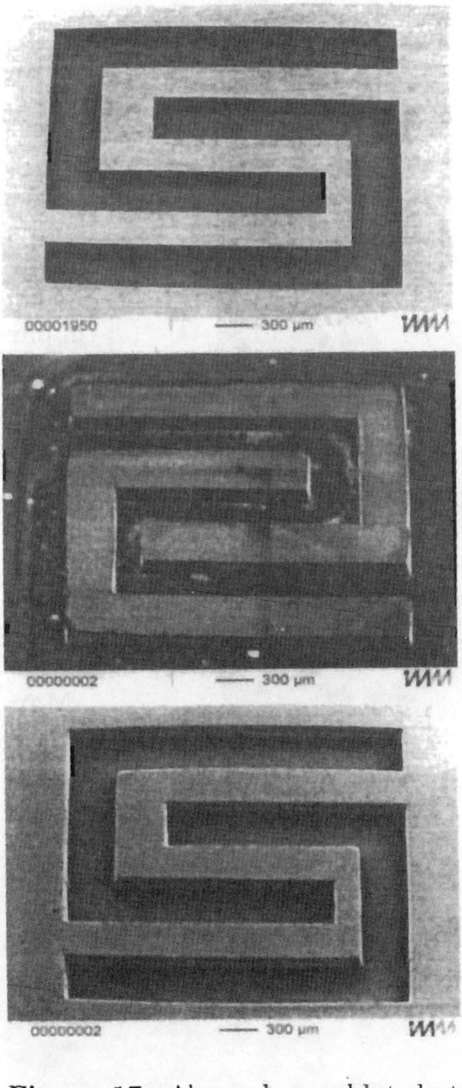

Figure 17 Above: laser-ablated structure in PMMA. Middle: inverse nickel replica. Below: injection molded product. (Courtesy of Lambda Physik.)

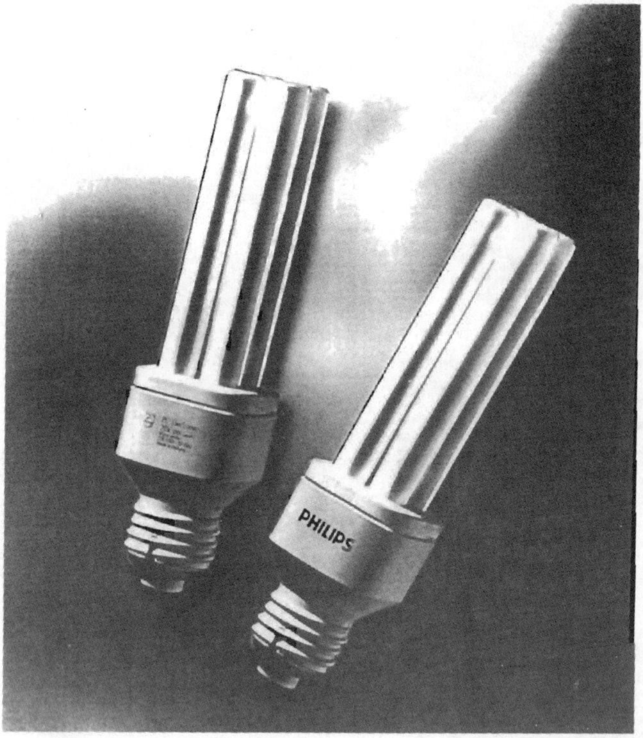

Figure 18 Laser surface modification: laser power 9 W, pulse length 120 ns. (Courtesy of Philips Lighting, Terneuzen.)

lasers. An example is given in Figure 18. A controlled carbonization causes the required grey color. The text is written with a speed of 1000 mm/s.

8.3.8 Laser Microcutting

Laser microcutting of metals and ceramics is a widespread application. The most commonly used laser is a low-power pulsed Nd:YAG type. Typical parameters for the cutting of 0.2-mm thick stainless steel are: pulse power 0.1 J, pulse length 0.1 ms, frequency 150 Hz, and cutting speed 5 mm/s. The cut

Figure 19 Laser fine cutting. (Courtesy of ILT, Enschede.)

width is less than the order of 100 μm, the roughness 2 μm, and an accuracy of about 5 μm dependent on the *xy*-table. An example is given by Figure 19.

NOTATION AND SYMBOLS

a	Thermal diffusivity ($a = \lambda/\rho c_p$) (m^2/s)
c_p	Specific heat capacity (at constant pressure) J/(kg·K)
c	Recondensation factor
D	Beam diameter (m)
E	Energy (photon) 1 eV = 1.6 10^{-19} J (eV)
E	Erosion resistance (kg/(m·s^5))
f	Focal length (m)
F	Fluence (energy density) (J/m^2)
I	Intensity (power density) (W/m^2)
I_a	Absorbed power density (W/m^2)
L_v	Latent heat of vaporization (J/kg)
m	Mass (kg)

M^2	Quality number of the laser beam
P_n	Pressure (N/m²)
P	Power (W)
R	Gas constant 8.314 J/(K·mol)
t	Time (s)
t_p	Time until plume formation (s)
T	Temperature (K)
T_c	Critical temperature (K)
T_n	Temperature of the normal boiling point (K)
T_v	Temperature of vaporization (K)
v	Velocity (of moving surface) (m/s)
w	Radius of the beam waist (m)
z	Coordinate (in direction of laser beam, depth) (m)
α	Penetration (or absorption) depth (m^{-1})
δ	Spot diameter in (focus) (m)
λ	Wavelength (m)
λ	Thermal conductivity (W/(m·K))
θ	Divergence angle (half cone) (rad)
ρ	Density (kg/m³)

REFERENCES

[1] D. Guillot, Microlasers. *Photonics Spectra* February, 143–146 (1998).

[2] F. W. A. Wisselink, Use of the plume initiation time in laser milling. Ph.D. Thesis, University of Twente (1996).

[3] A. H. Van Krieken et al., Laser micro-machining of material surfaces. *SPIE-The Int. Society for Optical Engineering* (1022), 34 (1988).

[4] M. von Allmen, *Laser Beam Interactions with Materials, Physical Principles and Applications*. Springer, Berlin (1987).

[5] V. N. Tokarev et al., Analytical thermal model of ultraviolet laser ablation with single photon analytical thermal model absorption in the plume. *J. Appl. Phys.* (78)2, 1241–1246 (1995).

[6] J. Meijer, F. W. A. Wisselink, and R. R. van Kessel, Process

control by plume registration in laser materials processing. In *Proc. ICALEO-Int. Congress on Applications of Lasers and Electro-Optics*, San Diego, November, 404–411 (1995).

[7] C. Z. Meijer and J. Meijer, Laser ablation rates of metals for 248 nm, experimental verification of an evaporation model. In *Proc. ICALEO-Int. Congress on Applications of Lasers and Electro-Optics*, San Diego, 130–139 (1997).

[8] B. N. Chichkov et al., Femtosecond, picosecond and nanosecond laser ablation of solids. *Appl. Phys. A* (63), 109–115 (1996).

[9] E. Ohmura and I. Fukumoto, Molecular simulation of laser ablation of fcc metal. *Int. J. Japan Soc. Prec. Eng.* (30)2, 128–133 (1996).

[10] X. Liu, D. Du, and G. Mourou, Laser ablation and micro-machining with ultrashort laser pulses. *IEEE J. Quantum Electronics* (33), 1707–1716 (1997).

[11] R. Windholz and P. Molian, Nanosecond pulsed excimer laser machining of CVD diamond and HOPG graphite. *J. Mater. Sci.* (32), 4295–4301 (1997).

[12] M. D. Shirk and P. A. Molian, Ultrashort laser ablation of diamond. *J. Laser Applications* (10)2, 64–70 (1998).

[13] M. D. Shirk and P. A. Molian. A review of ultrashort pulsed laser ablation of materials. *J. Laser Applications* (10)1, 18–28 (1998).

[14] G. Van der Meer, Micromachining with lasers in the electronics industry. *De Constructeur* (4), 26–30 (in Dutch) (1998).

[15] D. J. Elliot, *Ultraviolet Laser Technology and Applications*. Academic, San Diego (1995).

[16] E. K. Illy and J. A. Piper, Micromachining cuts costs for copper vapour lasers. *Photonics Spectra* March, 106–111 (1998).

[17] B. Lassiger, *Kontrollierter Formabtrag durch Sublimation mittels Laserstrahlung*. Verlag Shaker (in German) (1995).

9

Micromachining by Electrochemical Dissolution

Madhav Datta*
IBM Corporation, Yorktown Heights, New York

9.1 INTRODUCTION

Several nonconventional machining processes such as electrochemical machining and electropolishing are based on the principle of electrochemical metal removal [1,2]. These processes involve controlled metal dissolution from a workpiece that constitutes the anode in an electrolytic cell. Some of the unique features of these methods include their ability to machine complex features and complicated contours without machining marks, burrs, or surface stresses.

In electrochemical machining (ECM), a cathode tool is slowly advanced towards the workpiece anode but is separated

* *Current affiliation*: Intel Corp., Hillsboro, Oregon

from it by a small gap through which an electrolyte flows. ECM is characterized by a high current density (10 to 300 A/cm^2), a high electrolyte flow velocity (5 to 50 m/s) and a narrow interelectrode spacing (0.1 to 2 mm). A high electrolyte flow velocity is required to remove reaction products and to dissipate the generated heat. A close spacing is essential to reproduce the contours of the cathode onto the anode workpiece. Neutral salt solutions are generally used as electrolytes. The highest known rate for an electrochemical process is achieved in ECM where metal dissolution rates up to 10 mm/min are commonly attained.

Electropolishing is generally employed as a finishing operation to remove surface roughness from a workpiece, and therefore requires the removal of only small amounts of material. Electropolishing is usually carried out in concentrated acids with little or no electrolyte agitation, at current densities between 0.01 and 0.5 A/cm^2. The interelectrode spacing is not critical in electropolishing. Material removal rate in electropolishing is several orders of magnitude less than in ECM.

ECM and electropolishing processes are widely practiced in aerospace, automobile, and other heavy industries for shaping, milling, deburring, and finishing operations. Application of ECM in microfabrication and in the processing of thin films is referred to as electrochemical micromachining (EMM). EMM is now receiving considerable attention in the electronics and other high-technology industries, particularly as an alternative, "greener," method of processing advanced metallic parts.

In the electronics industry, subtractive methods of thin film processing by both dry and wet techniques are popularly known as etching technologies. These methods are widely employed in the fabrication of advanced components such as microelectronic packages, microengineered structures, sensors, and microelectromechanical systems (MEMS). Dry techniques for thin film etching include ion milling, sputter etching, reactive ion etching, and plasma etching [3]. These processes are particularly employed in the semiconductor industry for ultralarge scale integration (ULSI) because of their ability to remove material with precision. Wet etching methods are predominantly used in the processing of microelectronic components because of their selectivity, high etch rates, and relatively low capital investment costs. Although metal removal reactions in wet etching are generally electrochemical in nature, processes where the energy source for the dissolution reaction is derived from the etchant are popularly known as chemical etching. Electrochemical etching, on the other hand, relies on the passage of electric current for metal dissolution reaction to take place at the workpiece. Recent advances in the development of electrochemical metal removal technology for microfabrication are reviewed in this chapter.

9.2 MASKLESS AND THROUGH-MASK EMM

Microfabrication by EMM may involve maskless or through-mask material removal (Fig. 1). Microfabrication by maskless EMM requires highly localized material removal induced by capillary drilling or by impingement of a fine electrolytic jet [4,5]. High aspect-ratio holes are drilled by using a fine cathode tool in the form of a capillary that is advanced at a constant rate towards the workpiece. Thin film jet etching, on the other hand, uses a fine jet of electrolyte without advancement of the jet. Investigation of jet and laser-jet EMM demonstrates that neutral salt solutions can be effectively used for high-speed micromachining of many metals and alloys [5]. These studies demonstrate the feasibility of employing an electrolytic jet for generating complicated patterns in metallic foils and substrates. Other examples of maskless EMM include microfinishing of components and removal of unwanted layers of thin films by electromilling [6].

EMM in conjunction with a photoresist mask is of considerable interest in microfabrication of electronic components. Through-mask EMM involves selective metal dissolution from unprotected areas of a one- or two-sided photoresist patterned workpiece. Through-mask metal removal by wet etching is accompanied by undercutting of the photoresist and is generally

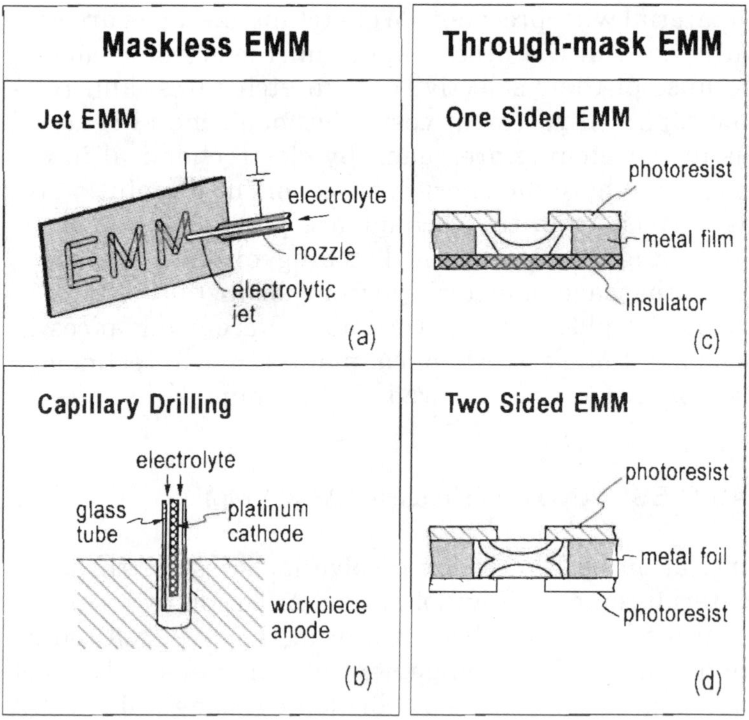

Figure 1 Different types of maskless (a,b) and through-mask (c,d) EMM applicable in microfabrication.

isotropic in nature. In isotropic etching, the material is removed both vertically and laterally at the same rate. This is particularly the case in chemical etching where the etch boundary usually recedes at a 45 degree angle relative to the surface [3]. In EMM, however, the metal removal rate in the lateral direction may be significantly reduced through proper considerations of mass transport and current distribution [7]. Photoresist undercut and etch factor are the key parameters that determine the suitability of an EMM process. Etch factor is defined as the ratio of the amount of straight-through etch to the amount of undercut [8]. For applications requiring high aspect-ratio, minimized undercutting of the photoresist and a high value of etch factor are desirable.

Production of the master artwork, surface preparation, choice of proper photoresist, and imaging are extremely important in the successful implementation of a through-mask etching process. Since parts produced by this process are a direct reflection of the master artwork, it is essential that all aspects of preparing the artwork are understood. These include a priori knowledge of the metal removal rate and etch factor. Imaging is another important step, the objective of which is to reproduce the artwork features as closely as possible onto the workpiece. The imaging process capability is measured by its resolution. In EMM, careful design of the walls and height of the photoresist provide opportunities to alter current distribution that reduce the photoresist undercutting.

9.3 BASIC PRINCIPLES

Atom-by-atom removal of metal by anodic dissolution is the basic principle underlying electrochemical metal removal processes. The workpiece is made an anode in an electrolytic cell in which a salt solution is used as an electrolyte and controlled metal removal takes place by an external current. The anodic reaction is:

$$M \rightarrow M^{n+} + ne^- \tag{1}$$

where n is the valence of metal dissolution or the number of electrons removed from dissolving metal atoms by anodic oxidation. At the cathode, electrolytic reduction of water takes place resulting in the formation of hydrogen gas and hydroxyl ions:

$$2H_2O \rightarrow H_2(gas) + 2OH^- \tag{2}$$

The cathode, therefore, remains unaltered. Depending on the nature of the material being machined and the pH of the electrolyte, metal ions and hydroxyl ions may combine to form hydroxide precipitates:

$$M^{n+} + n\mathrm{OH}^- \rightarrow M(\mathrm{OH})_n(s) \tag{3}$$

The hydroxide precipitates remain in suspended form in solution and can be filtered.

Depending on the metal–electrolyte combination and operating conditions, different anodic reactions take place at high current densities. In the ECM literature, electrolytes are generally categorized into two types: passivating (electrolytes containing oxidizing anions such as nitrates and chlorates) and nonpassivating (electrolytes containing aggressive anions such as chlorides, bromides, iodides, and fluorides). In passivating electrolytes, the current efficiency for metal dissolution is often lower than 100%. This condition arises as a part of the applied current and is consumed by an oxygen evolution reaction simultaneously occurring at the anode:

$$2\mathrm{H_2O} \rightarrow 4\mathrm{H}^- + \mathrm{O_2} + 4e^- \tag{4}$$

The metal removal rate, microfeature profile, surface finish, and uniformity of metal removal are some of the performance criteria that determine the technical feasibility of a metal removal process. In EMM, these criteria are dependent on the ability of the system to provide desired mass transport rates, current distribution, and surface film properties at the active surface (Fig. 2). An understanding of the metal–electrolyte interaction under high-rate anodic dissolution conditions is a prerequisite for choosing the optimum process parameters such as electrolyte composition and voltage/current. The development of precision tools requires an understanding of the influence of hydrodynamics, current distribution, and process parameters on the EMM performance. A precision tool should provide conditions of desired current distribution and a high rate of uniform mass transport at the dissolving surface [7]. In through-mask processes, additional issues related to lithog-

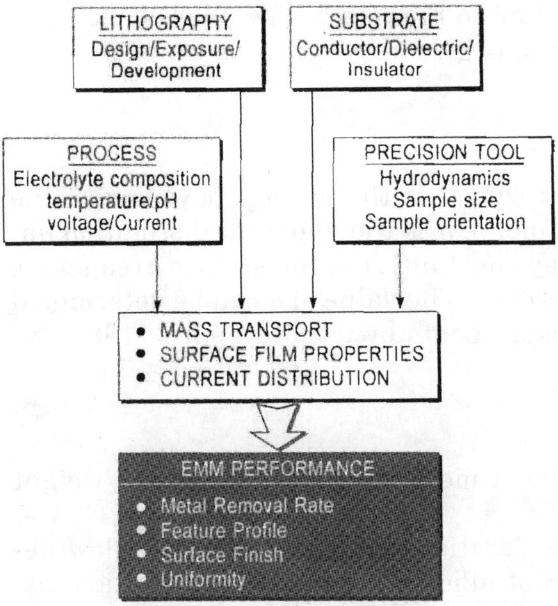

Figure 2 Dependence of EMM performance on processing and tool parameters.

raphy processing are critical to achieving desired performance. Careful design of the walls and height of photoresist masks provides opportunities to alter current distribution that reduces the photoresist undercutting. The conductivity of the substrate material is also important in influencing the current distribution of a dissolving thin film.

9.4 MATERIAL REMOVAL RATE

In an electrochemical dissolution process, the material removal rate depends on the specific electrochemical behavior of the metal/electrolyte system and is determined by the applied current density according to Faraday's Law [9]. The material removal rate, r in cm/s, is given by

$$r = \frac{IM}{nFA\rho} \tag{5}$$

where I is the current (A), M is the molecular weight of the dissolved material (g/mole), n is the apparent dissolution valence, F is the Faraday constant, A is the surface area (cm^2), and ρ is the density (g/cm^3). The value of n can be determined from weight loss measurements by use of Equation (6):

$$n = \frac{ItM}{\Delta WF} \tag{6}$$

where t is the dissolution time (s) and ΔW is the anodic weight loss (g).

With proper considerations of high electrolyte flow velocities and high current efficiency for metal dissolution, ex-

Table 1 Metal Dissolution Valence in Different Metal Electrolyte Systems

Metal	Electrolyte	Dissolution Valence
Ni	NaCl	2
Fe	NaCl	2 and 3
Ni	$NaNO_3$	2*
Fe	$NaNO_3$	2*
Ni	$NaClO_3$	2*
Fe	$NaClO_3$	2*
Cr	$NaCl$	6
Cr	$NaNO_3$	6
Cu	KCl	1 and 2
Cu	KNO_3	2 and 1
Cu	K_2SO_4	2 and 1
Ti	NaCl	4
Ti	NaBr	4
Mo	KOH	6
Mo	K_2CO_3	6

* Accompanied by oxygen evolution.

tremely high metal removal rates can be obtained. A knowledge of the dependence of n on the applied voltage/current density is essential in determining the operating conditions. The literature data on experimentally determined dissolution valences for different metal–electrolyte systems are summarized in Table 1 [9].

In ECM, the distribution of the metal dissolution rate on the workpiece determines its final shape in relation to the tool [10]. The machining performance is, therefore, influenced significantly by the current density dependence of anodic reactions. Passivating metal–electrolyte systems are known to give better ECM precision because of their ability to form oxide films and evolve oxygen in the stray current region [10]. Similar results have been confirmed during electrolytic jet EMM, where passivating electrolytes have been found to yield minimized stray cutting [5].

9.5 MASS TRANSPORT EFFECTS

Mass transport processes influence the EMM performance in several ways. First, they influence the maximum rate of an electrodissolution reaction, thus giving rise to a so-called limiting current; second, mass transport-controlled anodic reactions affect the morphology of dissolved surfaces; and finally, mass transport processes influence the macroscopic and microscopic current distribution on the workpicce. An understanding of mass transport effects are, therefore, a prerequisite for the development of EMM processes. In the following, a simple description of mass transport in electrochemical systems is presented with special reference to an anodic dissolution process.

During anodic dissolution, the concentration at the anode surface can be significantly different from that of the bulk. Since these concentrations are mainly determined by the rate of mass transport, transport mechanisms and diffusion layer thickness play an important role in high-rate anodic dissolution processes. Metal ions produced at the anode are transported into the solution by convective diffusion and migration. To maintain electroneutrality, electrolyte anions accumulate near the anode, causing the rate of convective diffusion away from the anode to be compensated by the rate of migration toward the anode. The extent of ion build-up depends on the current density, metal dissolution efficiency, and hydrodynamic conditions.

The Nernst diffusion layer concept has been used frequently to obtain a simplified description of mass transport effects in high-rate anodic dissolution of metals [9,11]. A stagnant diffusion layer of thickness δ is thus assumed to exist at the anode as shown in Figure 3. Inside the diffusion layer, a concentration gradient exists and the transport occurs exclusively by diffusion. Outside the diffusion layer, transport oc-

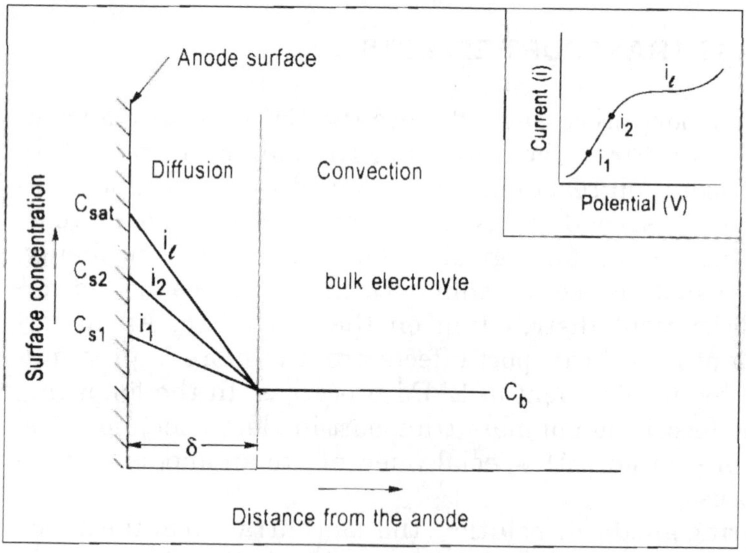

Figure 3 Nernst diffusion layer concept describing mass transport at the dissolving anode surface. A typical anodic polarization curve exhibiting a limiting current is also included (inset).

curs by convection and the electrolyte concentration is assumed to be constant. The thickness of the anodic diffusion layer depends on hydrodynamic conditions and is given by

$$\delta = \frac{L}{Sh} \tag{7}$$

where L is a characteristic length and Sh is the Sherwood number that represents the nondimensional mass transport rate. The Sherwood number can also be regarded as the normalized or nondimensionalized diffusion layer thickness. An exhaustive list of derived expressions describing $Sh = f(\text{Re}, Sc)$ are available in the literature for various flow situations and geometries [12]. For an unsubmerged circular impinging jet, the mass transport rate in the impingement region is given by [13]

$$Sh = 0.9\ \text{Re}^{1/2}\ Sc^{1/2} \left(\frac{h}{d}\right)^{-0.09} \tag{8}$$

while for a flow channel cell under laminar conditions [9]:

$$Sh = 1.85 \left(\text{Re}\ Sc\ \frac{Dh}{L}\right)^{1/3} \tag{9a}$$

and under turbulent conditions [9]:

$$Sh = 0.22\ \text{Re}^{7/8}\ Sc^{1/4} \tag{9b}$$

where Re ($= D_h\, v\, /\nu$) is the Reynolds number that characterizes the flow and Sc ($= D\, /\nu$) is the Schmidt number; D_h is the hydraulic diameter, v is the flow velocity, ν is the kinematic viscosity, h is the nozzle height, and d is the nozzle diameter.

The effect of fluid flow on the convective mass transport in a photoresist patterned workpiece is expressed by the Peclet number Pe, which is defined as follows,

$$Pe = \frac{\nu L}{D} \tag{10}$$

The influence of the Peclet number on the average mass transfer rate in a cavity has been correlated based on experimentally measured average mass transfer coefficients during etching of patterned Cu samples [14]. The following empirical correlation was obtained,

$$Sh_{av} = 0.3\left(\frac{L}{H}\right)^{0.83} Pe^{0.33} \tag{11}$$

where L and H are, respectively, the width and height of the cavity.

Equation (11) has been found to be in agreement with the average mass transfer rates calculated by solving the equations for Stokes flow by the finite element method and by a combination of the boundary integral method and Lighthill boundary layer analysis [14].

Investigation of anodic dissolution behavior of several ECM and electropolishing systems has indicated that polarization of a workpiece at anodic high potentials leads to a limiting current plateau for metal dissolution reaction [9,11,15]. These studies have confirmed that the limiting current density is controlled by convective mass transport [9,11,15].

For an anodic reaction that is controlled by convective mass transport, the anodic current density, i, is given by

$$i = nFD\,\frac{C_s - C_b}{\delta} \tag{12}$$

where D is the effective diffusion coefficient that takes into account the contributions from transport by migration [9], C is the surface concentration, C_b is the bulk concentration, and δ is the diffusion layer thickness.

An increase in current density leads to an increase in the rate of metal ion production at the anode. When the metal ion concentration at the surface exceeds the saturation limit, precipitation of a thin salt film occurs. This leads to an increase in the surface resistance thereby increasing the anode potential. The polarization curve under these conditions exhibits a limiting current plateau (Fig. 3). At the limiting current, i_l, Equation (12) becomes:

$$i_l = nFD \frac{C_{sat}}{\delta} \qquad (13)$$

where C_{sat} is the saturation concentration of the precipitating salt at the surface. The electrolyte is assumed to contain a negligibly small concentration of metal, hence C_b is taken to be zero.

Interestingly, in many systems the limiting current plateau does not correspond to the maximum machining rate; it merely signifies the limiting rate of a dissolution reaction for a given oxidation state [9]. The maximum rate of machining is limited by the ability of the system to eliminate reaction products and Joule heating. Nevertheless, the limiting current has a profound influence on the EMM performance since it influences the dissolution stoichiometry, surface finish, and current distribution as described below.

The formation of salt films on the anode influences the surface morphology of a dissolved workpiece. Different studies have conclusively demonstrated that two distinctly different surface morphologies result from dissolution [9,11,15]. At low current densities, surface etching is observed which, depending on the metal electrolyte combination, reveals crystallographic steps and etch pits, preferred grain boundary attack, or finely dispersed microstructure, leading to extremely rough surfaces. On the other hand, electropolished surfaces are obtained from dissolution at, or above, the limiting current density. Under these conditions, formation of salt films at the surface suppresses the influence of crystallographic orientation and surface defects on the dissolution process, thus yielding microfinished surfaces. The presence of salt films may increase the anode potential to such high values that dissolution reactions involving different oxidation states or the onset of oxygen evolution may become possible [9,11,15].

Several investigations have demonstrated the importance of salt films on the current distribution during EMM of photoresist masked anodes [16,17]. Below the limiting current, widely spaced patterns were found to dissolve more rapidly than the closely spaced patterns because the current distribution under those conditions was dependent mainly on the electrical field in the interelectrode gap [16]. In the presence of salt films, the nonuniformity of the electrical field plays a minor role, the rate of dissolution being governed by local hydrodynamic conditions. Therefore, at the limiting current, at which the current distribution is dependent on the rate of transport of dissolution products from the anode into the bulk, uniform dissolution occurred independently of pattern spacing [16]. The important role of surface layers of dissolution products in influencing the shape profiles during EMM has been emphasized in several publications [7,8,10,14,16,17].

From the above discussion it is apparent that the operating current density in EMM should be equal to, or higher than, the limiting current density to obtain dissolution uniformity and electropolished surfaces. The operating current density should also be high enough to yield desired directionality of dissolution [7,17]. On the other hand, application of high current density in the processing of thin films leads to an uncontrollable fast process and issues related to nonuniformities, particularly at the point of electrical contact. In order to alleviate these problems in an EMM process, it is desirable to choose conditions that provide a low rate of metal removal and hence a low operating current density. These two opposing requirements are best met by using pulsating current, which permits the application of extremely high peak currents (voltage) and, in addition to giving directionality, permits the metal dissolution to take place with high efficiency. Pulsed current EMM has, indeed, been demonstrated to be extremely effective in the patterning of thin films and foils [7,17].

9.6 PULSED EMM

The uses and advantages of pulsating current in electroplating have been well documented in the literature [18]. Several studies have shown that a pulsating current can also be advantageously employed in EMM, ECM, and finishing operations [7,17,19–21]. An important aspect of using a pulsating current is the possibility of varying instantaneous mass transport conditions at the anode by independently varying the pulse parameters [19]. In the following section, a short discussion on mass transport during pulsed anodic dissolution is presented. In principle, either current or voltage pulses of any shape can be applied. The discussion is restricted to rectangular current pulses separated by zero currents.

For a pulsating current, the average current density i_a is given by

$$i_a = i_p \frac{t_p}{t_p + t_p'} = i_p \gamma \tag{14}$$

where i_p is the peak current density, t_p is the pulse on time, and t_p', is the pulse off-time. The ratio $tp/(tp + t_p')$ is defined as the duty cycle γ.

Mass transport in pulse electrolysis is a combination of steady-state and nonsteady-state diffusion processes. In pulsed ECM (PECM), mass transport has been characterized by a duplex diffusion layer model [19]. According to the model, a stagnant diffusion layer of thickness δ exists at the dissolving anode, the value of δ depending on the hydrodynamic conditions that apply. The presence of a pulsating current leads to the formation of a time-dependant diffusion layer of thickness δ_p close to the anode surface within which the dissolved metal ion concentration is a periodic function of time. The diffusion layer thickness δ_p is predicted to be

$$\delta_p = \left(\frac{4}{\pi} D t_p (1 - \gamma) \right)^{0.5} \tag{15}$$

The pulse-limiting current density is defined as the peak current density at which the surface concentration reaches C at the end of the pulse. Its predicted behavior is given by the expression [19]

$$i_{pl} = i_l \left\{ \frac{\delta_p}{\delta} (1 - \gamma) + \gamma \right\}^{-1} \tag{16}$$

Thus by suitable choice of magnitude of pulse parameters, it should be possible to achieve high instantaneous mass transport rates even at low electrolyte flow rates. An analysis aimed at predicting electrolyte heating in narrow gaps under pulsed ECM conditions has indicated that the instantaneous and average electrolyte heating can be minimized by working with a small "pulse-on" time and a low duty cycle [19]. Experimental results obtained under well-controlled hydrodynamic conditions in a flow channel cell using passivating and nonpassivating systems have indicated that a good surface finish and high current efficiency for metal dissolution can be obtained even at low average current densities [19]. Figure 4 compares the current efficiency for nickel dissolution in a nitrate electrolyte as a function of current density, obtained by direct current and pulsed current dissolution [19]. It is interesting to note that at a current density at which most of the current is consumed for O_2 evolution in d.c. operation, use of a pulsed current yields nearly 100% metal dissolution. This behavior allows the disso-

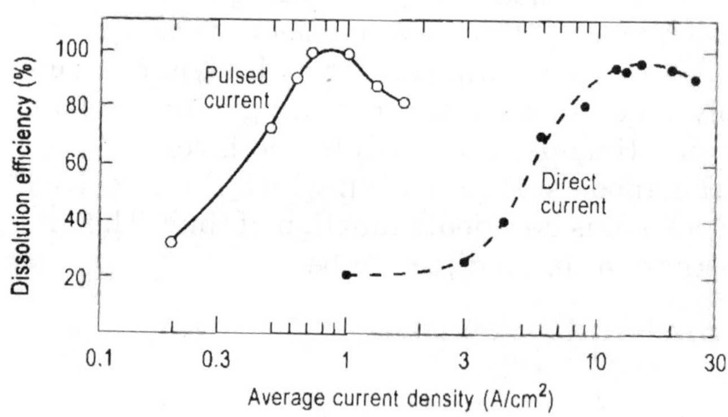

Figure 4 Current efficiency for nickel dissolution in 6-M sodium nitrate electrolyte under pulsed and direct current conditions. (From Ref. 19.)

lution of passivating metals under transpassive conditions with a high current efficiency without the need for a high average current density and high electrolyte flow.

The above concepts have been used to develop an electrochemical saw for maskless metal cutting [20]. Recently, pulsed EMM has been used to fabricate an array of high precision, microsmooth nozzles in metal foils for use in inkjet printers [17]. Pulsed dissolution has been found to be particularly suitable for through-mask micromachining of thin films, an application in which low dissolution rates are desirable for better control over the machining process [17].

9.7 SHAPE EVOLUTION AND CONTROL

The prediction of shape evolution during high-rate anodic dissolution requires solving the current distribution at the anode along with a moving boundary algorithm. The current distribution at the anode depends on the geometry, anodic reaction kinetics, electrolyte conductivity, and hydrodynamic conditions. For simulation of shape evolution during EMM, the current distribution is solved for an initial anode profile. The surface of the anode is then moved proportionately to the current distribution by use of Faraday's law. This process is repeated at several time-steps to predict the shape evolution at the anode. A wide variety of numerical techniques, such as the finite element (FEM) and boundary element methods (BEM) are available for solving fluid flow, mass transport, and current distribution problems. These numerical techniques have been extensively used to investigate shape evolution during maskless and through-mask EMM under primary and tertiary current distribution conditions [22,24].

In a recent investigation of maskless electrolytic jet-EMM, the finite element method was used for modeling the system [22]. The modeling was performed by means of FIDAP, the Fluid Dynamics Analysis Package by Fluid Dynamics International. The influence of flow velocity and interelectrode gap on the shape evolution was investigated by examining flat and cavity impingement surfaces. The mass transport rate at the side walls of the cavity was found to be higher than at the bottom or a flat surface. The high machining rate at the side walls led to lateral expansion of the machined microfeatures, leading to tapered profiles, as shown in Figure 5 [22].

The shape evolution during through-mask EMM has been modeled under the conditions of primary current distribution through the boundary element method [23,24]. Figure 6(a) shows a schematic of a photoresist-patterned metal film supported by an insulating film or substrate. During EMM, the current density along the exposed metal surface is a function of the cavity geometry and hence the undercut and the shape of the evolving surface is governed by the (i) aspect-ratio

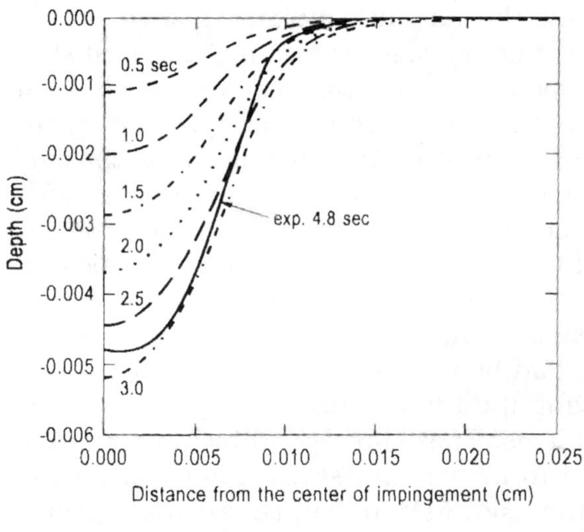

Figure 5 Numerically modeled profiles of holes at different times in a jet-EMM process, and an experimentally determined profile (solid line) of a hole in tungsten foil formed by an electrolytic jet of 3 M NaOH using parameters: nozzle diameter 100 μm; electrolyte velocity 1000 cm/s; interelectrode spacing 2 mm; cell voltage 150V; and dissolution time 4.8 s. (From Ref. 22.)

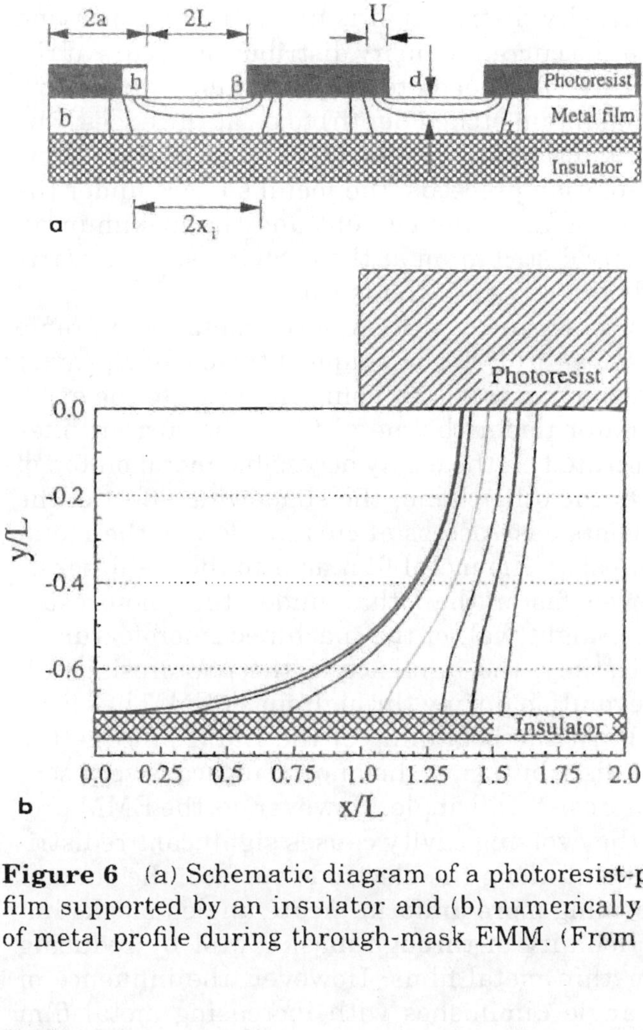

Figure 6 (a) Schematic diagram of a photoresist-patterned metal film supported by an insulator and (b) numerically modeled shape of metal profile during through-mask EMM. (From Ref. 23.)

(h/L), (ii) spacing-to-opening ratio (a/L), and (iii) film thickness ratio (b/L).

For aspect-ratios greater than 0.5, the maximum vertical displacement of the metal surface is at the center and the shape of the evolving cavity can be fit by an ellipse. The etch factor is independent of the spacing-to-opening ratio and decreases as the cavity evolves. For very low aspect-ratios, the initial current density distribution is highly nonuniform and the maximum in the current density distribution occurs at the edges of the feature. This leads to the maximum vertical displacement (or maximum etched depth) to be at the edges. The shape of this evolving cavity cannot be fit by an ellipse. However, as the dissolution proceeds, the metal surface under the photoresist draws some of the current and the maximum in the current density distribution at the evolving surface starts to move towards the center of the feature.

Figure 6(b) shows the evolution of the metal wall profile during through-mask EMM. The shape of the evolving cavity has been shown starting from the point just prior to the exposure of the insulator under the metal film. The current lines are more concentrated on the easily accessible metal on top of the insulator. On the other hand, the shadowing effect of the photoresist prevents easy access of current flow to the metal under the photoresist. The metal film near to the insulator is, therefore, removed faster than that under the photoresist. This leads to a straight wall of the machined microfeature.

In a recent study, the influence of the photoresist wall angle on shape evolution during through-mask EMM has been investigated [23]. At the beginning of the EMM process, the primary current distribution at the anode surface is very sensitive to the photoresist wall angle. However, as the EMM process continues, the evolving cavity causes significant redistribution of the current along the electrode surface, the current distribution becoming more uniform. Acute-angled masks improve the directionality of through-mask EMM by reducing the undercut for thin metal films. However, the influence of the mask wall angle diminishes with increasing metal film thickness [23].

9.8 EMM TOOLS

Development of an effective EMM process requires careful design and fabrication of a tool that provides desired current distribution and mass transport conditions at the dissolving surface. The electrolyte delivery system is one of the main considerations in the design of a precision tool. Different electrolyte delivery systems that are applicable in EMM include channel flow, electrolytic jet, slotted jet, and multinozzle systems [5,11,13,25,26]. Sample orientation, electrical contact, and provisions for filtration are some of the other important design aspects that need to be taken into account.

9.8.1 Jet-EMM Tool

In jet-EMM, a stream of electrolyte flows through a nozzle and impinges on a workpiece which is made an anode while the nozzle acts as the cathode. This leads to an extremely localized dissolution on the workpiece at the impingement region. Mass transfer studies at a submerged jet as well as a free-standing jet impinging on a barrier substrate have been reported in the literature [13,25]. For a free-standing circular jet impinging on a flat workpiece surface, there is a stagnation area in the impingement region where the mass transport boundary layer thickness is relatively independent of radial position. The diameter of this region of uniform mass transfer is roughly equal to twice the jet diameter [13].

A high-precision jet-EMM tool for maskless patterning is shown in Figure 7 [7]. The tool consists of a movable sample anode assembly, a jet assembly, and an electrolyte reservoir. All of these components are assembled on a vibration-free X–Y table. The sample holder, mounted on the arm of the X–Y table, moves in a plane perpendicular to the electrolyte jet. The nozzle, which also acts as a cathode, is a metal-shanked quartz capillary (microglass) mounted on stainless steel. The nozzle assembly is fitted to a rail-table which can be moved in the Z-direction thus allowing the adjustment of the interelectrode distance. The electrolyte is pumped through a fine filter (mace) for corrosion product removal and the flowrate is measured with a flowmeter. A three-way valve directs the flow either directly in the electrolyte tank or through the nozzle.

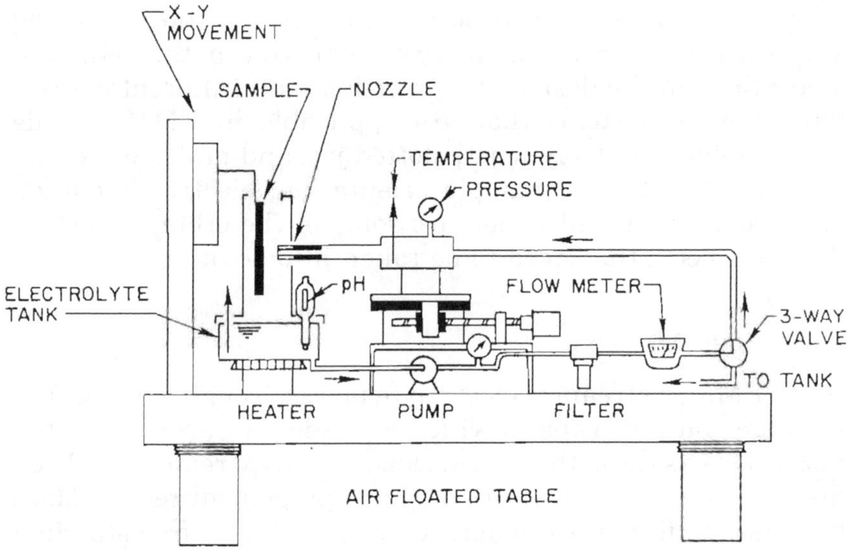

Figure 7 Schematic diagram of a high-precision jet-EMM tool for maskless patterning. (From Ref. 7.)

The temperature of the electrolyte is measured in the tank and at the nozzle. The hydraulic pressure is measured at the pump exit and just before the nozzle. A pH-meter is dipped in the electrolyte tank. Pipes and connectors are made of stainless steel. A high-voltage (400 V) power supply provides the required current or cell voltage and a high-speed multimeter is used to accurately measure the current.

9.8.2 One-Sided Through-Mask EMM Tool

Figure 8(a) shows a one-sided EMM tool [26]. The tool consists of a driving mechanism (*XYZ* table) for sample movement, and an electrolyte delivery system in the form of a multinozzle assembly. Other accessories (not shown in Fig. 8) include an electrolyte reservoir and electrolyte pumping and filtration units. The multinozzle assembly also acts as the cathode. The sample is held in a sample holder which is attached to the *XYZ*

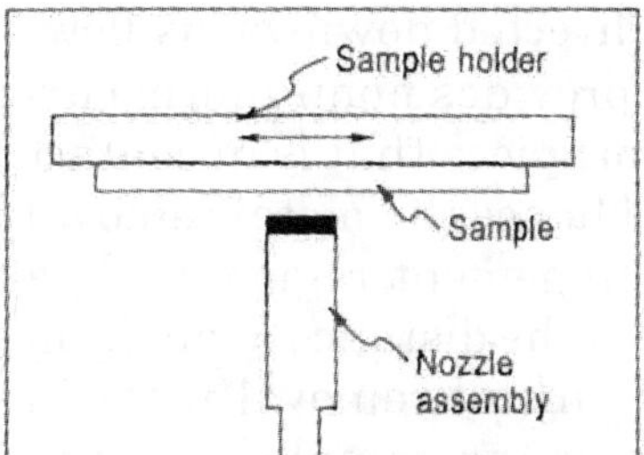

Figure 8 Schematic diagram of (a) one-sided and (b) two-sided through-mask EMM tools. (From Refs. 26 and 27.)

table and is moved at a constant speed over the multinozzle cathode. The interelectrode spacing is kept constant between 1 to 3 mm. A 25-mm wide multinozzle flow assembly provides high-speed electrolyte impingement at the dissolving surface thus permitting effective removal of the dissolved products and of the heat generated by joule heating. The tool can be used for EMM of samples of different sizes. The active area at a given time during micromachining is defined by the electrolyte in contact with the sample. Within the interelectrode gap, the electrolyte emanating from the multinozzle cathode flows towards the workpiece and is directed downwards flowing on the sides of the cathode. This provides nonuniform current distribution at the part of the workpiece that is in contact with the electrolyte. The current, and hence the metal removal rate, attains a high value in the impingement region while a gradual drop of current as a function of the distance away from the impingement region leads to low metal removal rates in these regions. Scanning the cathode or the sample serves to equalize the distribution of the metal removal rate by compensating for the stray current effect since every part of the sample undergoes both high current and stray current regions in cycle.

9.8.3 Two-Sided Through-Mask EMM Tool

Figure 8(b) shows a two-sided EMM tool that was recently developed for high-speed fabrication of molybdenum masks [27]. The tool uses a novel concept of localized dissolution induced by scanning two cathode assemblies over a vertically held workpiece providing movement of the electrolyte. Highly localized dissolution by using a small cathode width and an extremely small interelectrode spacing provides directionality of metal removal and uniformity of current distribution. The electrolyte flows through the cathode body at between 0.8 and 3 gpm. It then flows across the cathode surface between the cathode and the mask anode. Two flow types were investigated in this study: shearing flow from top to bottom, and impinging flow directed into the mask anode. The cathodes were scanned back and forth across the vertically held anode mask at rates between 0.5 and 7 cm/sec by use of an Anorad linear motion tool. Unlike many spray systems that are pressurized and involve solution spilling, the EMM tool is a nonpressurized system with electrolyte flowing in the downward direction. The tool is extremely flexible; it can handle different sample sizes and it can employ different interelectrode spacings and electrolyte flow.

9.9 APPLICATIONS OF EMM

Some selected examples of maskless and through-mask EMM processes are described in the following paragraphs. These examples demonstrate the opportunities offered by EMM in the fabrication of microcomponents, particularly as a high-speed, environmentally friendly, and cost-effective process.

9.9.1 Maskless EMM

In maskless EMM, one of the main considerations of process feasibility is its ability to localize the dissolution (machining) process so that "selective" dissolution takes place from the desired area of the workpiece. Selectivity in maskless EMM is governed by the current density dependence of anodic reactions [7,9]. In the machining area where the workpiece directly faces the electrolytic jet, the anodic reaction rate is constant. Away from the machining area, current density on the workpiece decreases to zero. In a passivating metal electrolyte system, in which oxygen evolution takes place at low current densities, current efficiency for metal removal varies as a function of the distance away from the machining area. In nonpassivating systems, on the other hand, current efficiency may remain independent of current density. Therefore, a better selectivity is expected in a passivating system because of its ability to form oxide films and evolve oxygen in the stray current region. In the following, electrochemical drilling and electrolytic jet etching methods are described as maskless EMM processes of microfabrication. Electrochemical drilling is generally used for microfabrication of high aspect-ratio holes in thick plates, which involves advancement of the tool into the workpiece similar to that in conventional ECM. On the other hand, in jet-EMM, the interelectrode gap does not change significantly during machining, since the thickness of material to be removed is very small, thus eliminating the need to move the cathode towards the workpiece.

Capillary Drilling of Cooling Holes

The jet engine industry requires drilling of high aspect-ratio, fine diameter holes through relatively thick blades [4]. These fine holes provide cool air passages to turbine blades that are vulnerable to overheating. A large number of fine holes in tough cast alloys are required to depths between 6 and 16 mm and their diameters between 0.25 and 0.4 mm. The process must produce the holes to a consistent and acceptable quality, and satisfy necessary production targets for reliability and cost. Electrochemical capillary drilling has proved to be the only machining method that fully meets these criteria. Electron beam (see Chapter 11) and laser drilling, as alternatives, do not fully satisfy the stringent surface quality standards. Electrodischarge drilling is inconsistent in performance and is not economical above the 10:1 depth-to-diameter geometry (see Chapter 7).

A schematic diagram of a capillary drilling process is shown in Figure 1(b). The drill tube is a glass capillary through which flows a dilute solution of nitric acid as the electrolyte. The cathode is a platinum wire sized to suit the fine tube bore without restricting the electrolyte flow through the tube. The wire is positioned 2 mm back from the tube tip. The process operates at 700 kN/m^2 and 100 V. The tooling features a multiple tube drilling assembly with a simple snap-on location to the machine feed and acid supply. By this technique, holes are drilled on production components within 0.05 mm of nominal position and to a dimetral tolerance of 0.05 mm. Electrochemical drilling is a cost-effective method capable of drilling fine holes above a 10:1 aspect-ratio [4].

Maskless EMM by Electrolytic Jet

Several investigations of electrolytic jet-EMM for microfabrication have been reported in the literature [5,7,22,28]. A systematic study was conducted to determine the influence of applied cell voltage, nozzle diameter, and dissolution time on metal removal rate and shape evolution of different through-

and blind-patterns [28]. A 5-M sodium nitrate solution was used as the electrolyte. Experimental data consisted of micromachining arrays of holes and slots and a preprogrammed pattern. Samples consisted of foils of copper, molybdenum, and stainless steel of thickness varying between 50 and 250 μm. The interelectrode gap was kept constant at 3 mm. The nozzle dimensions varied between 50 and 175 μm, applied voltage varied between 100 and 300V for a constant interelectrode gap of 3 mm, and dissolution time varied between 1 and 8 s for EMM of holes. Results are summarized in Table 2 [7,28], which shows that an increase in cell voltage leads to an increased micromachining rate but has little influence on the patterned diameter/width. The pattern diameter/width is approximately twice the nozzle diameter indicating that machining is mainly concentrated in the stagnation area of the impingement region.

An investigation of jet and laser-jet EMM in which sodium chloride and sodium nitrate solutions were used indicated that a laser beam helped in focusing the applied current into the machining area thereby increasing the effective machining rate and precision in a chloride solution [5]. An electrolytic jet of nitrate solution gave high current efficiency for metal dissolution and minimized stray cutting at high current

Table 2 Rate and Precision of Jet EMM

Material	Nozzle dia. (μm)	Voltage (V)	Rate (μm/s)	Hole dia. (μm)
	50	300	12.5	100
	100	100	—	173
302 SS	100	200	12.5	195
	100	300	16.7	200
	175	300	16.7	304
Copper	100	100	08.3	200
	100	200	16.7	210
	100	100	04.5	194
Moly	100	200	08.3	204
	100	300	11.0	205

densities. However, the incorporation of a laser beam in a jet of nitrate solution was found to be undesirable since it promoted oxygen evolution.

Figure 9 shows a photograph of a micromachined pattern in a stainless foil produced with a high-speed maskless jet-EMM tool. The jet-EMM technique has also been extended to many other materials including conducting ceramic films. It is expected that jet-EMM will find applications in many specialized processes such as circuit repair and micromachining of hard to machine materials. In these applications jet-EMM is particularly attractive since it eliminates the need for expensive photolithographic steps.

9.9.2 Through-Mask EMM

Application of through-mask EMM in microfabrication requires an understanding of some of the complexities and challenges associated with the process. The most important of

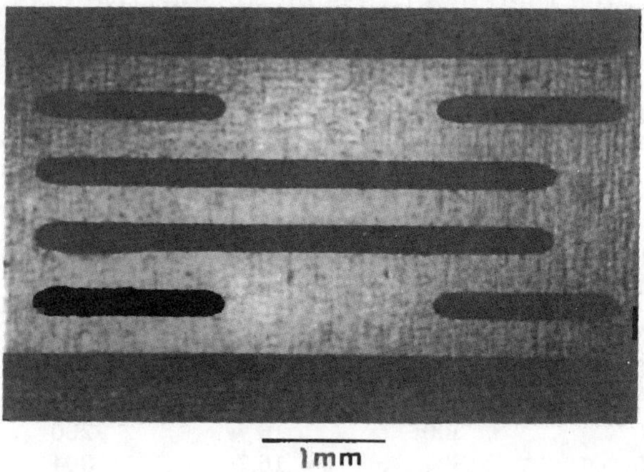

Figure 9 Through-slots micromachined in a 50-μm thick stainless steel sheet using a 5-M sodium nitrate electrolyte jet. The nozzle diameter is 200 μm. (From Ref. 5.)

these is the elimination of loss of electrical contact in the case of one-sided EMM. Other challenges include the ability of EMM to provide: uniformity of metal removal on the sample scale and on the feature scale, straight and smooth walls, and minimized undercutting for fine features. These factors are primarily governed by current distribution and mass transport conditions on the dissolving sample.

In the following, several different applications of through-mask EMM are presented. Two examples show the use of one-sided through-mask EMM to obtain patterns with precise angular walls and conical structures. The third example shows the application of two-sided through-mask EMM in the fabrication of metal masks.

Inkjet Nozzle Plates

Inkjet printing technology relies on the basic principle of forcing ink through a nozzle in the printer head. Inkjet printers used for computer output typically have nozzle diameters in the 40-to 100-μm range [29]. The print quality is very sensitive to small differences in nozzle shape and dimensions. Electroformed nozzles are currently used in inkjet products manufactured by Siemens, Dataproducts, and Hewlett-Packard [29]. Electroformed nozzles are produced by plating nickel onto a mandrel (mold) which defines the image of the nozzle, and then removing the finished part [29,30]. A thin protective film of gold is often used in cases where a particular ink otherwise might corrode the nickel. The electroforming process, however, is limited to materials that can be electroplated and is relatively expensive for low-end applications. In a recent publication, a cost-effective, high-speed process for the fabrication of precision nozzles using through-mask EMM has been described [17]. The process involves fabrication of a series of flat-bottomed V-shaped nozzles in a metal foil. The process is applicable to a variety of materials including high-strength corrosion-resistant materials such as conducting ceramics.

Through-mask EMM, therefore, provides the possibility of fabricating high nozzle density plates employing mechanically stable foil materials.

The fabrication of nozzle plates by EMM involves the following steps [17]. A cleaned metallic foil is laminated with photoresist on both sides of the foil. The photoresist on one side is then exposed and developed to define the initial pattern, consisting of an array of circular openings. A controlled EMM process is employed to fabricate flat-bottomed V-shaped (frustum of right circular cone) nozzles on the sample. The photoresist is then stripped and the sample is inspected for entry and exit holes. A sample typically consists of a series of photoresist-patterned nozzle plates each containing thousands of exposed vias to be micromachined. A 25-μm thick stainless steel foil is laminated with 25-μm thick photoresist on both sides. The photoresist on one side is exposed and developed for patterning while the blanket photoresist on the back side of the foil serves as a protective insulating layer. The photoresist pattern consists of an array of circular openings, 55 μm in diameter.

Direct and pulsed voltage experiments were performed with a neutral salt solution of sodium chloride and glycerol mixture as the electrolyte. The results demonstrated the importance of the role of mass transport-controlled-limiting current and surface films in the fabrication of precision nozzles with smooth surfaces. By controlling the extent of micromachining, nozzles of desired shapes could be fabricated. An array of nozzles fabricated by this method is shown in Figure 10 [17]. The final nozzle shape was determined by several factors that included undercutting, etch factor, dissolution time, and dissolution conditions.

Pulsating voltage EMM was found to be effective in providing dimensional uniformity of an array of nozzles. This was due to the possibility of applying extremely high peak currents (voltage) which, in addition to giving directionality, enabled breakdown and elimination of inhibiting layers thus facilitating activation of all the openings at the same time.

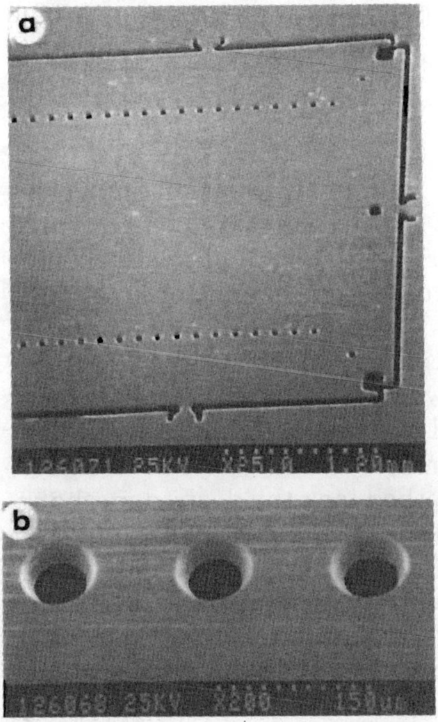

Figure 10 SEM photographs of nozzles fabricated in a 25-micron thick stainless steel foil using one-sided through-mask EMM technique: (a) part of the nozzle plate showing an array of nozzles; and (b) details of nozzles showing precision and surface smoothness. (From Ref. 17.)

The feasibility of fabricating an array of hundreds to thousands of precision nozzles with microsmooth surfaces in copper and stainless steel foils was thus demonstrated. On a 25-μm thick stainless foil patterned with a photoresist opening of 55-μm diameter, a specified exit hole dimension of 55 μm was achieved with a standard deviation below 2.0. A desired nozzle angle of 27° was produced with an etch factor of 2 (etch factor was defined earlier in Section 9.2). The results of these studies conclusively demonstrated the effectiveness of through-mask EMM in the fabrication of precision nozzle plates for inkjet printers [17].

Cone Connectors

Cone connectors represent a new generation of connectors with low total load which finds application in pad-on-pad cable connectors for flex, chip burn in pads, high performance boards, and the like [31]. Effective cone connector structures are characterized by small tips, tall cones, and strong material of fabrication. At present the cones are fabricated by laser ablation of polymeric films followed by metallization. This technique produces relatively good quality cones but involves several steps thus making the process expensive. Furthermore, the cones fabricated by this method lack the desired mechanical strength. A high-speed process of fabricating cones by through-mask EMM has been developed and patented [31]. The process is applicable to a variety of metals and alloys and is independent of the hardness of the material.

A photoresist pattern in the form of evenly spaced dots is generated on the metallic material that is suitably selected for the pad-on-pad connector. During EMM, the anode material dissolves in those areas that remain unprotected by the photoresist. As anodic dissolution continues, the removal of material between the dots leads to formation of cavities and finally leads to the formation of cones as shown in Figure 11. Preferential dissolution in the desired direction is achieved by employing a multinozzle assembly in which extremely high impinging electrolyte flow can be applied. Cones on copper and hardened stainless steel (Fe–13Cr) sheets have been generated by this method. Figure 11 shows SEM microphotographs of the cones fabricated on a hardened stainless steel sheet. The desired size and shape of cones could be obtained by proper design of the photoresist dimensions and by properly controlling the amount of charge passed during EMM. By choosing proper electrolyte and machining conditions many other metals and alloys can be used to fabricate such cones.

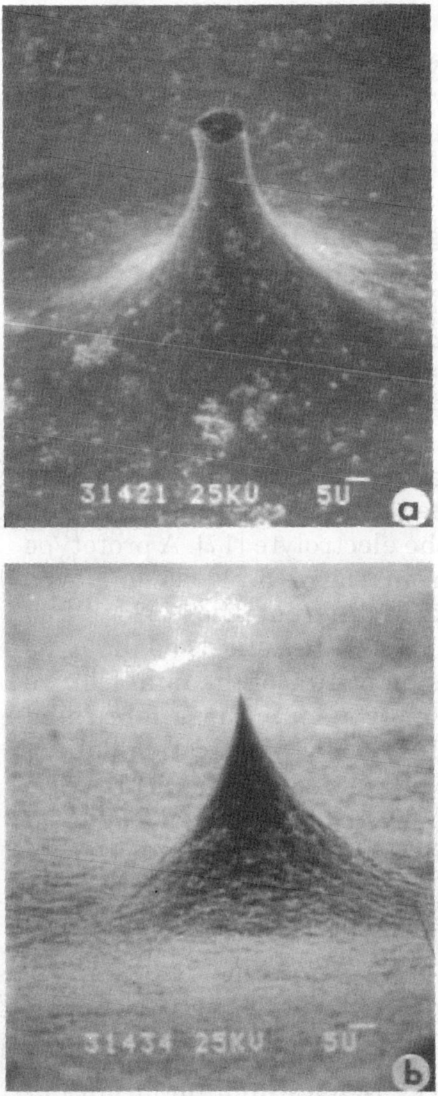

Figure 11 SEM photographs of cones fabricated by one-sided through-mask EMM; cones with (a) flat tip, and (b) extremely fine tip. (From Ref. 31.)

Metal Masks

Fabrication of metal masks involves "through" patterning by etching of a foil that is coated with perfectly aligned patterned photoresist on two sides. In a typical present-day processing, molybdenum masks are etched in a spray etcher using heated alkaline potassium ferricyanide solution. The solution will lose its etching activity as a larger quantity of the ferricyanide is reduced to ferrocyanide. Larger volume users regenerate the etchant electrochemically or by using a chemical oxidizer such as ozone. The spent etchant must be disposed of as hazardous waste. In addition, the rinsewater from the etching operation must also be segregated and treated as a hazardous waste stream.

As a greener alternative, a novel EMM process has been developed recently for high-speed fabrication of molybdenum masks using a salt solution as the electrolyte [17]. A prototype precision tool of the type shown in Figure 8(b) was employed to fabricate molybdenum masks of different sizes (225 × 225 mm and 250 × 250 mm). Features on the sheet, as many as 120,000, were etched to a precision of 10% of the total feature size. An SEM photograph of a molybdenum mask fabricated by EMM is shown in Figure 12. The microfabrication data of EMM obtained in salt solution at ambient temperature were compared with those obtained by a conventional chemical etching process with ferricyanide solution. Performance criteria included machining rate, surface finish, aspect-ratio, and simplicity of operation. The metal removal rate in EMM was found to be orders of magnitude higher than that in chemical etching. Operating EMM at, or higher than, the limiting current density provided conditions for microsmooth surface and patterning uniformity. Ability to maintain a thin layer of salt film at the surface was a key to obtaining uniformity of etching and a high aspect-ratio. A higher aspect-ratio was achieved by increasing electrolyte impingement at the surface [17].

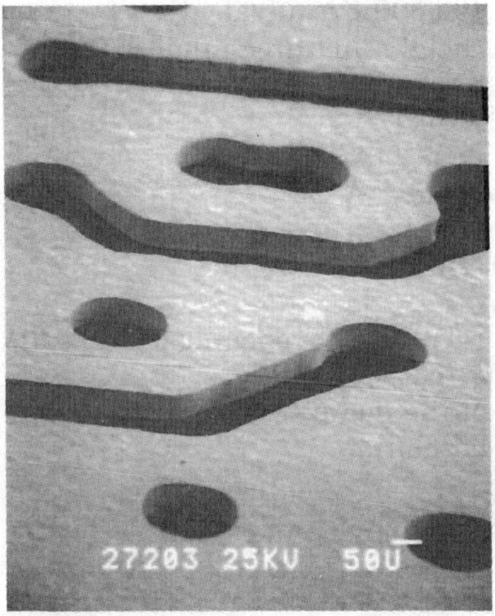

Figure 12 SEM photograph of a metal mask fabricated by two-sided through-mask EMM. Note that smooth surfaces and straight walls are obtained by EMM. (From Ref. 27.)

SYMBOLS AND NOTATION

a	Half width of photoresist covered spacing (cm)
A	Surface area (cm^2)
b	Thickness of metallic film (cm)
C_s	Concentration of metal ions at the surface (mol/cm^3)
C_{sat}	Saturation concentration of the precipitating salt (mol/cm^3)
C_b	Concentration of metal ions in the bulk solution (mol/cm^3)
d	Nozzle diameter (cm)
D	Effective diffusion coefficient of metal ions (cm^2/s)
D_h	Hydraulic diameter (cm)

F Faraday constant (96,500 C/equivalent)
h Photoresist mask height in through-mask EMM (cm)
i Current density (A/cm^2)
i_a Average current density (A/cm^2)
i_l Limiting current density (A/cm^2)
i_p Peak current density (A/cm^2)
i_{pl} Pulse limiting current density (A/cm^2)
I Applied current (A)
L Characteristic length or feature opening width (cm)
L' Final feature width (cm)
M Molecular weight of the metal (g/mol)
n Valence of metal dissolution
Pe Peclet number
r Material removal rate (cm/s)
Re Reynolds number
Sc Schmidt number
Sh Sherwood number
t Time (s)
t_p Pulse-on time (s)
t_p Pulse-off time (s)
v Flow velocity (cm/s)
w Half-width of the feature (cm)
ΔW Amount of material removed (g)

GREEK

β Photoresist angle
δ Diffusion layer thickness (cm)
δ_p Pulsating diffusion layer thickness (cm)
ν Kinematic viscosity of electrolyte (cm^2/s)
ρ Density of material (g/cm^3)
γ Duty cycle

REFERENCES

[1] J. A. McGeough, *Advanced Methods of Machining*. Chapman & Hall, New York (1988).

[2] W. J. McTegart, *The Electrolytic and Chemical Polishing of Metals*. Pergamon, London. (1956).

[3] D. P. Seraphim, R. C. Lasky, and C. Y. Li, Editors, *Principles of Electronic Packaging*. McGraw-Hill, New York (1989).

[4] D. A. Glew. In *Proceedings of ISEM-6*, Cracow, Poland, 309 (1980).

[5] M. Datta, L. T. Romankiw, D. R. Vigliotti, and R. J. von Gutfeld, Jet and laser-jet electrochemical micromachining of nickel and steel. *J. Electrochem. Soc.* (136) 2251 (1989).

[6] C. van Osenbruggen and C. de Reg. *Phillips Tech. Rev.* (42) 22 (1985).

[7] M. Datta, Electrochemical micromachining. In *Electrochemical Technology: Innovations and New Developments*. N. Masuko, T. Osaka, and Y. Ito, Editors, Kodansha/Gordon and Breach, Tokyo, 137 (1996).

[8] M. Datta, *The Electrochem. Soc. Interface* (4) 2, 32 (1995).

[9] M. Datta, *IBM J. Res. Dev.* (37) 2, 207 (1993).

[10] M. Datta, R. V. Shenoy, and L. T. Romankiw, *J. Engineering for Industry, Transactions of the ASME* (118) 29 (1996).

[11] M. Datta and D. Landolt, *Electrochim. Acta* (25) 1255, 1263 (1980).

[12] J. R. Selman and C. W. Tobias, *Adv. Chem. Eng.* (10) 211 (1978).

[13] D. T. Chin and K.-L. Hsueh, *Electrochim. Acta* (31) 561 (1986).

[14] R. C. Alkire, H. Deligianni, and J.-B. Ju, *J. Electrochem. Soc.* (137) 818 (1990).

[15] M. Datta and D. Vercruysse, *J. Electrochem. Soc.* (137) 3016 (1990).

[16] E. Rosset, M. Datta, and D. Landolt, *J. Appl. Electrochem.* (20) 69 (1990).

[17] M. Datta, *J. Electrochem. Soc.* (142) 3801 (1995).

[18] J. C. Puippe and F. Leaman, *Theory and Practice of Pulse Plat-*

ing. American Electroplaters and Surface Finishers Society, Florida, (1986).

[19] M. Datta and D. Landolt, *Electrochim. Acta* (26) 899 (1981); (27) 385 (1982).

[20] M. Datta and D. Landolt, *J. Appl. Electrochem.* (13) 795 (1983).

[21] E. Rosset, M. Datta and D. Landolt, *Plat. Surf. Finish.* (71) 60 (1985).

[22] S. J. Jaw, J. M. Fenton, and M. Datta, Electrochemical microfabrication 11. In *ECS Proceedings*, Vol. 94-32, M. Datta, K. Sheppard, and J. Dukovic, Editors, The Electrochemical Society, New Jersey, 217 (1995).

[23] R. V. Shenoy and M. Datta, *J. Electrochem. Soc.* (143) 544 (1996).

[24] R. V. Shenoy, M. Datta, and L. T. Romankiw, *J. Electrochem. Soc.* (143) 2306 (1996).

[25] R. C. Alkire and T.-J. Chen, *J. Electrochem. Soc.* (129) 2424 (1982).

[26] M. Datta and L. T. Romankiw, U.S. Patent No. 5,284,554 (1994).

[27] M. Datta and D. Harris, *Electrochimica Acta* (42) 3007 (1997).

[28] C. Clerc, M. Datta, and L. T. Romankiw. In *Patterning Science and Technology*, R. Gleason, G. J. Haffron, and L. K. White, Editors, Electrochemical Society, New Jersey, PV 90-1 (1990).

[29] W. J. Lloyd and H. H. Taub, Ink jet printing, in *Output Hard Copy Devices*, R. C. Durbeck and S. Sherr, Editors, Academic Press, New York (1988).

[30] *Hewlett Packard Journal*, (36) 5, entire issue (1985).

[31] M. Datta, D. E. King, A. D. Knight, and C. J. Sambucetti, U.S. Patent No 5,105,537, April 21 (1992).

10

Ion Beam Machining

Joseph McGeough
The University of Edinburgh, Edinburgh, Scotland

10.1 INTRODUCTION

Ion beam machining takes place in a vacuum chamber, with charged atoms (ions) fired from an ion source towards a target (the workpiece) by means of an accelerating voltage. Ion beam machining (IBM) is associated with the "sputtering" phenomenon first reported by Grove in 1852 (see Carter and Colligon [1]). While investigating the electrical conductivity of gasses, Grove discovered that metallic substances had become deposited on the glass walls of the glow discharge tube that he was using. He inferred that metal atoms had been removed from the surfaces of the electrode, and subsequently had adhered to the walls of the glass tube. Later the mechanism underlying

Grove's finding was established as the ejection of atoms from a surface when it is bombarded by other ions.

10.2 ION BEAM MACHINING SYSTEM

An ion beam machine has several main components:

1. A plasma source that generates the ions;
2. Extraction grids for removing the ions from the plasma and accelerating them towards the substrate (or specimen); and
3. A table for holding the specimen.

10.2.1 Ion Source

For the removal of an atom from the surface by impingement of an ion, or ions, a source of ions is required that should produce a sufficiently intense beam with an acceptable spread in its energy. Jolly et al. [2] have reviewed suitable ion sources for IBM equipment. They draw attention to the "electron-bombardment" ion source, or "Kaufman" system. Its main characteristics have been summarized by Spencer and Schmidt [3] and are presented in Table 1.

10.2.2 Plasma Source

As indicated in Figure 1, a heated filament, usually tungsten, acts as the cathode, from which electrons are accelerated by means of a high voltage (above 1 kV) towards the anode. During the passage of the electrons from the cathode to the anode, they interact with argon atoms in the plasma source (which is sustained by keeping the gas pressure at about 10^{-4} torr). The following reaction then occurs,

$$Ar + e^- \rightarrow Ar^+ + 2e$$

Argon ions are thereby produced. A magnetic field, obtained from an electromagnetic coil or a permanent magnet, is often

Table 1 Characteristics of Ion Beam Sources

Ion gun	Beam current (mA)	Beam current density ($mAcm^{-2}$)	Beam voltage (kV)	Beam diameter (cm)
Kaufman	10–50	0.85 at 1 kV	0.5–2.0	5.0
Duoplasmatron	10	10^3	0–25	Focus to 0.3 mm

Source: Ref. 3.

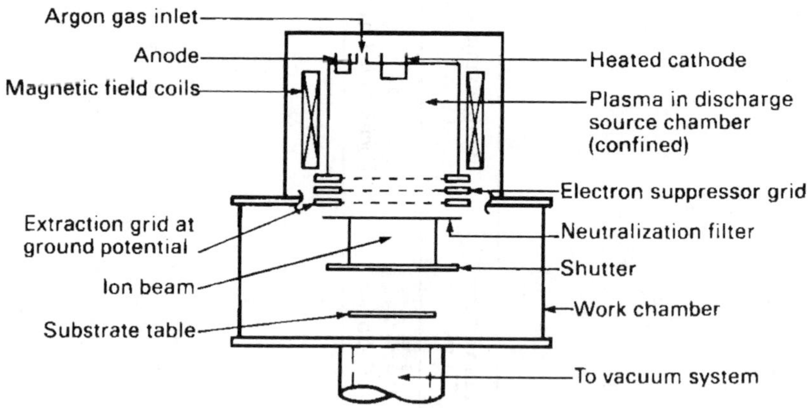

Figure 1 Main features of ion beam machine. (From Ref. 4.)

applied between the anode and cathode to make the electrons spiral. Spiralling increases the path length of the electrons and hence increases ionization.

10.2.3 Extraction Grids

The ions are removed from the plasma by means of extraction grids. The grids are normally made of two or three arrays of perforated sheets of carbon or molybdenum; these materials can withstand erosion by ion bombardment. The perforations in each of the sheets are aligned above one another. The shape of the holes and the spacings of the grids are significant elements when ion source systems are being designed to give the best conditions of ion current and grid erosion.

The outer grid is usually kept at ground potential, which is a more negative level than that of the anode. This grid therefore provides the negative field that is needed to remove the ions from the plasma. The second grid is held at a negative potential below the ground value. The escape of electrons from the plasma is thereby prevented, as is their diffusion back from the work chamber. A third grid, which is maintained at the anode potential, is sometimes added—placed between the plasma and the electron suppressor grid—to improve the performance of the source.

Extraction voltages of 0.5 to 2 kV with associated current densities of approximately 2 mA cm^{-2} are used with Kaufman ion sources.

10.2.4 Substrate Mounting

When the ions have been removed from the source, they "drift" in a field-free region to the component, specimen, or substrate which is to be machined or milled.

As shown in Figure 1, the specimen is usually mounted on a water-cooled table that can be tilted through an angle of 0 to 90°. The specimen is separate from the plasma. Machining variables such as acceleration, flux, and angle of incidence can all be independently controlled.

The highest currents are obtained at the lowest spacing between the grids, and for grids carrying the largest number of holes of the smallest size.

In Figure 2, an ion source from which is produced a 25-mm diameter beam of argon (Ar^+) ions, of current 110 mA is shown striking an alumina target (which is in the lower center of the photograph). Surfaces mounted on a water-cooled rotating holder are coated with the sputtered alumina (top right). Figure 2 also includes a 15-cm source (left-hand side) which is producing up to 0.5 A of oxygen ions at 500 eV to mill the surface of the alumina specimen.

Figure 3 shows the profile of the beam from this source. An industrial version of an ion beam machining system is shown in Figure 4.

10.3 COLLISION MECHANISM

The interactions of the ions with atoms are now considered.

The size of an ion is normally comparable to that of an atom. When an ion strikes the surface of a material its colli-

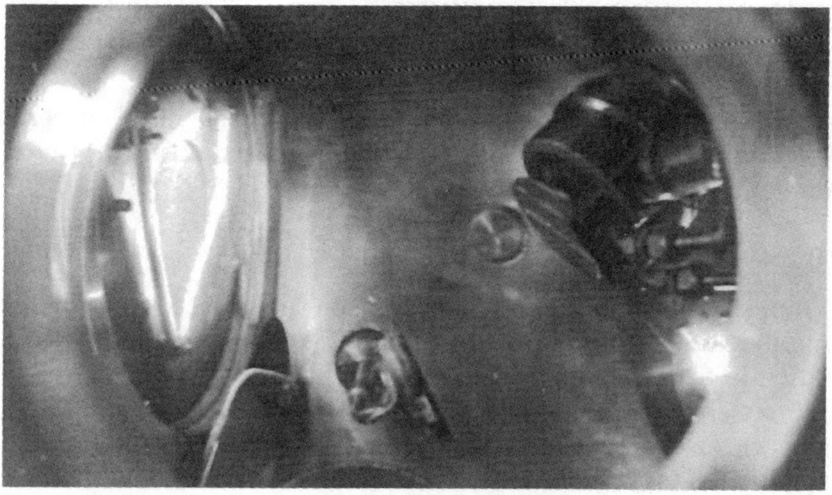

Figure 2 Argon ion beam striking alumina target. (Courtesy of Oxford Applied Research.)

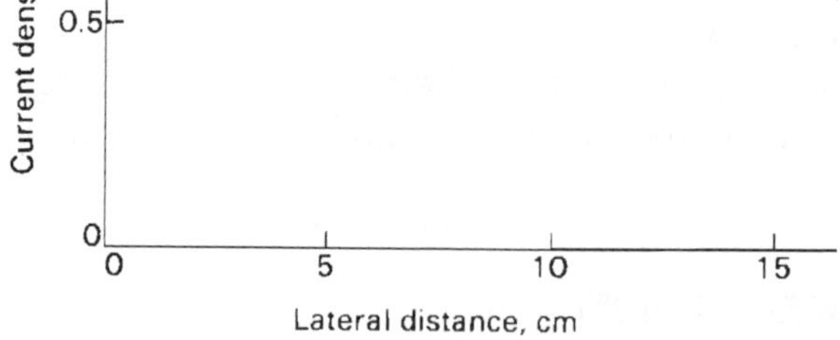

Figure 3 Profile of ion beam from 15-cm source. Source-probe distance: 10 cm; beam energy: 500 eV; beam current: 202 mA; gas flow (STP): 2.4 ml min^{-1}. (From Ref. 2.)

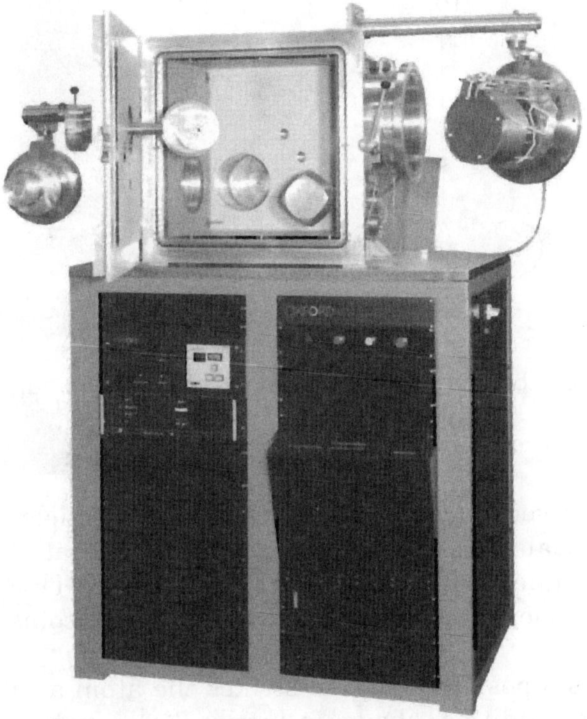

Figure 4 Elements of industrial ion beam machine. Upper part shows vacuum chamber with specimen holder on the right. Left-hand side shows control cabinet and power system. (Courtesy of Oxford Instruments Ltd.)

sion with an atom there often occurs in a direction that is normal to the surface. If the mass of the ion is less than that of the atom of the surface, the former will bounce back, away from the surface, and the atom will be driven in a direction farther into the material. Figure 5(a) illustrates this condition.

If the mass of the incident ion is greater than that of the surface atom, after collision, the ion and atom move from the position of collision towards the interior of the surface of the material, irrespective of the angle (i.e., head-on or glanc-

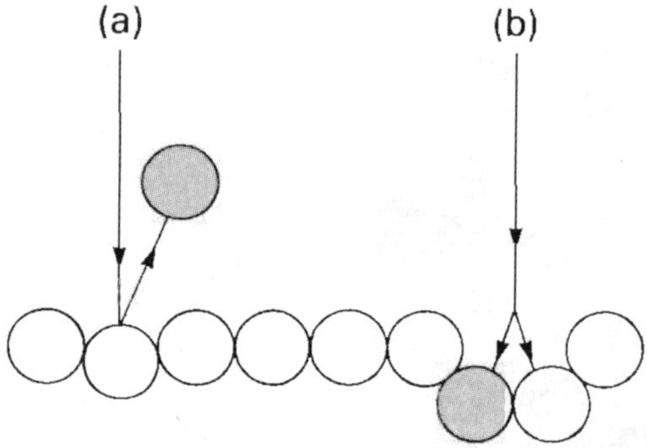

Figure 5 Ion (shaded) bombardment at normal incidence to surface. (From Ref. 5.)

ing) of the collision; see Figure 5(b). Usually both particles move into the material at energies that are less than that of the incident ion, yet much greater than the lattice energy. This type of primary collision does not lead to removal of atoms from the surface.

Alternatively, suppose that an ion strikes the atom at a glancing angle. One condition, shown in Figure 6(a), involves the incident ion impinging on a stationary surface atom at an angle of 90°. This atom does not obtain a velocity component in a direction away from the surface as a direct consequence of this primary collision. However, the primary collision will cause at least one, and often two, further secondary "binary" collisions just below, and very close to, the surface. Figure 6(b) illustrates this case. After the collision between ion and atom either particle should be able to leave the point of impact at more than 45° to the original direction of velocity of the ion.

Then, a secondary collision should be possible in the same plane of motion, to result in a lattice atom leaving the point of secondary impact at an angle greater than 45°. That is, a total angle of more than 90° is obtained. With an angle of this size the lattice atom has a velocity component which is in a

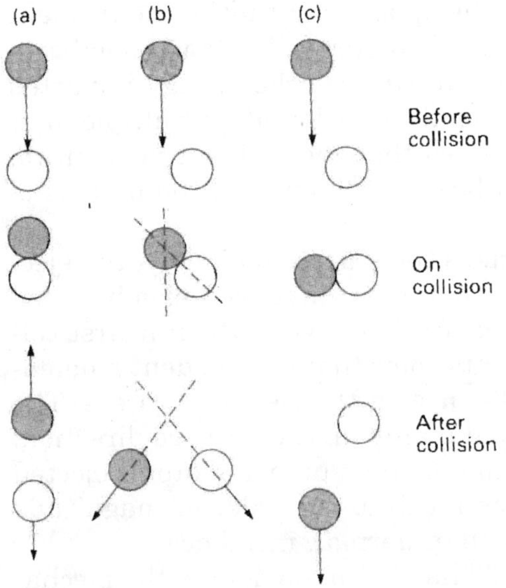

Figure 6 Three types of collision between ion (shaded) and atom (mass of ion less than that of atom). (From Ref. 5.)

direction outward from the surface. Therefore the atom has the capacity to be ejected.

However, the atom cannot be ejected in a direction parallel to the normal to the surface, that is, in a direction opposite to that of the incident ion. That movement would require two 90° deflections. At least one of these deflections would involve the deflection of a lattice atom through 90°, during which it would acquire zero velocity. An atom with zero velocity cannot be ejected. For the same reason the atom cannot cause ejection. The final set of conditions is illustrated in Fig. 6(c).

Although ejection might then be expected to occur most commonly in directions away from the normal to the surface, for ion impingement on the normal direction to the surface, atoms have been observed to be ejected from the surface in a cosine distribution. That is, ejection is most likely to occur in a direction that is exactly the opposite to that of the incident ion (it should be noted that this occurs only for ion bombardment at normal incidence). Apparently the energy imparted by the incident ion is so randomly distributed by multiple collisions before the atom is ejected that the incident momentum vector is fully lost. It then has no influence on the process of ejection.

If the incident ion strikes the surface obliquely, the ejection is very likely to result from the primary collision between the incident ion and the surface atom with which it first collides. In this case experiments show that the incident momentum vector has a great influence on the ejection process. The atoms are found to be ejected mainly in the forward direction.

The sputtering yield, that is, the number of atoms ejected per incident ion, may be as much as an order of magnitude greater for oblique, rather than normal, incidence.

Spencer and Schmidt [3] have also considered the mechanisms at work in IBM. They explain material removal in terms of the transfer of momentum from the incident ions to atoms on the surface of the material. Thus in Figure 7 an atom is removed from the surface, and the ion is also deflected away from the material. Spencer and Schmidt propose that energies greater than the binding energy of 5 to 10 eV are needed to achieve removal of atoms.

As indicated also in Figure 7, at higher energies sufficient momentum may be transferred by the impinging ions for several atoms to be removed from the material in a cascade-type effect near its surface. The higher the energy of the incident ion, the more deeply this cascading effect occurs into the material. Several atoms, or molecules, ionized or neutral, are likely to be ejected from the material. The incident ion will become implanted deep into the material, damaging it, by displacement of atoms.

Various theoretical expressions have been reported for the yield, that is, the number of atoms removed per incident ion. See, for example, Spencer and Schmidt [3] and Somekh [6]. These workers all confirm that the yield depends on the

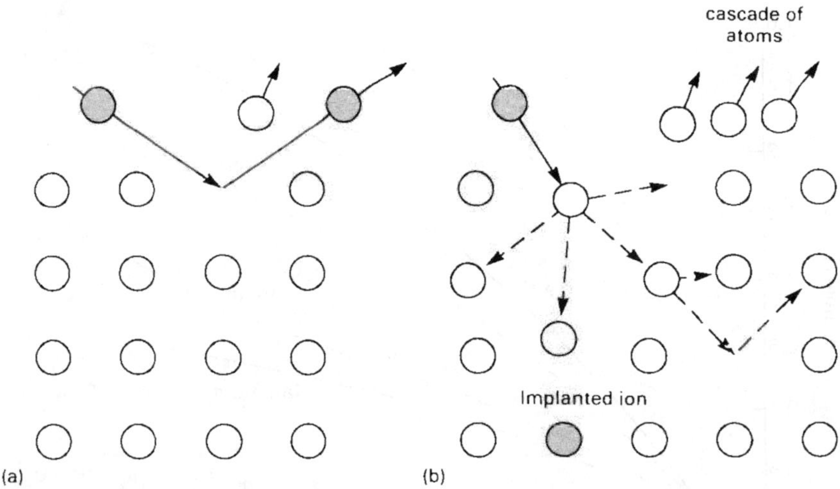

Figure 7 Effects of low and high energies on atom removal: (a) low energy case; (b) high energy case. (From Ref. 3.)

material being treated, the type of atoms and their energy, the angle of incidence, and in some cases, the gas pressure. The relationships that are presented depend often in a complicated way on other fundamental quantities. Despite the undoubted values of these expressions, the more practical aspects of the process can be readily assessed from experimentally obtained values.

10.4 RATES OF MATERIAL REMOVAL IN IBM

Typical experimental results in which yield increases nonlinearly with incident ion energy, for argon ions impinging on a range of materials, are shown in Figure 8. For higher ion energies a yield of approximately 0.1 to 10 atoms per incident ion is representative of IBM. Spencer and Schmidt [3] discuss the effects on yield of a wide range of process conditions, not only

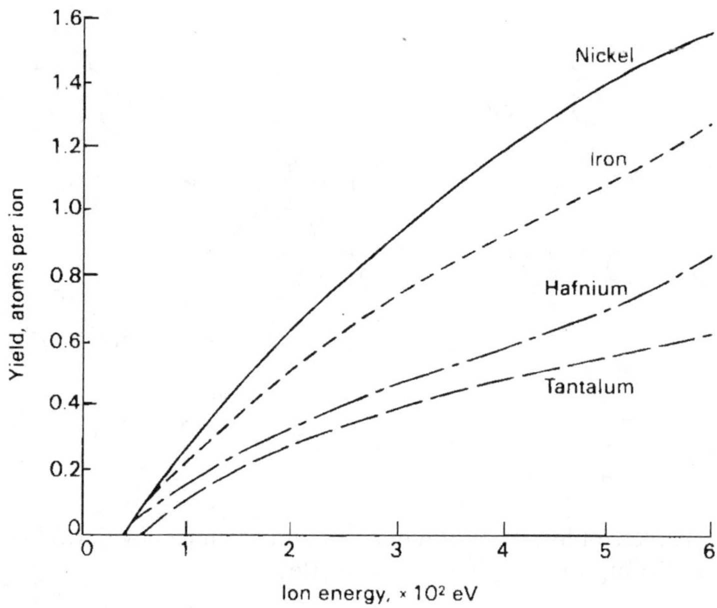

Figure 8 Variation of yield with ion energy. (From Ref. 3.)

incident ion energy but also the angle of incidence, ionic and atomic periodicity, and heat of sublimation. They point out, for example, that the rate of material removal (or etching) can also be enhanced at grain boundaries and can be affected by crystallographic orientation.

An alternative way of recording rates of removal is in terms of the milling rate, which typically varies between 1 and 2 μm s^{-1} for, respectively, normal and 50° angles of incidence. Some typical milling rates for a range of materials that have been ion beam machined are presented in Table 2.

Useful experimental information is available on these results which supplements that given in Table 2. The results were obtained with apparatus that included a low voltage gun, in which a hot filament was utilized as the source of electrons to ionize the argon gas. The hollow anode source used as the

Table 2 Typical Removal (Milling) Rates by IBM (Hollow Anode Source)[a]

Material	Removal (milling) rate ($\mu m\ hr^{-1}$)
Quartz	2
Garnet	1
Ceramic	1
Glass	1
Gold	2
Silver	3
Photo resist material (KTFR)	1
Permalloy	1
Diamond	1
GaAs (500-μA current)	10
GaP (500-μA current)	125

[a] Data: argon ion beam 60 to 70° from normal; pressure = 3×10^{-4} Torr; voltage = 6 kV; current = 100 μA; current density = 1 mA cm^{-2} over 1-cm diameter area.
Source: Ref. 3.

ion gun was about 5 cm in diameter and 15-cm long. The anode and cathode had matching arrays of 300 holes, each of 0.3-mm diameter, and spaced apart such that uniform material removal or milling, accurate to within about 3% over the machining area, could be obtained. With this anode–cathode configuration an area of 300-cm diameter could be treated. Each beam was capable of 100 μA with only 1 kV voltage applied; that is, the total beam current was 30 mA.

Further experimental results for an argon ion beam at normal incidence are presented in Tables 3 and 4.

Relevant data concerning IBM can now be summarized. For a typical IBM operation on a material of area 1 cm^2, a depth of approximately 10^{-4} mm is removed with an ion beam of 100-μA current. At a removal rate of one atom per incident ion, 1 $\mu m hr^{-1}$ of material would be machined over that area; that is, about 3 atoms s^{-1} or one atomic monolayer per second per square centimeter of area. The removal rate can be increased by use of greater beam fluxes or higher energy ions,

Table 3 Removal Rates by IBM[a]

Material	Removal (milling) rate ($\mu m\ hr^{-1}$)
Silicon	2.0
GaAs	15.9
Silica (ceramic)	2.3
KTFR photoresist	2.3
Silver	18.0
Gold	2.6

[a] Data: incident beam normal to surface; pressure = 3×10^{-4} torr; voltage = 1 kV; current = 0.85 mA cm^{-2}; beam diameter = 5 cm.
Source: Ref. 3.

Table 4 Removal (Milling) Rates Obtained with 500 eV Argon Ions Incident Normally on Target[a]

Material	Removal (milling) rate ($nm s^{-1}$)
Carbon	0.07
Aluminum	1.2
Silicon	0.62
Chromium	0.83
Manganese	1.5
Silver	2.5
Gold	2.4
SiC	0.52
SiO_2	0.67
Fe_2O_3	0.78
AZ 1350 (photoresist)	0.50
PMMA (photoresist)	0.93

[a] Current density 1 mA cm^{-2}.
Source: Ref. 2.

although lower energy conditions are preferable. For most practical circumstances, IBM can be regarded as a sequential operation in which each ion can be considered to act alone. As single ions impinge on the material each one can be regarded as removing an atom.

Somekh [6] confirms that the yield depends on the material being etched, the type of atoms and their energy, the angle of incidence, and, in some cases, the gas pressure. He draws attention to the dependence of yield on the binding energy of the atoms in the material being etched. The amount of yield can therefore be varied by the introduction of reactive gasses.

These reactive gasses can react with the surface of the material and vary its binding energy, and hence the rate of material removal (i.e., the etch-rate). For example, a flux of reactive species containing fluorine is known to react with materials such as titanium or silicon to form loosely bound or volatile compounds, which increase the etch-rate.

Jolly et al. [2] discuss further aspects of reactive IBM. They have investigated conditions in which a flux of a reactive species such as methane (CH_4), fluorine (CF_4), rather than inert argon ions is directed at the specimen target.

The free radicals chosen should react with the material at its surface, forming volatile products or those which can be readily milled (removed) by the effect of the kinetic energy of the bombarding ions.

10.5 ACCURACY AND SURFACE ROUGHNESS IN ION BEAM MACHINING

Jolly et al. [2] report that dimensions as small as 100 nm should be possible by IBM, and that features of size less than 10 μm are obtainable. The slope of the walls of the machined surface and its surface finish are determined by the angle of incidence of the ion beam which is fully controllable. They claim that the process variables can be monitored to an accuracy of ±1.0%, with a repeatability of ±1.0%.

Surfaces can be textured by IBM, which can produce cone- and ridgelike configurations, on the order of 1 μm in size. Smoothing of a surface to a finish of less than 1 μm can also be achieved by IBM. Jolly and coworkers also propose that an already smooth surface can undergo ion beam machining without significant increase in its roughness. They report typical results, in which a depth of 100 μm can be machined from a surface; an initial surface roughness of 1.5 nm was found to be altered by less than ±1 nm.

10.6 APPLICATIONS

10.6.1 Smoothing

The use of IBM for smoothing of laser mirrors and for modifying the thickness of thin films and membranes without affecting the surface finish was also reported by Jolly et al. [2].

10.6.2 Ion Beam Texturing

Hudson [7] has shown that an ion beam source is a controlled method for texturing surfaces. Hudson proceeded to investigate 26 materials, including stainless steel, silver, and gold. He has also reported a number of other investigations in which a microscopic surface texture created by sputter-etching performed simultaneously with the sputter-deposition of a lower yield material onto the surface.

Applications of ion beam texturing have been discussed further by Jolly et al. [2]. They include enhanced bonding of surfaces, increased surface area capacitors, and surface treatment of medical implants.

10.6.3 Ion Beam Cleaning

Atomically clean surfaces can be produced by IBM. This technique can be preferable to electron beam and electrical discharge methods which can damage the surface. Harper et al.

[8] discuss in detail this well-established application of ion beam technology. For example, they report substantial improvements in the adhesion of gold films to silicon and aluminium oxide Al_2O_3 substrates by use of argon or oxygen ion beam sputter-cleaning of the substrate, prior to evaporation. The cleaning consisted mainly of removal of absorbed water and hydrocarbons. When a layer surface oxide has to be removed in order to clean a surface, higher ion energies, of several hundred eV, are needed. Damage to the substrate material may then arise: in this case a reactive gas with high selectivity of oxide etching can be used.

10.6.4 Shaping, Polishing, and Thinning by IBM

Thinning by use of oblique incidence argon ions has been used to enhance polishing (see Harper et al. [8]). Macroscopic thinning and shaping of materials can be applied to the fabrication of magnetic heads and surface acoustic wave devices.

The polishing and figuring of optical surfaces has also been reported by Harper et al. [8] and by Taniguchi [9]. The latter gives a valuable summary of the various applications of IBM including aspherizing of lenses, sharpening of diamond indentors, and cutters and cutting tools.

Taniguchi [9] points out that these operations are performed by the direct sputtering of preforms in glass, silica, and diamond. Unlike conventional technology involving cutting, grinding, lapping, and polishing, the ion beam process has no inherent polishing surface (e.g., guideways), the reference being the preform or patterning mask.

Thinning of samples of silicon to a thickness of 10 to 15 μm has been obtained by argon ions impinging at normal incidence. The production of samples for transmission electron microscopy (TEM) is another widespread practice; two opposing beams thin a circular region on a rotating sample until the center etches through, leaving thin fringe areas suitable for TEM.

10.6.5 Ion Milling

Earlier work revealed encouraging signs for ion milling. For example, Jolly and coworkers [2] reported that ion milling is especially useful for the accurate production of shallow grooves, such as that illustrated in Figure 9. Milling through masks, to produce regular arrays of pits with widths of 5 to 200 μm and depths of up to 1 mm for enhanced bonding, has also been achieved. These authors also pointed out that pillarlike configurations useful in the manufacture of precision electrical resistive and fiber-optic arrays can be produced by ion beam methods.

Other workers have also confirmed the usefulness of the ion milling technique as an alternative to the fabrication by chemical etching of devices of fine geometry. IBM is limited only by masking capabilities. For example, line widths of 0.2 μm have been achieved by IBM in the fabrication of bubble memory devices; depth-to-width ratios 2:1 have been achieved. Problems associated with chemical etching, such as

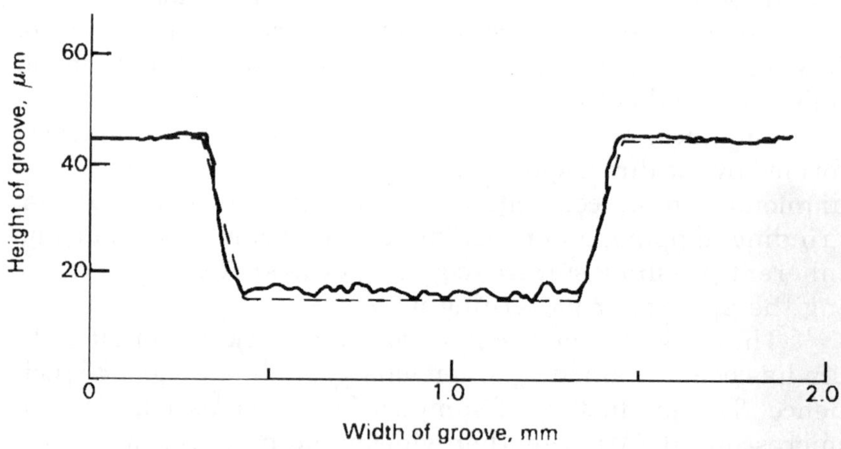

Figure 9 Groove machined in refractory material: solid line-profile obtained; dashed line-required profile. (From Ref. 2.)

lack of line delineation owing to failure of resist adhesion and undercutting of layers, are avoided since masking is only needed to shadow the beam (see Bollinger [10]). Typical results from Bollinger's detailed account of the manufacture of solid-state devices of fine geometry are given in Figure 10, which shows the fine etching by IBM of near vertical walls on a GaAs substrate. The milled gold structure accurately holds the pattern. No undercutting occurs at the gold–photoresist interface and the channel floor is flat right up to the wall interface. Further discussion of patterned microfabrication by IBM is given by Harper et al. [8]. They describe the fabrication of line widths as narrow as 80 A° in 200-A° thick carbon membranes by argon ion beam etching.

Most recent applications have been limited to milling of multilayered structures, and require techniques that will mill through elements such as gold [11].

Other applications for ion beam technology are known in the III–IV semiconductor industry, for etching facets in semiconductor lasers. Ion beam technology is used in the manufacture of hard disk heads and for etching metallic tracks on ce-

Figure 10 Ability of ion milling to etch near-vertical walls. (From Ref. 10.)

ramic substrates. Figuring of large optical surfaces has been reported in which a 3-cm ion source is moved above the surface [12].

10.7 CONCLUSIONS

Much work has been published on the use of microfocused ion beams for machining, although this technique is almost entirely confined to academic research. For production applications laser ablation is usually preferred for small spots, and plasma etching for large areas. Much of future ion beam technology is likely to deal increasingly with deposition [11]. In this method, the material to be deposited is ionized inside the source and then directed in the form of an ion beam to the surface on which it will condense. The technique is used to study growth of thin films for deposition of metals such as pure iron and semiconductor layers of silicon. An ion beam sputter deposition technique is also reported, the material being sputtered from a target by an energetic ion beam and collected onto the substrate. Films of oxides and nitrites can be deposited by reactive ion beam sputtering.

New developments on radio frequency or microwave ion guns continue. Most trends are concentrated on automatic control of the source for utilization in complex processes, neutralization of the ion beam in reactive gas environments, increases in ion density, and very low levels of contamination [12].

REFERENCES

[1] G. Carter and J. S. Colligon, *Ion Bombardment of Solids*. Heinemann, London, Chaps. 7 and 9 (1968).

[2] T. W. Jolly, R. Clampitt, and P. Reader, Ion beam machines and applications. In *Proc. 7th Int. Symposium on Electromachining*, Birmingham, April 12–14, 201–210 (1983).

[3] E. G. Spencer and P. H. Schmidt, Ion beam techniques for device fabrication. *J. Vac. Sci. Technology* (8) 5, S52–S70 (1972).

[4] C. M. Melliar-Smith, Ion etching for pattern delineation. *J. Vac. Sci. Technology* (13) 5, 1008–1022 (1976).

[5] R. V. Stuart, *Vacuum Technology Thin Films and Sputtering—An Introduction*. Academic, London, 92–97 (1983).

[6] S. Somekh, Introduction to ion and plasma etching. *J. Vac. Sci. Technology* (13) 5, 1003–1007 (1976).

[7] W. R. Hudson, Ion beam texturing. *J. Vac. Sci. Technology* (14) 286–289 (1977).

[8] J. M. E. Harper, J. J. Cuomo, and H. R. Kaufman, Technology, and applications of broad-beam ion sources used in sputtering. Part II Applications. *J. Vac. Sci. Technology* (21) 3, 737–756 (1982).

[9] N. Taniguchi, Current status in and future trends of ultraprecision machining and ultrafine materials processing. *Annals of the CIRP* (32) 2, 1–8 (1983).

[10] L. D. Bollinger, Ion milling for semiconductor production processes. *Solid State Technology* (20) 11, 66 (1977).

[11] T. Jolly, Personal communication (1999).

[12] L. Wartski, C. Schwebel, and J. Aubert. Radio frequency, microwave and electron cyclotron resonance ion source for industrial applications. A review (invited). *Rev. Sci. Instrum.* (6) 3, March, 895–900 (1996).

11

Electron Beam Machining

Joseph McGeough
The University of Edinburgh, Edinburgh, Scotland

11.1 INTRODUCTION

This chapter deals with the phenomena arising when electrons are generated within a vacuum chamber. Electron beam machining depends for its operation upon such effects. Its principles form the bases for the technique of electron beam lithography discussed in the next chapter. Some of the earliest work on the utilization of the electron beam for material removal can be attributed to Steigerwald, who designed a prototype machine in 1947. Modern electron beam machines work on the same principles.

11.2 BASIC EQUIPMENT

The main components of an EBM installation are shown in Figure 1. They are housed in a vacuum chamber, evacuated to about 10^{-4} torr. The source of electrons is an "electron gun," which is basically a triode consisting of a cathode, a grid cup negatively biased with respect to the cathode, and an anode at ground potential. The cathode is usually made of a tungsten filament, which is heated to between 2500 and 3000°C in order to emit the electrons. A measure of this effect is the emission current, the magnitude of which varies between 25 and 100 mA. Corresponding current densities lie between 5 and 15 Acm^{-2}. This quantity, however, is determined by a range of

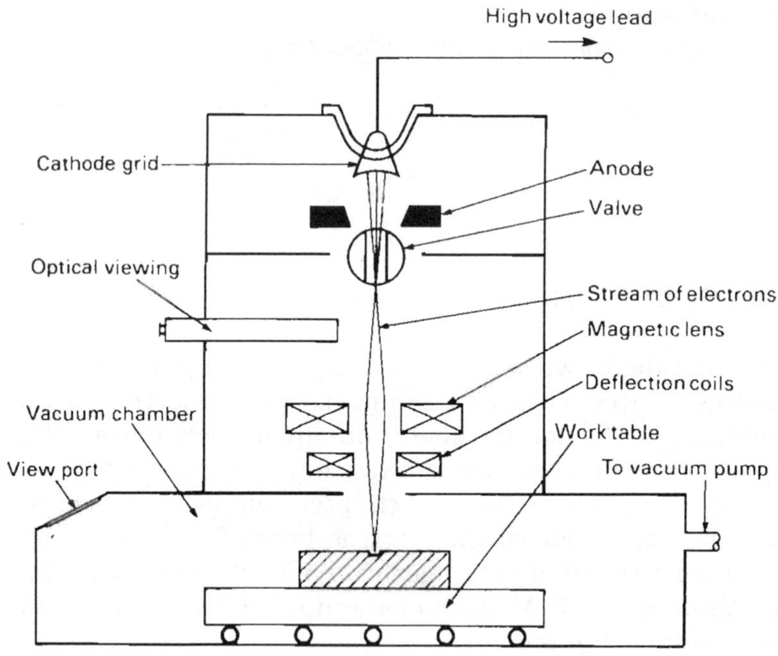

Figure 1 Components of electron beam machine. (From Ref. 1.)

factors, including the type of cathode material and its temperature (see below). The size of the emission current is also influenced by a high voltage, usually about 150 kV, which is applied between the cathode and anode in order to accelerate the electrons in the direction of the workpiece. Figure 2 illustrates how the emission current increases with cathode temperature and accelerating voltage, when that electrode is made of a tungsten filament.

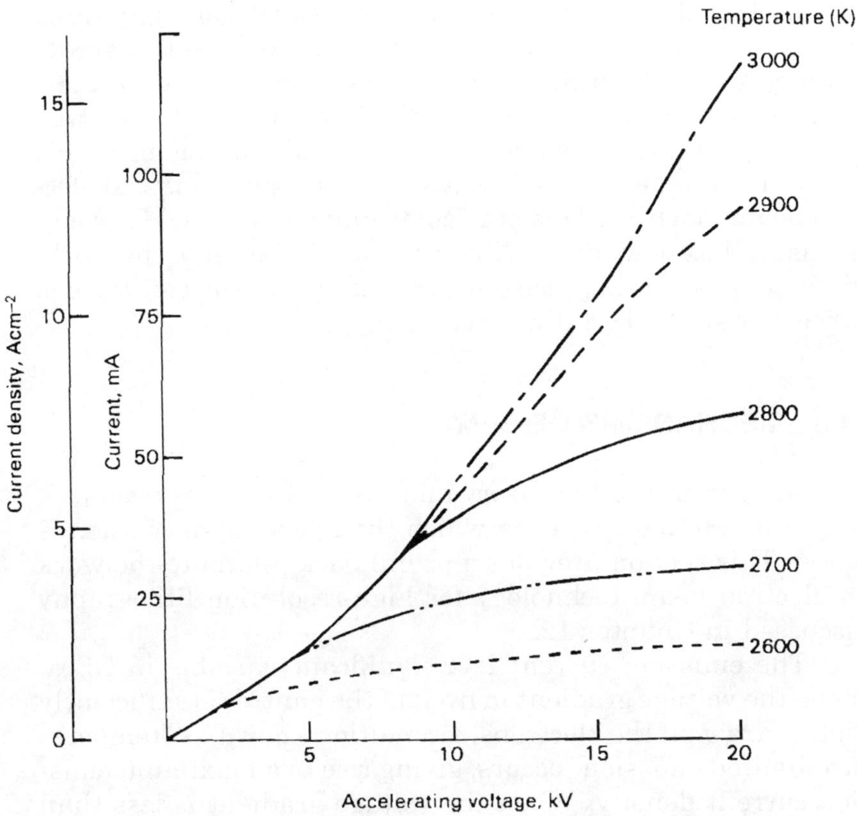

Figure 2 Variation of emission current with cathode temperature and accelerating voltage. (From Ref. 2.)

After acceleration, the electrons are focused by the field formed by the grid cup so that they travel through an aperture in the anode. On its exit from the anode cavity, the electron beam is refocused by a magnetic or electrostatic lens system. By this means, the beam has its direction towards the workpiece kept under control. The electrons maintain the velocity imparted by the accelerating voltage, until they strike the workpiece specimen, over a well-defined area, typically 0.025 mm in diameter. There the kinetic energy of the electrons is rapidly translated into heat, causing a correspondingly rapid increase in the temperature of the workpiece, to well above its boiling point. Material removal by evaporation then occurs. With power densities on the order of 1.55 $MWmm^{-2}$ involved in EBM, virtually all engineering materials can be machined by this technique. Figure 3 shows details of an industrial electron beam machine. This particular unit can be used for welding as well as machining. Accurate manipulation of the workpiece coupled with precise control of the beam can yield a process that can be fully automated.

11.3 MECHANISMS OF EBM

A useful account of the theory underlying EBM has been presented by McGeough [1] on which the following discussion is based. This section provides a useful background to the work on electron beam technology for high-resolution lithography discussed in Chapter 12.

The emission current is a significant variable in EBM. When the voltage gradient in front of the emitter is sufficiently high to draw off the electrons, a condition known as "temperature-limited emission" occurs, giving rise to a maximum emission current density. When the voltage gradient is less than that above, the emission of electrons is hindered by a negative space charge in the vicinity of the cathode. Mutual electron repulsion takes place there. The emission current is then known as space-charge limited.

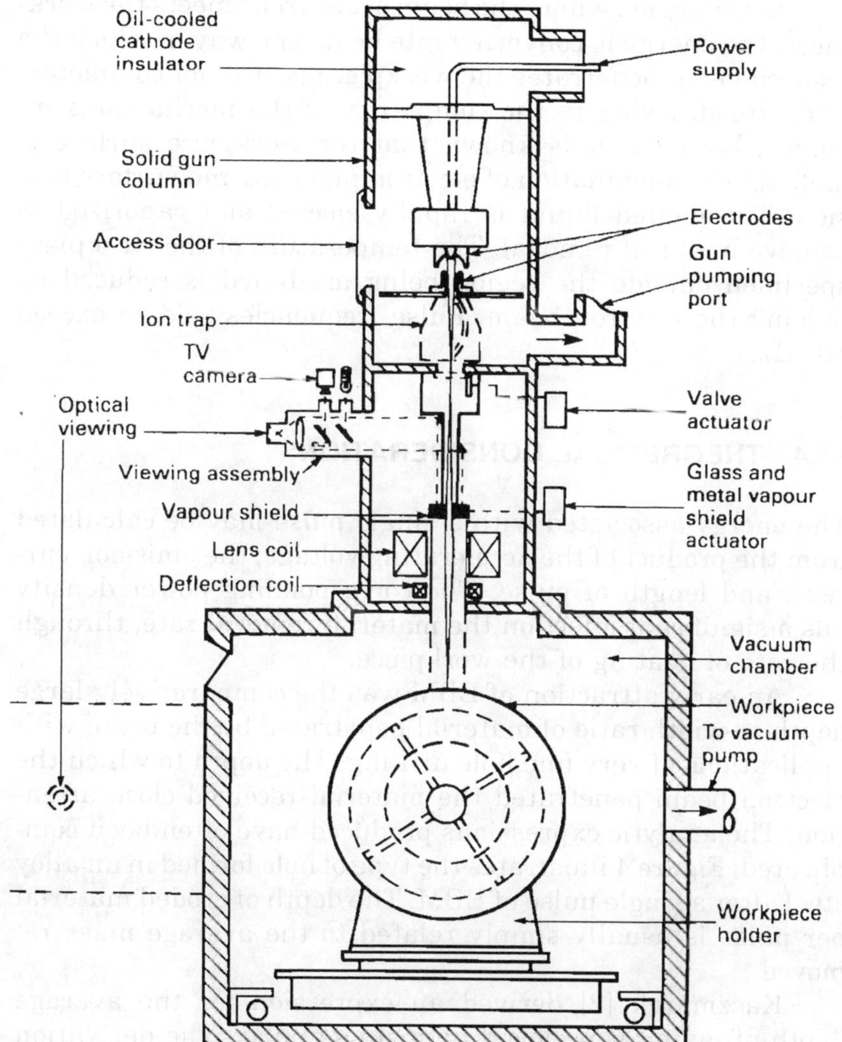

Figure 3 Industrial electron beam machine.

In the region where the beam of electrons meets the workpiece, the energy is converted into heat. The way in which the focused beam penetrates the workpiece is still not completely understood, owing to the complexity of the mechanisms involved; however, it is known that the workpiece surface is melted by a combination of electron pressure and surface tension. The melted liquid is rapidly ejected and vaporized to achieve material removal. The temperature of the workpiece specimen outside the region being machined is reduced by pulsing the electron beam. Pulse frequencies seldom exceed 10^4 Hz.

11.4 THEORETICAL CONSIDERATIONS

The energy associated with a single pulse may be calculated from the product of the accelerating voltage, the emission current, and length of pulse. The corresponding power density has a significant effect on the material removal rate, through the rate of heating of the workpiece.

An early attraction of EBM was the comparatively large depth-to-width ratio of material penetrated by the beam with applications of very fine hole drilling. The depth to which the electron beam penetrated the material received close attention. The analytic expressions produced have often been complicated. Figure 4 illustrates the type of hole formed in an alloy steel after a single pulse of EBM. The depth of eroded material per pulse is usually simply related to the average mass removed.

Kaczmarek [2] derived an expression for the average depth of material removed by a single pulse. The derivation is complicated, as indeed is the equation itself. Kaczmarek's analysis reveals the existence of an optimum value of the accelerating voltage for which the average depth, eroded by a single pulse, reaches a maximum value. From his work the number of pulses required to erode a hole of depth can also be obtained. The graphical interpretation of his results shown in Figure 5 indicates that there is a minimum number of pulses

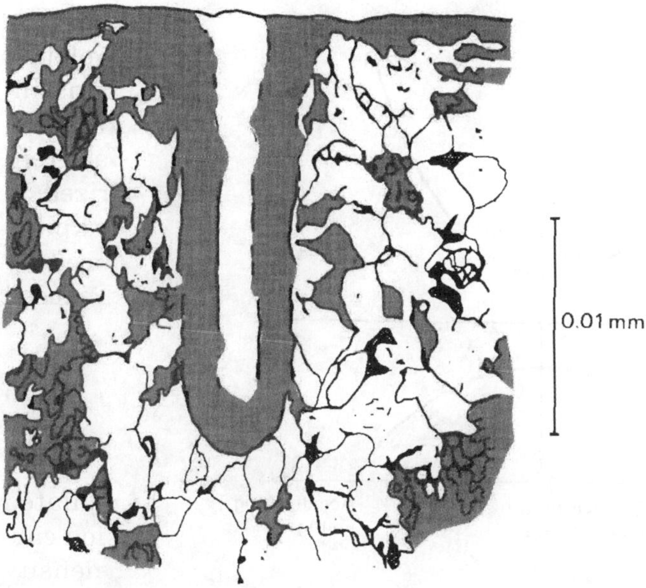

Figure 4 Cross-section of a cavity in chromium–molybdenum steel formed by a single pulse. (From Ref. 2.)

associated with an optimum accelerating voltage. In practice the number of pulses needed to produce a given hole depth is usually found to decrease with an increase in accelerating voltage. His analysis also reveals that, for a fixed set of process conditions, the number of pulses required increases hyperbolically as the depth increases. In practical terms this result means that when a certain depth has been reached, any further EBM to deepen the hole would require a very large increase in the number of pulses.

11.5 RATES OF MATERIAL REMOVAL IN EBM

Electron beam machining rates are usually evaluated in terms of the number of pulses required to evaporate a particular amount of material. Two types of pulse numbers are used. The

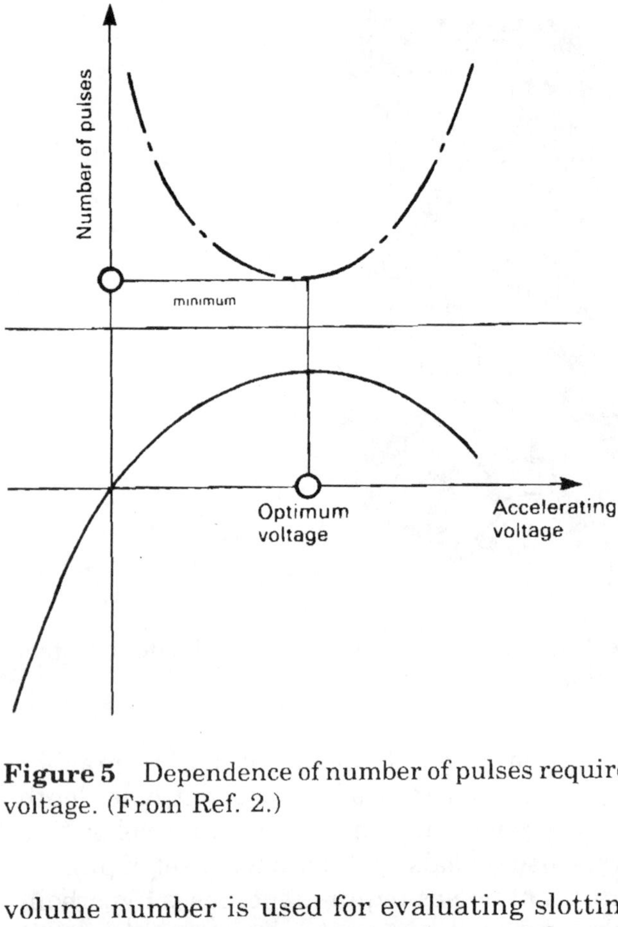

Figure 5 Dependence of number of pulses required on accelerating voltage. (From Ref. 2.)

volume number is used for evaluating slotting by EBM, that is, "cutting-off" or "cutting-into" materials, and is given by the ratio of the mass to be removed to that due to a single pulse. A linear number is adopted for hole sinking, and is the ratio of the depth of the hole required to that sunk by a single pulse. Studies of the EBM of different metals have revealed that their boiling point and thermal conductivity play a significant role in determining how readily they can be machined. Properties such as electrical conductivity of the material are additional factors. The complexity of the relationship between the material properties and machinability renders difficult any fully quantitative analysis of removal rates in EBM.

Table 1 Removal Rates in EBM[a]

Material	Volumetric removal rate (mm^3s^{-1})
Tungsten	1.5
Aluminum	3.9

[a] Power 1 kW.
Source: Ref. 3.

Experiments still have to be performed to obtain representative values of machining rates. Table 1 shows typical results for tungsten and aluminum.

Experimental results on the dependence of pulse numbers on accelerating voltage given in Figure 6 show that increasing the hole depth requires a much greater rise in the number of pulses at low voltage, due mainly to a relative rise in heat losses resulting from conduction and melting of the adjacent metal layers. For a given number of pulses little im-

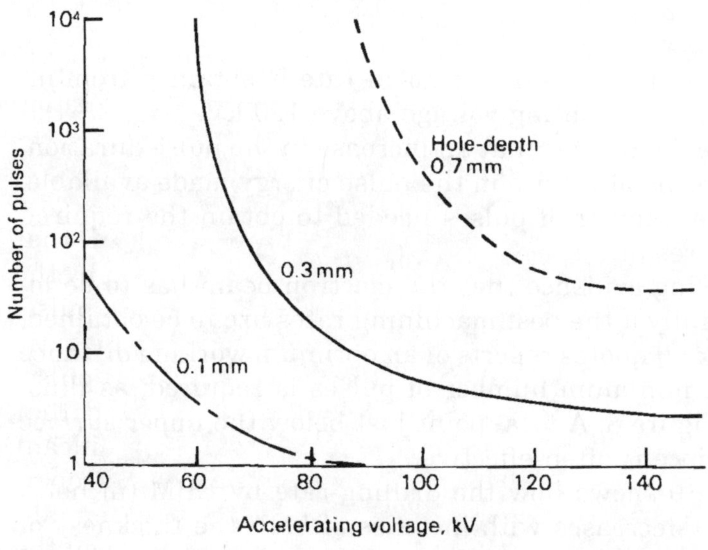

Figure 6 Effect of accelerating voltage on number of pulses required. (From Ref. 2.)

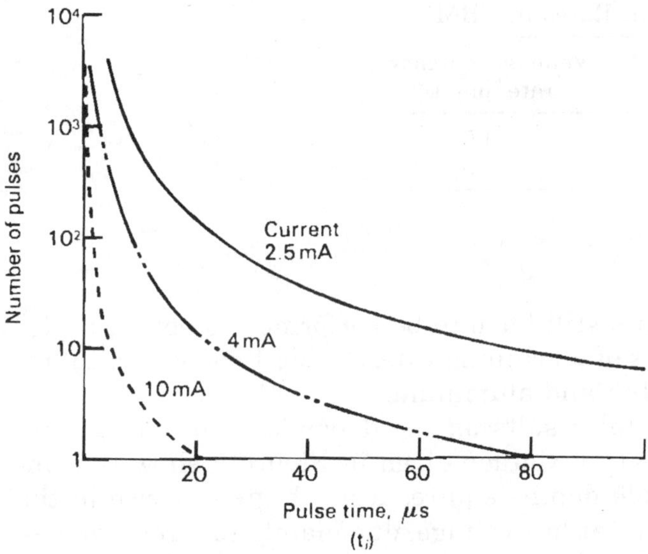

Figure 7 Dependence of number of pulses needed on pulse duration. (From Ref. 2.)

provement in the material removal rate is obtained from increasing the accelerating voltage above 120 kV.

Figure 7 indicates that an increase in the pulse duration, with a corresponding rise in the pulse energy made available, reduces the number of pulses needed to obtain the required machining result.

Following evidence that the electron beam has to be focused carefully if the best machining rates are to be obtained, Kaczmarek [2] quotes reports of an optimum working distance at which a minimum number of pulses is required, as illustrated in Figure 8. A focal point just below the upper surface of a workpiece is often effective.

Figure 9 shows how the drilling-rate by EBM (in holes per second) decreases with increase of both the thickness of the workpiece and the diameter of the hole to be produced. The relevance of these results to micromachining of thin materials is discussed below.

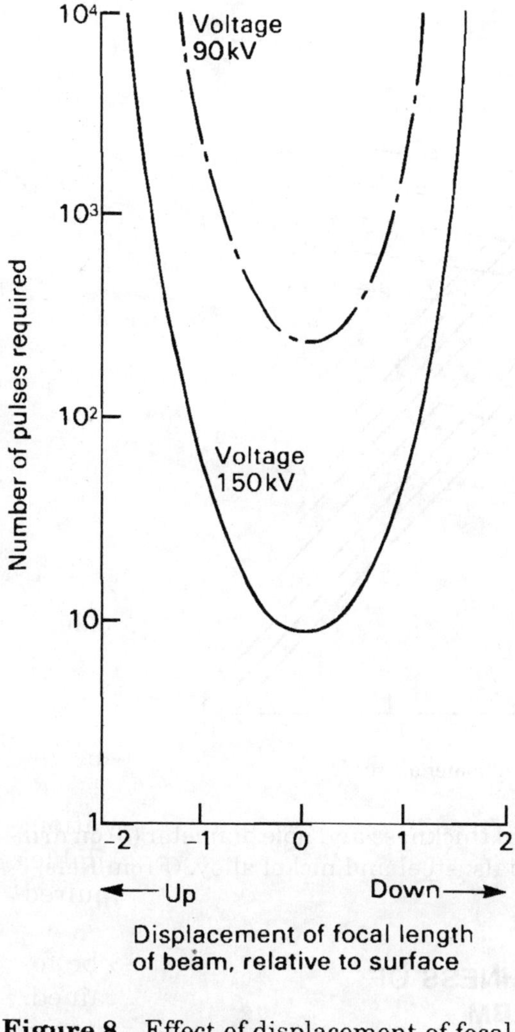

Figure 8 Effect of displacement of focal length of beam relative to upper surface of workpiece on number of pulses. Thickness of specimen is 2 mm. (From Ref. 2.)

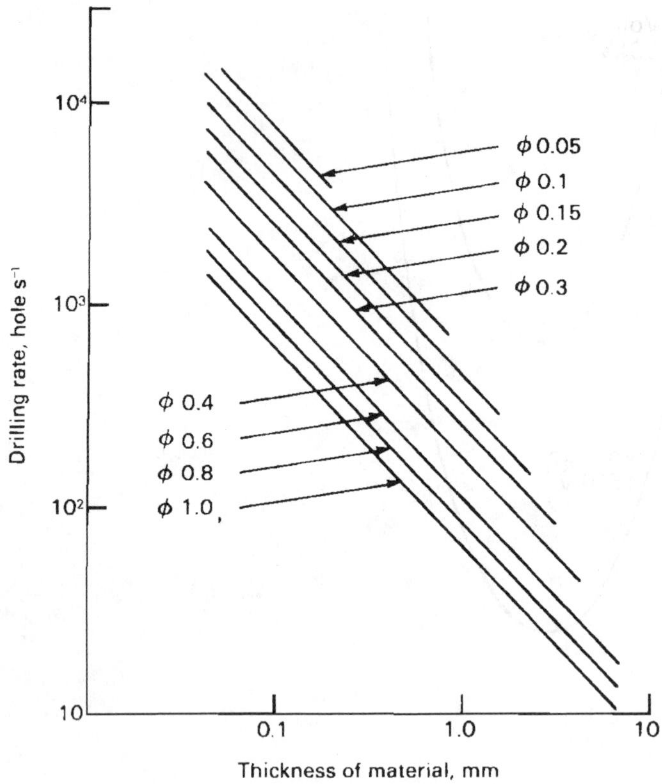

Figure 9 Effect of material thickness and hole diameter (ϕ) on drilling rate. Workpiece materials: steel and nickel alloy. (From Refs. 3 and 1.)

11.6 SURFACE ROUGHNESS OF WORKPIECE IN EBM

The quality or surface roughness of the edges produced in EBM depends greatly on the type of material. Local pitting of the surface is a common occurrence, the extent of which is influenced by the thermal properties of the workpiece, and by the pulse energy or charge. Figure 10 illustrates how surface roughness can increase with pulse charge for a range of com-

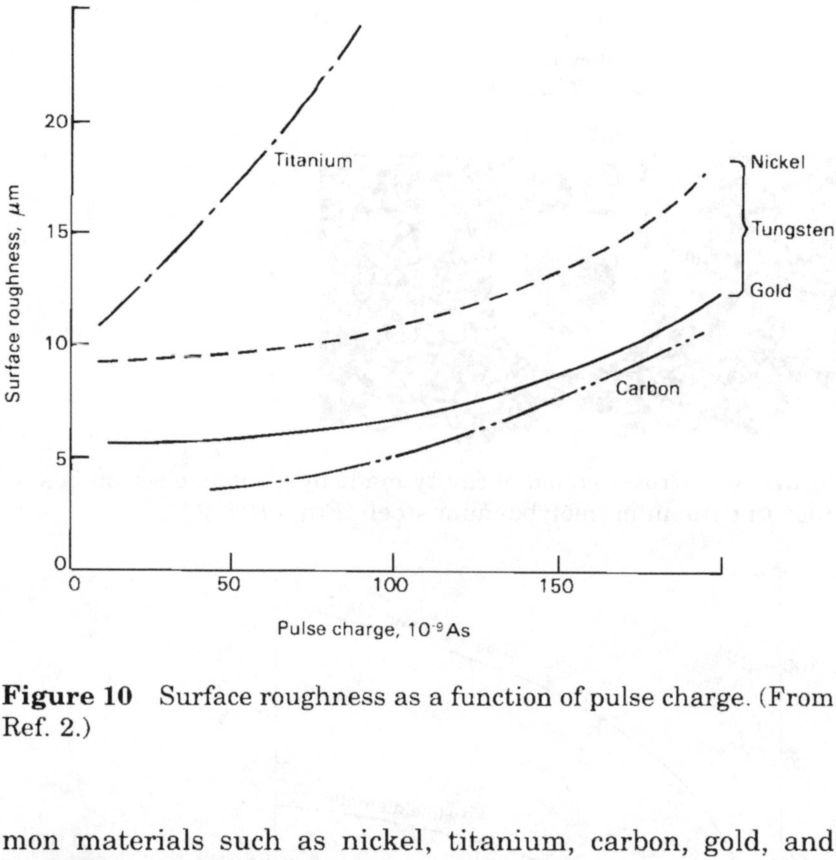

Figure 10 Surface roughness as a function of pulse charge. (From Ref. 2.)

mon materials such as nickel, titanium, carbon, gold, and tungsten.

11.7 HEAT-AFFECTED ZONE

As the evidence in Figure 11 indicates, the surface layers of materials treated by EBM are affected by the high temperatures of the focused beam, illustrated by the white ring surrounding the hole. Further evidence is given in Figure 12, which shows how the diameter of the damaged layer increases with pulse duration as well as hole diameter. The heat-affected zone can be as much as 0.25 mm in EBM.

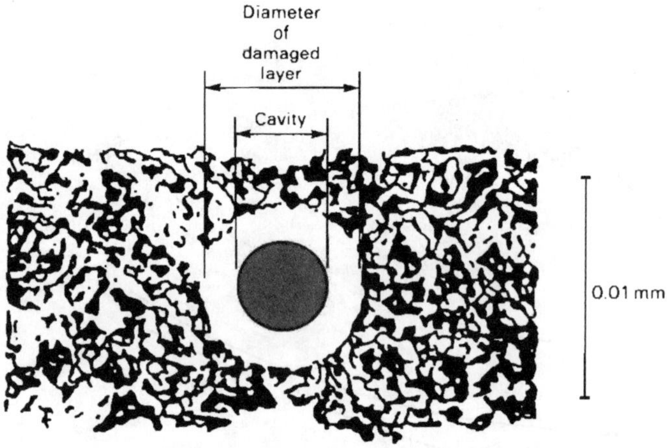

Figure 11 Cross-section of cavity made by a single electron beam pulse in chromium–molybdenum steel. (From Ref. 2.)

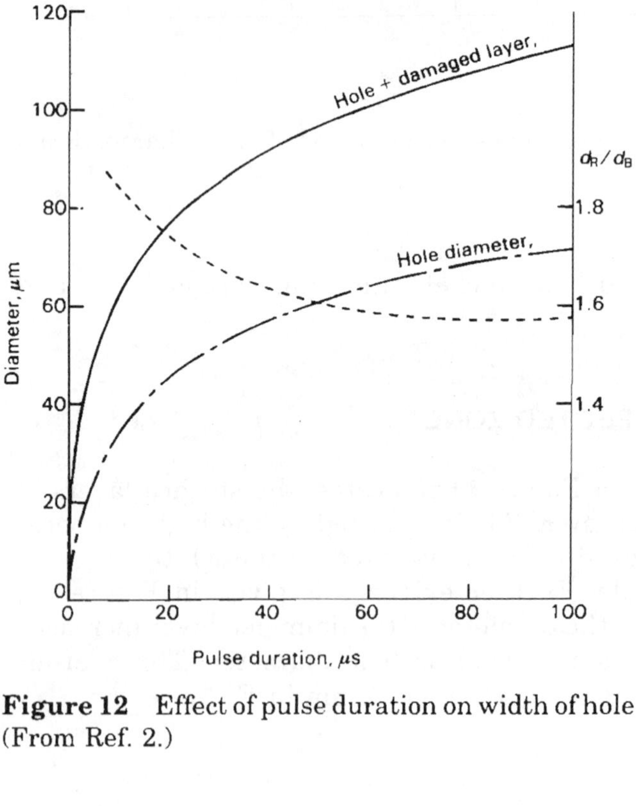

Figure 12 Effect of pulse duration on width of hole drilled by EBM. (From Ref. 2.)

11.8 APPLICATIONS OF EBM

Applications lie in the following main areas: (a) drilling, (b) perforating of sheet, (c) pattern generation associated with integrated circuit fabrication (with which milling is also associated), and (d) texturing.

11.8.1 Drilling

Steigerwald and Meyer [4] gave early consideration to EBM for hole-drilling. They concluded that improved reproducibility, greater working speeds, and deeper holes of accurately controlled shapes were needed. Later Boehme [5] discussed drilling applications with electron beam machines fitted with systems for numerically controlling the beam power, focus, and pulse duration. As a result cylindrical and other configurations, such as conical- and barrel-shaped holes, of various diameters were drilled with consistent accuracy at rates of several thousand holes per second. Drilling of inclined holes, at an angle of 15°, was also investigated.

At that time Boehme [5] reported that the largest diameters and depths of holes that could be accurately drilled by EBM were, respectively, 1.5 and 10 mm, and that the ratio of depth-to-diameter was normally in the range 1:1 and 1:15. For example, Binnie and Champney drilled stainless steel plate 0.25-mm thick with 0.2-mm holes [6].

For deeper holes, in the range 2.5 to 7.5 mm, Steigerwald and Meyer [4] emphasized the need for a stable power supply that can emit the required groups of pulses and for a well-controlled beam of closely defined diameter, the angle of aperture of which has a strong bearing on the shape of the hole produced. Under laboratory conditions, holes of about 19 mm were achieved by their team. Figures 13(a) and (b) show samples of drilling obtained by them for different materials. A further useful summary of the characteristics of EBM of various materials is given in Table 2. Figure 14 shows the cross-section of holes drilled by EBM under the results (a) of Table 3.

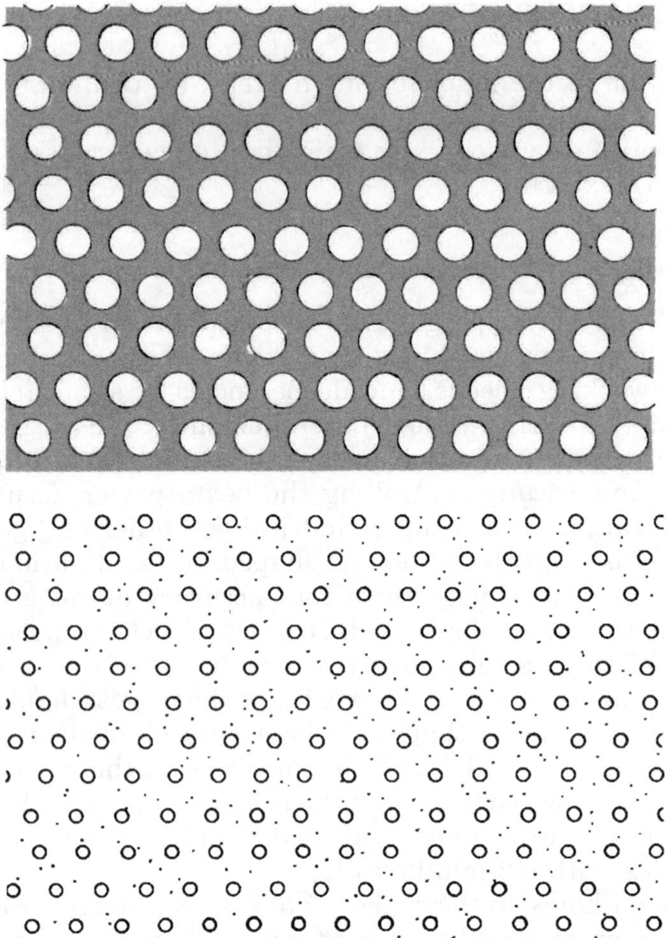

(a)

(b)

Figure 13 Pattern of holes drilled by EBM. (a) Workpiece material: stainless steel; thickness: 0.2 mm; diameter of holes: 0.09 mm; density of holes: 4000 per cm^2; distance between holes: 0.16 mm; distance between rows: 0.16 mm; time required to drill one hole: 10 μs. (b) Workpiece material: synthetic fabric; thickness: 0.012 mm; diameter of holes: 0.006 mm; density of holes: 20,000 per cm^2; distance between holes: 0.07 mm; distance between rows: 0.07 mm; time required to drill one hole: 2 μs. (From Ref. 4.)

11.8.2 Perforation of Thin Sheet

The sheet to be perforated is usually lined with an auxiliary material. The electron beam first penetrates the sheet forming a vapor channel within the fused material, and then enters the auxiliary lining. An eruption of vapor occurs, causing ejection of molten material [5]. For perforation by EBM to be economically acceptable, 10^4 to 10^5 holes per second have to be produced. Thus single pulses lasting only a few μs are needed. In some applications the sheet or foil metal is stretched on a rotating drum, which is simultaneously shifted in the direction of its axis. Rows of perforations following a helical line are thereby produced. Manipulators capable of linear and rotary movement in as many as four axes are used, especially for perforation by EBM of jet engine components.

Foil made of a synthetic material has been perforated with 620 holes per square millimeter for filter applications at a rate of one hole every 10 μs. Some melting does occur, especially with plastics, as occurred with the fabric shown in Figure 13(b) where foil with 205 perforated holes per square millimeter was produced.

Finally, Figure 15 shows a cross-section of a hole 0.125 mm in diameter drilled in 30 μs through a sheet of nickel alloy 0.4-mm thick. This technique can be applied to the production of filters and masks for color television tubes. Other applications for perforation lie in sieve manufacture, for sound insulation, and in glass fiber production.

11.8.3 Pattern Generation for Integrated Circuit Fabrication

This section provides useful background to the work on electron beam technology for high-resolution lithography discussed in Chapter 12.

Birnie and Champney [6] were among the first to draw attention to the use of electron beam technology in scribing thin film circuits for the electronics industry. An attraction of the former is the wavelengths that are some orders of magni-

Table 2 Applications of EBM

Material	Voltage (kV)	Current (μ A)	Pulse width (μ s)	Pulse frequency (Hz)	Beam focus condition	General comments
Aluminum wafers 0.25 mm gauge	90	150	80	150	Focused	0.10-mm slot cut at 300-mm min^{-1}
0.25-mm gauge	125	60	80	50	Focused	0.075-mm hole cut in 10 s
0.75-mm gauge	150	200	80	200	Focused	0.10-mm slot cut at 610-mm min^{-1}
0.75-mm gauge	125	60	80	50	Circle generator	0.30-mm hole cut in 30 s
Sapphire crystal 0.65-mm gauge	110	20	9	50	Beam circular	0.064-mm hole- <30 s
Ferrite wafers (0.25-mm thick)	140	25	5	50	Focused	0.025-mm dia. holes drilled in < 1 s
Molybdenum shim (0.25-mm thick)	140	20	20	50	Focused	<0.050-mm dia. holes drilled in <1 s on 0.075-mm centers

Microdiodes scribing	110	7	12	50	Focused	Scribed to ≈0.025-mm depth Scribing rate ≈1500-mm min^{-1} (Maximum rate not evaluated)
Silicon wafers (0.25-mm thick) (gold deposited)	130	70	4	3000	Focused	Scribed to 0.05-mm depth at 127 mm min^{-1}
Thin film register (Tantalum, 100 Å)	100	20	9	1000	Focused	Cut by manual programming
Mylar tape (0.038 mm)	110	600	Continuous beam	Continuous beam	Focused	117–200 mm min^{-1}
Steel drill (0.36-mm dia.)	140	200	80	50	Focused circular deflected	Drilled in ≈3 min

Source: Ref. 6.

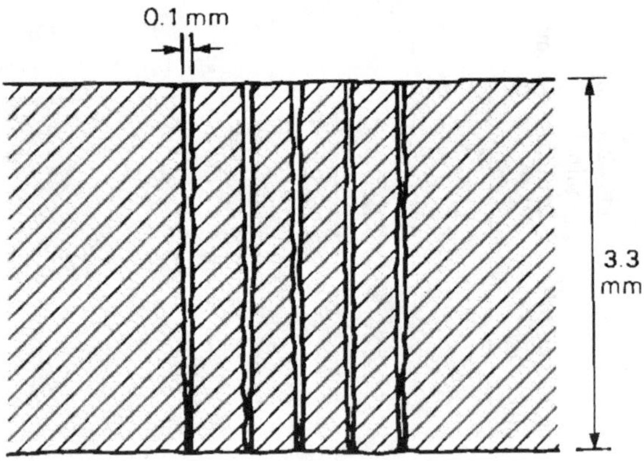

Figure 14 Cross-section of holes drilled by EBM. Workpiece material: high temperature nickel alloy. (From Ref. 4.)

tude shorter than those of light systems, which were used before integrated circuits became so complex.

A detailed early account of EBM for the manufacture of integrated circuits has been presented by Yew [7]. The beam is positioned accurately by means of deflection coils at the location where a pattern is to be written, by exposing a film of electron resist coated on either a chrome mask blank or a wafer, for the production of the lithographic definition required. (Although they are discussed more fully in Chapter 12, it is

Table 3 Drilling of Nickel Alloy[a]

	(a)	(b)	(c)	(d)
Thickness (mm)	3.3	3.3	3.3	1.6
Length of drilled hole (mm)	3.3	5.2	5.2	3.1
Diameter of drilled hole (mm)	0.1	0.7	0.2	0.3
Taper (°)	0	2	1	0
Angle of drilling (°)	90	35	35	30
Time of drilling (μs)	1	15	3	10

[a] Accelerating voltage (130–150 kV).
Source: Ref. 4.

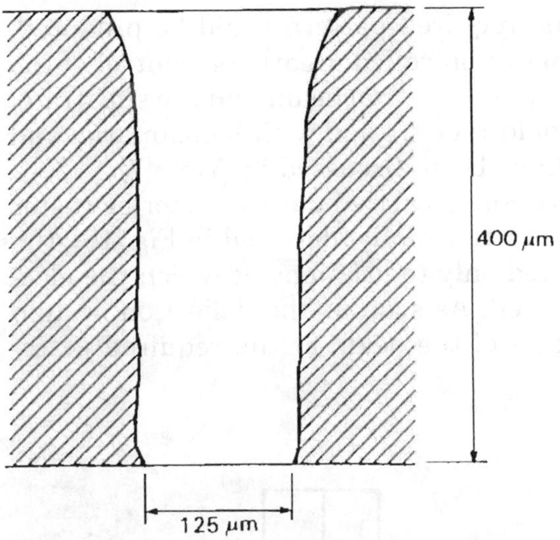

Figure 15 Hole drilled in metal alloy sheet.

noted that electron resists are polymeric materials, similar to photoresists used in lithography, except that the former are sensitive to exposure to electrons rather than ultraviolet light, as is the case with the latter.)

An electron beam of energy about 10 to 20 kV can readily break the bonds between polymer molecules in the case of positively acting electron resists. It can also cause crosslinking in the polymers for negatively acting electron resists. With the onset of either of these conditions, the solubility changes when the resist film is immersed in the developer, usually a solvent for the resist. Due to the difference in solubility between the original and exposed resist polymers, differential material removal occurs. A fine pattern of polymer is thus obtained. This pattern is then used as an active mask to avoid unwanted etching of the integrated circuit mask or wafer.

For integrated circuits produced on silicon wafers 75 mm or greater in diameter, a moving worktable was employed to position precisely each area of the chip under the electron beam, in order that the required pattern could be produced. The accuracy of this operation relied greatly on control of the relative position between the electron beam and the substrate. Two systems, field-to-field registration with benchmarks and laser interferometry, have been discussed by Yew [7].

The pattern generation is carried out by vector or raster scanning. With the former technique (depicted in Fig. 16), the electron beam is deflected only to locations at which the electron resist is to be exposed. As soon as the deflection system completes the positioning of the beam at the required place,

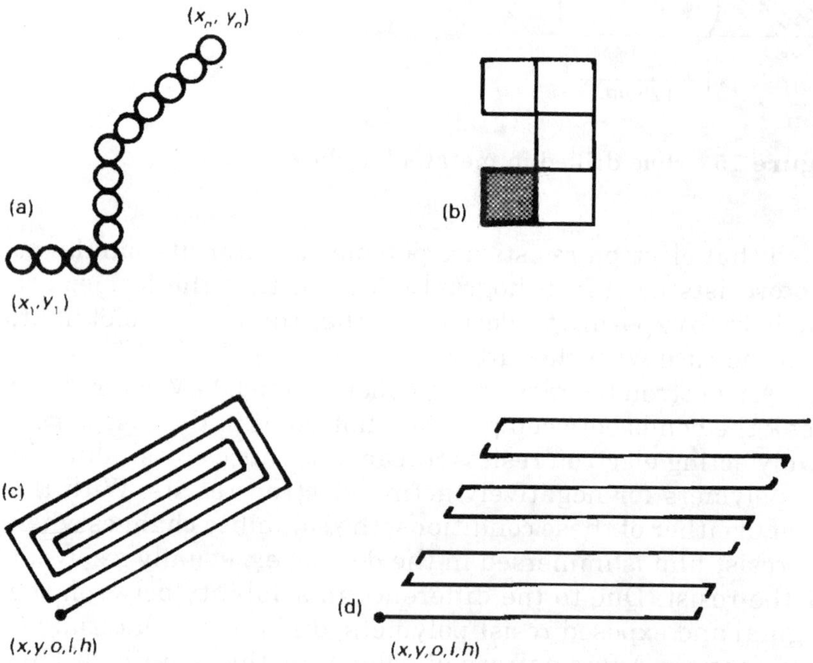

Figure 16 Vector scan writing method. (a) Simple: each dot has (x, y) address output from a computer through a D/A converter to a beam deflection and blanking system; (b) one spot: 2.5 μm × 2.5 μm; minimum positioning step: 0.25 μm; and (c), (d) deflected beam: chip 6 × 6 mm (max). (From Ref. 7.)

the electron beam action is started. As illustrated in Figure 17 with the raster scan system, the chip pattern is first divided into subfields. Each subfield is scanned by the electron beam in a raster, like that employed with television. The electron beam is turned on and off along each raster line as needed. The required pattern is fully formed by the combined effects of electron beam exposure and subsequent resist development. With the raster scan method, the electron beam has to cover most of the mask or wafer area and is therefore less attractive than vector scanning when low density patterns have to be produced.

The EBM system can produce primary chrome masks in a single step, since the electron beam is extremely fine and can have its position very accurately controlled. When the electron beam is used to manufacture primary chrome masks, conventional chrome and resist-coated glass blanks are used, except that electron, instead of photo, resists are employed. The elec-

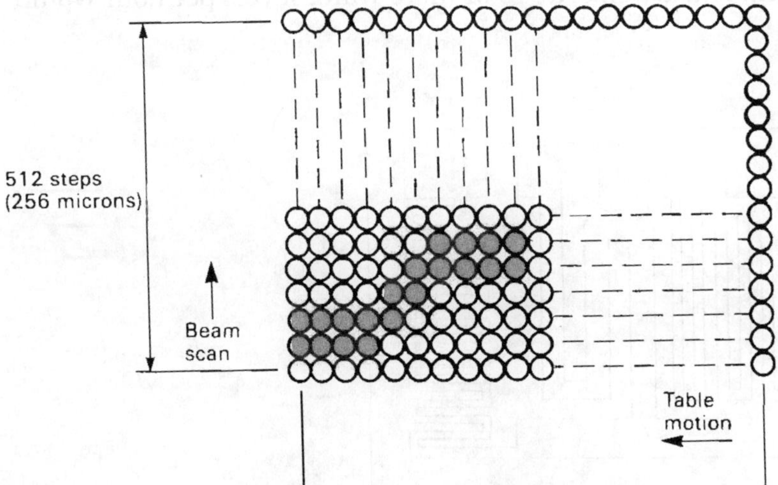

Figure 17 Raster scan method. (From Ref. 7.)

tron resist is exposed until complete exposure is achieved on all the chips. The exposed blank is then withdrawn from the work chamber, and developed by a conventional spray process. After development subsequent etching of the chrome is undertaken in a fashion similar to common photolithographic techniques. The mask should then be ready for use.

Line widths as small as several hundred could then be written with electron beam techniques, and writing speeds up to about 20 MHz are obtainable. Across a 125-mm mask an accuracy of 0.125 μm could be achieved, and a mask of this size could be manufactured in about 60 minutes. Direct wafer processing electron beam systems were built to produce up to 22 wafers per hour.

In his report on electron beam lithography, Richman [8] described a machine capable of producing masks or wafer levels up to 150 mm across, with a 0.125-μm address size, and figure placement accuracies better than 0.02 μm. He comments that industrial requirements for machines capable of 0.25 μm feature size, at 10 or more wafer levels per hour would

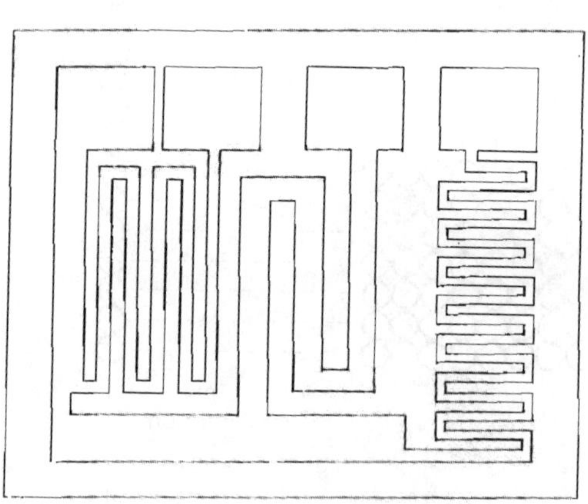

Figure 18 Hybrid circuit engraved with 40 μm traces machining speed greater than 5 m/s (Courtesy of H. Elhofy.)

be achieved in the 1990s. Advancements made since that time are discussed in Chapter 12.

In Figure 18 a miniature electronic hybrid circuit engravement produced by EBM with a 40-μm wide trace is shown.

11.8.4 Electron Beam Texturing (EBT)

Although this topic is not directly related to micromachining, recent accounts show how electron beam technology, among other new methods, has superseded shot blasting in producing required surface textures on steel mill rolls made of hard alloys in order to meet industrial demands for consistent and reproducible steel sheet [9].

REFERENCES

[1] J. A. McGeough, *Advanced Methods of Machining*. Chapman & Hall, London (1988).

[2] J. Kaczmarek, *Principles of Machining by Cutting Abrasion and Erosion*. Peter Peregrinus, Stevenage, 514–528 (1976).

[3] G. Bellows, *Non-Traditional Machining Guide—26 Newcomers for Production*. Metcut Research, Cincinnati, Ohio, 40, 41 (1982).

[4] K. H. Steigerwald and E. Meyer, New developments in electron beam machining methods, electrical methods of machining and forming. Institution of Electrical Engineers, Conf. Publ. no. 38, 252–258 (1967).

[5] D. Boehme, Perforation welding and surface treatment with electron and laser beam. *Proc. 7th Int. Symp. on Electromachining IFS* 189, 200 (1983).

[6] J. V. Binnie and M. A. Champney, The contribution of electron beams to machining and forming, in electrical methods of machining and forming. IEE, Cont. Publ. no. 38, 210–221 (1967).

[7] N. C. Yew, Electron beam—Nort, a practical LS1 production tool. *Solid State Technology*, August, W90 (1977).

[8] R. M. Richman, Precision engineering issues in electron beam lithography machines. In *Proc. 3rd Int. Precision Engineering Seminar*, Cranfield (1987).

[9] L. V. Van Hoye and C. de Mare, Inst. of Materials newsletter of steel division report on *Advances in Mill Roll Technology* (26) 3, 156 (1999).

12

High-Resolution Lithography

S. Thoms and D. Macintyre
University of Glasgow, Glasgow, Scotland

12.1 INTRODUCTION

12.1.1 Scope

In this chapter, optical, electron-beam, and X-ray lithographic techniques are described. Methods used to transfer lithographically defined polymer patterns into more substantial materials, such as metals, are also discussed. The chapter deals with the limits of lithography in terms of smallness and pattern placement accuracy. These techniques may be regarded as the ultimate form of micromachining in which a combination of lithographically large patterns, sacrificial layers, and selective etches are used to make structures such as cogs and pressure sensors, although such applications are not included here.

Optical lithography is capable of building very fine patterns on planar substrates with features as small as 0.18 μm. Such patterns are routinely mass-produced in quantities of more than 10^7 per circuit. To that end, electron beam lithography (EBL) is of particular significance as it provides the masks for most other lithographic techniques, playing a major part in defining their accuracy, and is capable of producing smaller features than any other "mature" lithography process.

As an example of the ultimate capability of EBL, Figure 1 shows a cross-section of part of a proposed 256-Mbit mem-

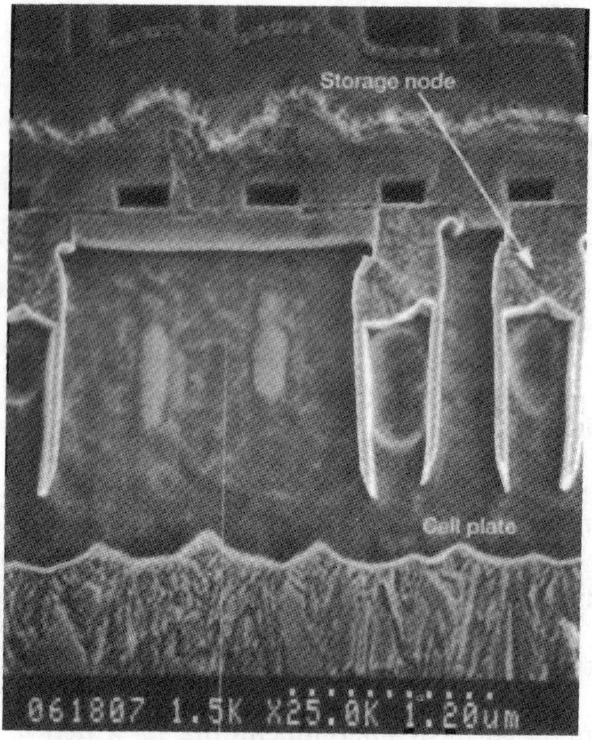

Figure 1 A device cross-section showing the three-dimensional nature of structures in an integrated circuit of a DRAM cell. (From Ref. 56.)

ory circuit, illustrating some trench capacitors. Their highly three-dimensional nature is evident, with submicron features in both the vertical and lateral directions. Many different materials are structured at these sizes to produce complete circuits, including silicon, silicon oxide (quartz), aluminum, and tungsten. The fabrication of such miniature geometries can be considered to represent the finest micromachining.

12.1.2 Steady Shrinkage: Moore's Law

In 1965, seven years after invention of the integrated circuits, Gordon Moore, cofounder of INTEL, observed that the number of transistors that could be mounted on a chip was doubling per year. This phenomenological trend, which became known as "Moore's law," continued until the late 1970s, when the pace decreased to a doubling of transistor numbers every 18 months, which quantity has been maintained to the present. Figure 2 shows that the increase stems from two sources: the shrinkage of lateral dimensions, every three years, from 10 to 0.25-μm design rules, over 1970 to 1999, and the doubling of chip area every three years, from $2 \times 3\ mm^2$ to $20 \times 30\ mm^2$, over 1970 to 1999.

Using Moore's law, semiconductor manufacturers can anticipate the technological development needed to maintain a competitive position.

12.2 PRINCIPLES

12.2.1 Lithography

Lithography is the process by which fine features are defined on a substrate. For conventional lithography the substrate needs to have a flat surface, and is typically a 200-mm diameter silicon wafer with a thickness of about 1.0 mm. Optical lithography is the most common form because it combines the accuracy required for current silicon circuits with considerable throughput capability. Electron beam lithography is capable of much higher resolution pattern definition but is a serial

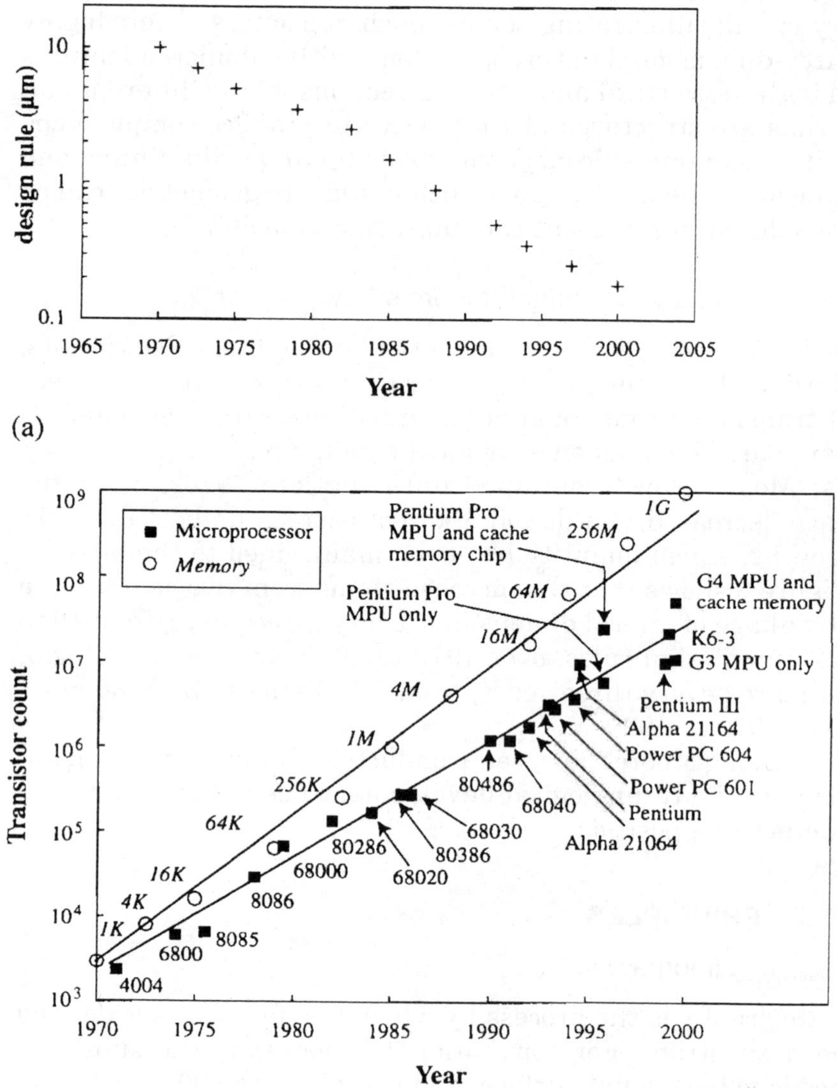

(b)

Figure 2 Moore's law showing (a) exponential decrease in minimum feature size per year (DRAM) and (b) corresponding exponential growth in transistor count per chip (IC). (From Ref. 57.)

process and therefore much slower than optical lithography, which is a parallel process. X-ray lithography combines the higher resolution of electron beam lithography with the high speed parallelism of optical lithography. However, it is not as technologically developed as optical lithography and is currently a research, rather than a production, tool.

The lithography process leaves a substrate coated with a finely patterned polymer film. Subsequent processing steps are needed to transfer this pattern to underlying materials that are usually metals, semiconductors, or dielectrics. The pattern transfer step is discussed in Section 12.2.3.

Optical Lithography

The process for optical lithography is shown in Figure 3. First the substrate is coated with a thin film of radiation-sensitive polymer known as a resist. Film thicknesses are typically on

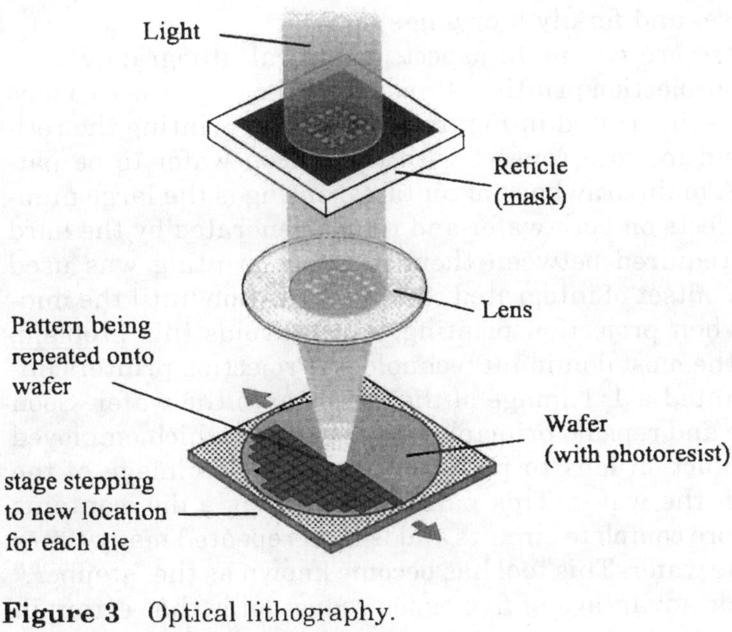

Figure 3 Optical lithography.

the order of 0.5 μm for 0.18-μm minimum feature sizes currently used. The second stage is the selective irradiation of areas of the substrate. Two avenues are possible: either irradiation of areas from which the resist is to be removed with the latter remaining elsewhere (*positive* tone resist); or the alternative (*negative* tone resist), irradiation of those areas that are to remain with all the other resist removed. A patterned reticle is needed for this process as shown in Figure 3. This typically consists of a chrome-coated quartz substrate with the desired pattern etched into the chrome. The optical system produces a reduced image of the reticle on the substrate accurately transferring the desired master copy of the pattern on the reticle to the wafer. The areas with chromium are opaque to the UV radiation whereas the quartz is transparent.

The final stage is resist development. Here the resist is treated to remove selectively exposed or unexposed areas of resist, depending on whether the resist is positive or negative tone. The simplest development process consists of simply immersing the wafer in a suitable developer solution, followed by a rinse, and finally a drying step.

There are two main aspects to optical lithography: contact and projection printing. Projection lithography uses a lens system as illustrated in Figure 3. In contact printing the reticle is held in close contact with the silicon wafer to be patterned. A main drawback of contact printing is the large number of defects on both wafer and reticle generated by the hard contact required between them. Contact printing was used from the outset of integrated circuit fabrication until the mid-1970s, when projection printing, which avoids this problem, became the most dominant technology. Projection printers initially printed a 1:1 image of the reticle onto the wafer. Soon the "step and repeat" principle was developed which employed a 5:1 reduction lens to produce a demagnified image of the reticle on the wafer. This pattern, known as a die, contains one or more complete circuits and is then repeated many times across the wafer. This tool has become known as the "stepper." The major advantage of five-times reduction is that errors in linewidth and positioning on the original reticle become less significant, making reticle production an easier and cheaper process. Further refinements have been made to this principle, largely through decreasing the wavelength from the visible Hg "G" line of 436 nm, through I line (365 nm) to the deep UV (DUV) wavelength of 248 nm generated by KrF lasers for present steppers. Shorter wavelengths are important because they allow smaller features to be resolved. Another refinement is the use of the "step and scan" principle in which the image area of the lens system is reduced to a rectangle or arc. Both the wafer and the reticle are then simultaneously scanned to create the complete image of the die, which, as before, is then stepped across the wafer. DUV (248 nm) steppers have typical scanning field sizes on the order of 5 by 22 mm. Shorter wavelength 193-nm (ArF laser) steppers are now commercially available but not used in production.

Electron Beam Lithography

The production of the mask for optical lithography is clearly an important issue. Generally in the lithographic process, it is the largest application of electron beam lithography. Other lithography techniques have been used for mask fabrication, most notably laser lithography, although this subject is not considered here. Electron beam lithography tools are also used for so-called "direct write" applications to pattern wafers directly. They are employed mainly when higher resolution is required than that obtainable from optical lithography.

The outcome of electron beam lithography is very similar to the concept implemented on personal computers, only on a much reduced scale. Computer-aided design packages are available to assist in the creation of complex designs which are then written onto substrates precisely as viewed on the computer screen, although on a submicron scale.

In electron beam lithography, the resist is exposed by an energetic beam of electrons, under the action of a voltage typically between 10 and 100 kV. Two main types of systems are in production, one with shaped beams; the other uses a Gaussian spot. Gaussian spot machines can be further subdivided into raster, such as the MEBES type and vector, scan machines. In these techniques, a finely focused beam of electrons is scanned across the surface of the substrate in order to expose the resist. This is termed a serial process. It contrasts strongly with optical lithography, a parallel process, in which the whole circuit is simultaneously exposed. As a consequence EBL is much slower than optical lithography, and therefore is generally used in production only when resolutions smaller than those available in optical lithography are required. Several methods of undertaking parallel electron beam lithography have been suggested, and currently SCALPEL (scattering with angular limitation in projection electron beam lithography) is under intense development at Bell Laboratories [1].

Gaussian spot machines are capable of the highest resolution whereas shaped beam machines are faster. In a raster scan system, the beam is scanned across the whole substrate and blanked or unblanked as appropriate to produce a pattern in a way analogous to that used in a television set. Vector systems only scan the beam where exposure is required which gives better-defined shape edges. It is also faster when the pattern density is low. Raster systems, however, can scan at much higher clock rates (typically 100 MHz in contrast to 10 MHz) which makes them faster for medium- to high-density patterns. Raster scan systems are utilized typically for maskmaking, whereas vector scan systems are employed in high-resolution direct write applications.

In all these tools the maximum possible movement of the electron beam by use of the deflection system is about 1 to 5 mm. This movement is limited by aberrations that affect both the spot size and the positional accuracy as the deflection increases. The largest area that can be written without movement of the substrate is called a "field." Larger patterns are made up by joining a number of fields in a similar way to the stepping action of an optical stepper. The major difference is that in an optical stepper a complete circuit is usually written with each step, whereas with electron beam lithography only a part of a circuit is written between steps. Where fields butt together there is inevitably some residual positional mismatch between the fields, known as "stitching error." The column deflection system is electronic and much faster than mechanical stage movements. To improve writing speed, writing strategies have been developed to minimize stage movement, and these involve a meander-style coverage of the entire substrate with as few movements as possible.

One key aspect of electron beam lithography is its flexibility. This characteristic stems partly from its much larger depth of focus, compared to optical lithography; also the beam focus can be varied during pattern writing. Thus EBL can be used to pattern highly topological surfaces provided they can be coated with resist. A further feature is that design changes can be rapidly implemented without the need to remake costly sets of reticles. EBL is therefore an invaluable tool in research and development, even when the feature sizes to be written could be patterned by means of optical lithography.

Other Lithographies

X-rays, ion beams, lasers [2], and scanning probes [3] can all be used to carry out lithography. The use of X-ray lithography was noted as early as 1972 [4], and has until recently been described as the "next generation tool" to be used when optical lithography reaches its limitations. The latter technique was expected to be replaced for dimensions smaller than 1 μm. However, recent reports (2001) show that circuits with minimum feature sizes of 0.13 μm are still being made by optical lithography, and sizes as small as 0.07 μm are also expected to be achieved by the same procedure.

The principles underlying X-ray lithography are similar to those of contact printing for optical lithography and therefore require a full-sized mask. The production of such masks for X-ray lithography remains a major technological difficulty.

Not only does mask production lack the benefit of a reduction system, but also writing on a 2-μm thick silicon membrane is needed, so that mechanical stability becomes a key issue.

Scanned laser lithography is used to make reticles, by a procedure similar to that of electron beam lithography except that a laser, rather than an electron beam, is scanned across the substrate. Scanning probes are currently capable of the ultimate lithographic limits with controlled atomic movement. This technique, however, is very slow, and technologically immature. Massive parallel arrays of scanning probes may be used for future lithography although much development is still needed.

Another new lithographic technique is "mechanical pattern transfer." Several modes are available, although in each a relief pattern on a master is replicated onto a substrate. The substrate to be patterned can either be a hard material, such as silicon, coated with a suitable polymer, or a solid piece of polymer. In the former, resists are often used as the polymer; this process is known as nanoimprint lithography [5]. By use of this technique, resist, and ultimately metal, patterns as small as 10 nm have been made on silicon. In the latter case both embossing [6] and injection molding [7] have been shown to be capable of sub-100 nm pattern transfer. [Recently 25-nm injection molding has been demonstrated (1999).] Mechanical pattern transfer for deep submicron features is at a very early stage of development; it has potential for the cheap mass replication of nanoscale structures.

12.2.2 Resists

Lithographic procedures generally involve the coating of energy-sensitive chemical substances called resists. The resist-coated substrate is then exposed by an energetic beam, which has been patterned in some way. Pattern transfer to substrates by mask technologies is well established for both optical and X-ray lithography [8]. However, high-speed scanning techniques are now becoming increasingly significant. In addition,

ion and electron beam lithography technology by both raster and vector beam scanning is well established. Recent developments in this field include projection ion beam and projection electron beam methods, such as SCALPEL. After patterning, resist-coated substrates undergo a development process that, depending on the resist used, selectively removes either exposed, or unexposed, areas of the resist to reveal the pattern.

The quality of pattern definition and the speed at which pattern transfer into the resist can occur is very dependent on the type of lithography tools and the resist used. A wide variety of resists is available; they are constantly evolving to satisfy demands for improved lithography tool performance, new fabrication processes, and legal requirements for use of environmentally safe solvent systems.

Resist Types and the Selection of Appropriate Resists

Resists are generally classified as either positive, such as the S1800 Shipley photoresists, or negative tone such as the SC Waycoat resist series. A number of image reversal resists exist. Energy patterns can be applied to these resists and the pattern reversed, normally by carrying out an additional baking step prior to development. One example is the Hoechst AZ5200 resist. An important measure of resist performance is the so-called contrast of the resist, which is a measure of the variation of the resist dissolution rate with exposure dose. High-contrast resists generally produce vertical resist profiles after development, whereas low-contrast resists produce sloped resist profiles. Resist sensitivity is defined as the energy required to produce complete solubility in the exposed region; it is a significant measure of resist performance.

Choice of resist depends on many factors. Key influences are the sensitivity of the resist to the energy being used for patterning, and the type of lithography. The suitability and thickness of the resist needed for subsequent processing stages such as metallization, wet or dry etching, electroforming, and the minimum feature sizes in the pattern must also be considered. Account must also be taken of the continuity and quality of supplies, cost, and the safety of solvents on which the resist is based.

The semiconductor industry now regularly uses 0.25-μm process technology. It has recently been demonstrated that semiconductor production technology can be used to print features less than 0.1 μm in size. This achievement has been developed on the design and application of a resist sensitive to 193-nm light sources. Certain photoresists are classified according to their sensitivity to major lines in the mercury spectrum since mercury lamplight sources are used in the longer wavelength lithography tools. Thus resists that are sensitive to light of wavelength about 436 and 356 nm are termed, respectively, "g-line" and "i-line" resists. Resists sensitive to 248-nm wavelength deep ultraviolet light are now established, while 193-nm DUV resists and those sensitive to extreme ultraviolet (EUV) are being developed to meet advances in lithography tool technology.

A less commonly used but nevertheless important group of photosensitive materials comes under the category of thick resists. They include epoxy-based negative tone resists such as SU-8 (first produced by IBM Corp.) which can be applied in thicknesses exceeding 100 μm, and photosensitive polyimides such as the Probimide series, which can be applied in film thicknesses greater than 30 μm. These materials have useful applications in the fabrication of high aspect-ratio three-dimensional features.

The volume market for electron beam sensitive resists is relatively small and so continuity of supplies can be problematic. Polymethyl methacrylate (PMMA) was first reported as a positive electron beam sensitive resist by Haller et al. in 1968, and it has since been well used for this purpose. PMMA is capable of patterning features less than 10 nm. It is relatively easy to process as it only needs to be applied to substrates, baked, exposed, and then developed. A drawback to PMMA is its poor dry etch resistance relative to Novolak-based photoresists, and its relatively poor sensitivity to electron beams. Typical resist sensitivity is 200 μC/cm^2 on silicon substrates. Thus PMMA often has to be used as a mask to produce more robust dry etch masks in other materials such as metal films; and pattern writing speeds are slower than machine capabilities. Improvement in the dry etch resistance of PMMA has been attempted through the formation of composite resist systems [9,10] by the addition of subnanometer etch-resistant particles such as ruthenium metal clusters and Buckminsterfullerene (C60). Copolymers of methyl methacrylate and methacrylic acid (PMMA–MAA) offer improved sensitivity and are commonly used in multilayer resist applications. Other positive tone copolymers used in electron beam lithography include ZEP-520 from Nippon Zeon (copolymer of chloromethacrylate and methylstyrene), the sensitivity of which is 10 times better than PMMA with comparable resolution capabilities. EBR-9 is a positive acrylate based (2,2,2-trifluoroethyl chloro acrylate) electron beam resist produced by the Toray companies for microfabrication in the VLSI industry. It has a sensitivity of 30 μC/cm^2 at 50 kV, can resolve 0.2 μm features, and has good adhesion to chrome-coated substrates. This resist is commonly used in the production of photomasks by electron beam lithography as is poly(butene-1-sulfone)–PBS which is more sensitive but involves more complicated processing.

Negative tone electron beam resists often suffer from resist swelling during development. Shipley SAL resists have been designed specifically for electron beam lithography and are claimed to be nonswelling. These resists are Novolak-based, have relatively good dry etch resistance, and reported sensitivities of approximately 7 μC/cm^2 at 20 kV.

It has been shown that certain positive and negative tone DUV photoresists can be used as sensitive (around 20 μC/cm^2 at 50 kV), electron beam resists [11] capable of defining pattern features less than 100 nm. These resists allow fast pattern writing. As they exhibit relatively good dry etch resistance, they can be used to design simpler fabrication procedures. Shipley UVIII, UV5, and UVNII are typical examples. They are in plentiful supply, owing to their increasing use in the semiconductor industry.

Resist Chemistry

This topic is discussed in detail in Moreau [12]. Many types of resists are available, and the processing conditions used greatly depend on the resist chemistry. In general, resists can be described as being one-, two-, or three-component resist material. A one-component positive resist material is typically a polymer whose molecular weight is reduced upon irradiation. An organic solvent or solvent mixture is used as developer and the dissolving power of the solvent varies strongly with the molecular weight of the polymer. PMMA is an example of a single-component positive resist, the sensitivity of which decreases slightly with increased molecular weight. Negative, single-component resists work by free radical-induced cross-linking of polymers which become insoluble above a gel point dose. These resists suffer from swelling during development. Typical two-component resists are i-line photoresists which are often based on diazonapthoquinone (DQN) dispersed in a Novolak resin [13]. The DQN has a dual role as a dissolution inhibitor and photosensitizer, while good dry etch resistance is attributed to the Novolak resin. Absorption of radiation destroys the dissolution inhibitor in patterned areas and aqueous alkaline developers are used to create the photodifferentiation. An increasing number of three-component resist systems are being produced. They offer good dry etch resistance and enhanced sensitivity from chemical amplification. Three-component positive resist systems often comprise a Novolak or acrylate matrix which provides good dry etch resistance and allows aqueous development, a dissolution inhibitor which is not photosensitive but which is decomposed in an acid-catalyzed hydrolysis reaction, and a radiation-sensitive component which generates an acid catalyst on exposure to radiation. The catalyst is not stoichiometrically consumed in the hydrolysis reaction and is regenerated. The original radiation event can thus lead to several catalytic cycles and the resists are thus termed "chemically amplified." The hydrolysis reaction is slow at room temperature and so chemically amplified resists are normally given a postexposure bake (PEB) prior to development in order to speed up this process. Reliable processing of these resists requires careful control of bake temperatures and times. In order to decrease the sensitivity of chemically amplified resists to airborne contamination, some of the more recent resists have been designed to have relatively high bake temperatures (approximately 135°C.) compared to i-line photoresists (80°C).

Resist Processing and Related Fabrication Issues

Both resist and substrate must be exceptionally clean since any particle contamination will result in poor pattern transfer and defects. Resists are commonly filtered to remove particles as small as 0.2 μm and relative humidity must be controlled below 50% RH (relative humidity) in order to obtain satisfactory resist adhesion. (Relative humidity is the ratio between the water vapor in the air and the amount of water vapor the air can actually hold.) Wafers must be completely dry prior to the application of the resist to substrates and adhesion promoters such as HMDS (hexamethyldisilizane) are often used to improve resist adhesion, particularly to silicon wafers. Resists are generally applied to wafers by spin coating and the process can be described by Newtonian models in which the film thickness t is given by

$$t = K\sqrt{\frac{C^2}{SS}} \tag{1}$$

where K is a constant depending on the type of resist and the casting solvent, SS is the spin speed, and C is the concentration of polymer in the solvent. The film thickness of the spun film is independent of the wafer diameter provided there is sufficient fluid present to cover about two thirds of the wafer. Resist spinning procedures normally produce highly uniform resist thicknesses over flat wafers. However, topographical features such as etched mesas or metal patterns can cause variations in resist thickness that can translate into variations in linewidth. These problems are sometimes eliminated by planarizing wafers with resists such as polyimide. The uniformity of resist films over entire wafers can be rapidly measured by equipment such as the Prometrix reflectometer and more detailed investigations into the surface roughness of resists is often carried out by atomic force microscopy (AFM) (see Chapter 2).

Resist-coated wafers are given a soft bake prior to exposure in order to drive off the casting solvent, and this is either effected in an oven or by use of a vacuum hot plate. In the semiconductor industry wafer coating and baking are carried out automatically on wafer tracks. After pattern exposure and possibly postexposure baking, resist-coated substrates are developed. The choice and concentration of developer largely depend on the resist used, but may also depend on whether contamination will arise from the presence of metal ions. Resist manufacturers generally recommend particular development techniques such as puddle development which give optimum resist performance. Bake temperatures and times, process delays, development temperature, and other conditions can influence the quality of lithography; design of experiment techniques is sometimes used to simplify process optimization.

Resist films are occasionally given a postdevelopment bake to enhance their resistance to dry etching. This procedure, combined with the etching process itself can lead to crosslinking in the resist which can render difficult removal of the resist at a later fabrication stage. A variety of specialized resist strippers is therefore manufactured to deal with this difficulty.

Fabrication techniques often exploit the performance of different resists. For example, in high resolution electron beam lithography, improved metal lift-off is achieved by using a bilayer of resists with different sensitivities to give sloped resist-profiles. The fabrication of T-shaped gate structures relies on the use of a highly sensitive resist such as copolymers of PMMA or UVIII on top of less sensitive resists such as PMMA.

12.2.3 Pattern Transfer Techniques

These techniques are usually divided into subtractive and additive processes which are illustrated in Figure 4. In the former the resist acts as a mask protecting the underlying material from the etch. In regions where resist is absent, the

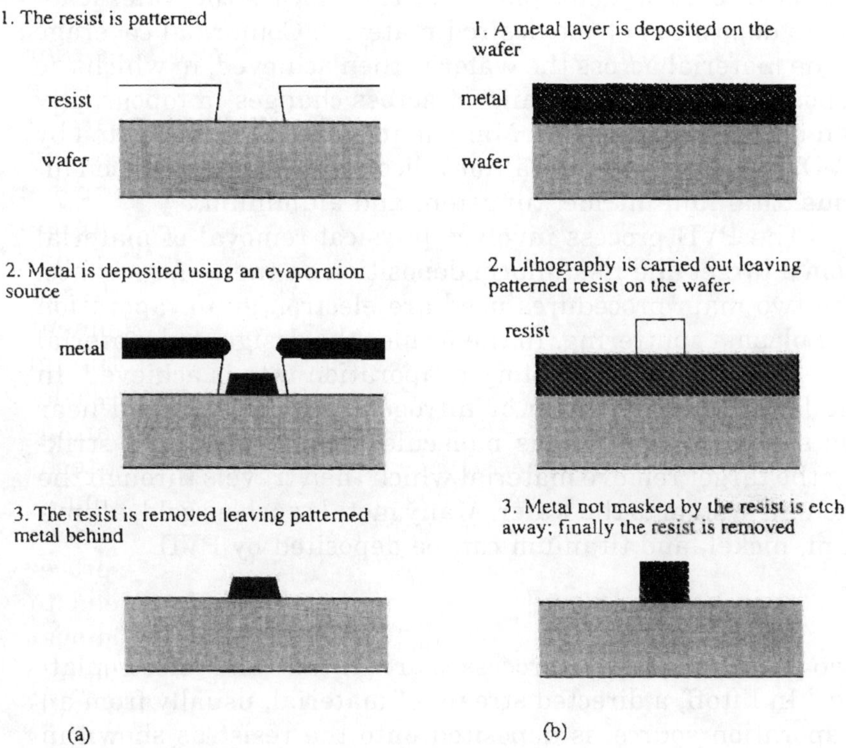

Figure 4 Pattern transfer from resist to hard material showing (a) additive process and (b) subtractive process.

material is removed. In the additive process, material is added in such a way that it remains only in those areas free of resist. In both types of process, on completion of the pattern transfer, the resist is removed, its function now having been fulfilled. These processes are described very briefly here. A fuller description is given by Chang and Sze [14].

Deposition

The two main classes are chemical vapor (CVD) and physical vapor deposition (PVD). In CVD reactant components are brought together in gaseous form in a suitable chamber. A chemical reaction takes place on the surface of the wafer leading to deposition of the desired material. Conformal coverage of the material across the wafer is then achieved, in which the deposited thickness is uniform across changes in topography and underlying material. Many materials can be deposited by CVD, including polycrystalline silicon, silicon oxide, silicon nitride, titanium nitride, tungsten, and aluminum.

The PVD process involves physical removal of material from a target and its uniform deposition on the wafer surface. The two main procedures used are electron gun evaporation and plasma sputtering. In the former the designated material is heated until a satisfactory evaporation rate is achieved. In the latter, a small argon or nitrogen plasma is formed near the source. Energetic gas molecules from the plasma striking the target remove material which then travels through the vacuum system to the wafer. Many metals such as gold, aluminum, nickel, and titanium can be deposited by PVD.

Additive Processes

Two principal additive processes are "liftoff" and "electroplating." In liftoff, a directed stream of material, usually from an evaporation source, is deposited onto the resist as shown in Figure 4. Conformal covering is unwanted in this case. After the deposition is complete, the resist and the metal coated on its top are removed by immersion in a suitable solvent, which undergoes agitation. Electroplating can also be used to add material in the gaps patterned into the resist. In this case a plating base is usually first needed.

Subtractive Processes

The subtractive process is illustrated in Figure 4. There are two main classes: wet etching, which occurs in solution, and dry etching, which takes place in a low vacuum and relies on either a plasma or a beam of atoms/ions. A decision that wet etching should be replaced by a dry etch process depends mainly on the verticality of the etch which can be achieved by the latter, as illustrated in Figure 4. This quality has become increasingly important with reduction of device dimensions below 2 μm. Wet etching makes use of chemicals such as hydrofluoric acid and potassium hydroxide to etch selectively one material, others being left untouched. Dry etching has both physical and chemical aspects. High energy atoms/ions striking the surface physically remove material by sputtering. If reactive species are present then a chemical reaction takes place that enhances the etch rate, and makes possible selective etching of one material in preference to others. Vertical sidewalls result partly from the directional aspect of the ions causing sputtering and also by deposition of reaction byproducts on the sidewalls, which inhibit lateral etching. As each different material requires etching, fresh chemical and plasma conditions have had to be developed.

Dry etching technology has many subclasses. In reactive ion etching (RIE) a plasma is struck between two parallel plate electrodes by applying radio frequency (RF) power causing a plasma to be struck between them. A self-bias is formed between the plasma and the substrate to be etched which is placed on the lower electrode. Ions and reactive radicals are accelerated across the "dark" space between the plasma and the substrate with energies typically in the range 20 to 400 eV. These basic operating conditions may be varied in order to increase the etch speed and reduce etch damage, the latter being caused by impingement on the substrate of high energy ions. In electron cyclotron resonance (ECR) etching electrons are confined by the interaction of a magnetic field with microwave radiation. The resulting plasma is of a very high density, approximately two to three orders of magnitude more than that in RIE, typically producing high etch rates. Inductively coupled plasma (ICP) etching uses a large coil external to the vacuum to couple RF power inductively into the plasma. The resulting plasma densities are also two to three orders of magnitude higher than those of RIE.

12.3 RATES OF MICROMACHINING FOR RELEVANT MATERIALS

Two issues are discussed in this section: the rapidity at which the lithographic process itself can be carried out (measured in area per unit time), and the rate of the pattern transfer (depth per unit time).

12.3.1 Lithography

Optical Lithography

For optical steppers the rate of area coverage is measured in wafers per hour. For 200-mm wafers typical values are between 30 and 60 per hour for 0.25-μm feature sizes.

Electron Beam Lithography

This procedure is much slower than optical lithography; typical times are measured in hours per wafer, and depend on the nature of the pattern to be written, the two main factors being the pattern density, or fill factor, and the ultimate resolution required. The required resolution determines the beam diameter which is typically set between two and five times smaller than the minimum feature size. An approximation for the writing-time T for a pattern can be expressed as

$$T = \frac{FF \cdot A \cdot S}{i} \tag{2}$$

where *FF* is the fill factor, *A* the total area to be covered, *S* is the sensitivity, and *i* is the beam current. The beam current *i* is related to the spot diameter *d* by

$$i = B\frac{\pi^2 d^2 \alpha^2}{4} \tag{3}$$

where *B* is the electron source brightness, and α is the beam semiangle at the substrate, typically 1 to 10 mrad. The beam brightness depends greatly on the electron source used. Two main types of sources are employed by EBL machines, namely, thermionic sources such as LaB6 and field emission sources, their respective brightnesses being about 10^6 and 10^8 Acm^{-2} sr^{-1}. Clearly the field emission sources have superior speed; they also provide smaller spot sizes giving improved resolution.

For a typical beam current of 5 nA and a resist sensitivity of 20 μC cm^{-2}, the time to write a square centimeter with 50% fill factor is 2000 s (or about one half hour). Practical fill factors are between 1 and 50% giving area coverage rates between 2 and 100 cm^2 per hour. Equation (2) considers only calculation of the beam time, that is, the time needed for the beam to scan the pattern. In calculation of writing time, account may also have to be taken of machine calibration, stage movement, and shape loading times. Although the beam time usually determines the writing time, in some cases these other factors may have a greater effect. In particular, the stage movement times required to cover a 200-mm wafer are noted to amount to about 2 hours for a typical vector scan machine, which is not taken into account in the determination of beam time.

12.3.2 Pattern Transfer

Deposition Rates

Typical rates are on the order 10 to 100 nm/min, and depend on the system and the materials being deposited.

Dry Etch Rates

Many etches have been developed for different applications with a large spread in etch rates. Nonetheless in the development of an etch, not only must the etch rate be considered, but also other aspects, notably selectivity and semiconductor damage. Selectivity is the differential etch rate of one material over another. Some selectivity is key to any etch process, as the mask must remain intact while the specified material is removed. Multilayer structures are common in lithographic applications. Etches are often required that etch through one layer and cease on another. A high degree of selectivity from one to the next material is thereby required. An etch is often required that removes at the same rate two or more different materials. Dry etch damage is inflicted on semiconductor materials by the rapidly moving ions in dry etching, which thereby degrades device performance. Such degradation is a vital consideration in development of any etch.

Table 1 cites different dry etches showing both the spread of different materials which can be microetched, and also the variation in etch rates obtained.

Wet Etch Rates

Most semiconductor etching is carried out by using plasma-based systems, although wet etches are still frequently used. Table 2 shows the characteristics of some common wet etches.

12.4 ACCURACY AND DIMENSIONAL CONTROL

12.4.1 Lithographic Tolerances

In lithography, size control for the smallest features is usually both crucial and difficult. The size of these smallest features is known as a critical dimension (CD); CD control is a major aspect of lithography, as the absolute positional fidelity with which patterns can be placed. In practical applications, the

Table 12.1 Etch Rates for Various Materials

Material to be etched	Mask	Etch gasses	Etch system	Rate (nm/min)	Depth (μm)	Feature size (μm)	Ref.
Si	600 nm photoresist	ICP	SF_6/CHF_3	680	0.5	0.5	34
Si	2.8 μm SiO_2	BCl_3/Cl_2	RIE	200	48	10.0	35
Si	15 nm Au	$SiCl_4$	RIE	140	0.06	0.015	36
Polycrystalline Si	tetraethoxysilane BPTEOS	HBr	ECR	230	0.4	0.3	37
GaAs	40 nm Cr	Cl_2/CH_4	ECR	460	2.3	0.2	38
AlGaAs	40 nm Cr	Cl_2/CH_4	ECR	280	1.4	0.2	38
InGaAs	100 nm NiCr	§	RIE	90	3	2	39
InGaAs	20/50 nm Ti/Ni	Cl_2/Ar	ICP	250	0.4	0.1	40
Au	500 nm SiO_2	Cl_2	ECR	250	0.4	1	41
AlCu (0.5% Cu)	photoresist	BCl_2/Cl_2	ICP	1000	0.7	0.4	42
W/WSi_x	resist	SF_6/Ar	RIE	100	0.4	0.3	43
W	45 nm Ni	SF_6	ECR	120	0.45	0.2	44
W	50 nm PMMA	SF_6/O_2	RIE	50	0.05	0.03	45
NiMnSb	photoresist	SF_6/Ar	ICP	>1000	0.5	1.2	46
$YBa_2Cu_3O_{7-x}$	50 nm C	ion beam milling	Ar	25	0.1	0.2	47
SiO_2	300 nm polycrystalline Si	$C_4F_8/O_2/Ar$	M-RIE	600	1.0	0.06	48
Si_3N_4	50 nm Ti	C_2F_6	ECR	25	0.1	0.03	49
C diamond	Al	CF_4/O_2	RIE	0.78	0.3	0.01	50
Polyimide	NiCr	O_2	RIE	??	0.6	0.1	51

Table 12.2 Various Wet Etch Rates

Material to be etched	Mask	Etchant	Rate (nm/min)	Comments
SiO_2	resist	HF	50	
Si_3N_4	resist	H_3PO_4	10	
Al	resist	HNO_3/ CH_3COOH/ H_3PO_4/H_2O	35	
Si	SiO_2	KOH	600	80°C, anisotropic. Stops on (111) plane.
GaAs	resist	H_2SO_4/H_2O_2/H_2O	800	anisotropic

From Ref. 52.

main positional parameter is the overlay accuracy. Semiconductor circuits are built by use of a large number, typically more than 20, of lithography and pattern transfer steps. Overlay error refers to the accuracy with which one lithographic step is aligned to the next, and the usual target is about 25% of the minimum CD.

Optical Lithography

The resolution of optical lithography is usually defined by the Rayleigh criterion discussed by Moreau [12],

$$R = \frac{k\lambda}{NA} \tag{4}$$

where R is the smallest printable feature, NA is the numerical aperture of the lens system, and λ is the exposing wavelength. For periodic patterns R is equal to one half of the pitch. For a given optical system, λ and NA are fixed and k is a factor that describes the modulation of the aerial image. The absolute minimum of the k-factor for periodic structures is 0.25 since at this stage the modulation becomes zero. In order to achieve the smallest possible features, k has to be made as small as possible. The minimum practical value of k is about 0.35, which is achieved by a combination of techniques including the use of phase-shifted masks and off-axis illumination [15]. For an assumed large *NA* of 0.7, 0.13-μm features can be printed with 248-nm steppers. Isolated (in contrast to dense) lines of 0.12 μm have been reported by use of a 248-nm stepper and phase-shifted masks [16].

Positional errors come from two main sources: the reticle and the lens system. For a SVG lithography DUV scanner lens, distortions are 35 nm, and tool-to-tool overlay errors amount to 90 nm [17].

Electron Beam Lithography

The smallest size of the features that can be written depends mainly on the spot size, beam energy, and the resist used. Control of this feature size across a wafer also depends on the pattern density and the nature of the substrate. In this procedure, impingement of the electron beam on a substrate is first considered. Figure 5 shows a Monte Carlo simulation [18] of electrons entering a resist-coated surface and illustrates the key

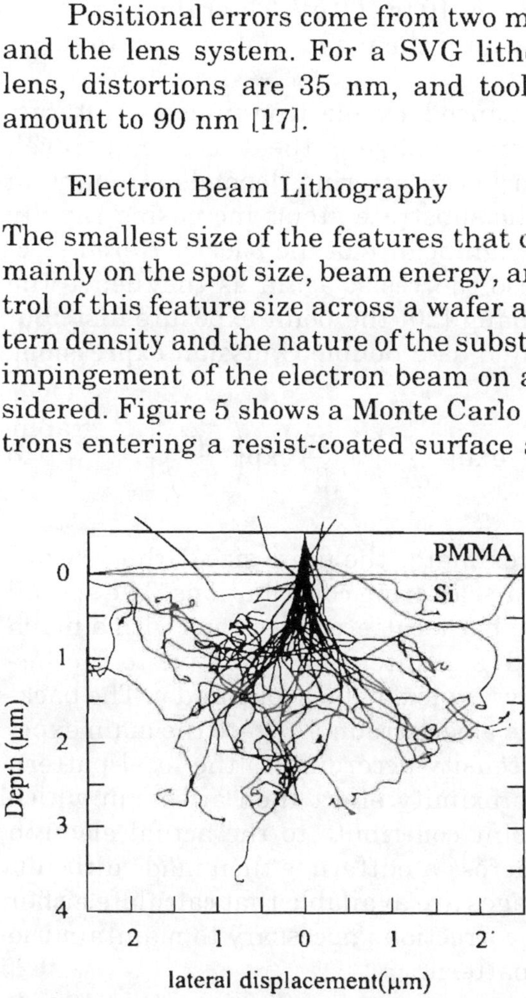

Figure 5 Monte Carlo simulation of electrons entering a resist-coated surface. The electron scattering within the resist and substrates is clearly visible. (From Ref. 18.)

processes that take place in lithography. The two main processes are forward- and backscattering. Forward-scattering is caused by electron–electron interactions and results in a gradual spread of the beam as it penetrates the resist. In addition, forward-scattering results in a shower of secondary electrons with energies on the order of 10 to 100 eV. Their range is about 3 nm in resist [19]. They form the major contribution to resist exposure.

Backscattering is caused by electron–nucleus interactions and can cause a large change in the direction of travel. The amount of electron backscattering depends strongly on the atomic number of the substrate atoms increasing rapidly for heavier atoms. The significance of the backscattered electrons is that they expose the resist again as they leave the substrate. Following Chang [20] the point exposure distribution is often approximated as a double Gaussian expression:

$$f(r) = \frac{1}{\pi(1 + \eta)} \left[\frac{1}{\alpha^2} \exp\left(\frac{-r^2}{\alpha^2} \right) + \frac{\eta}{\beta^2} \exp\left(\frac{-r^2}{\beta^2} \right) \right] \tag{5}$$

The first Gaussian approximates the finite size of the forward beam along with forward-scattering contributions. The second Gaussian represents the backscattered electrons with a much wider spread given by β, and intensity η relative to the forward beam. Table 3 shows typical values for β and η. The backscattered electrons act as a background "fog" to the main exposure which varies in intensity according to the local pattern density. The so-called proximity effect then arises, in which all shapes close to a point contribute to the actual electron dose there. CD control across a pattern is then made difficult, although software packages are available that calculate either the dose or feature size corrections necessary to maintain the desired CD across the pattern.

Even with zero beam size and sufficiently thin resist, which conditions should help to ensure that forward-scattering can be neglected, the minimum feature sizes would still

Table 12.3 Measured Electron Backscatter Characteristics for Different Substrates and Beam Energies

Substrate	Resist thickness (μm)	Energy (kV)	β (μm)	η	Ref.
Silicon	<0.5	25	2.81	0.92	53
Silicon	<0.5	50	8.80	0.75	
Si	10	25	5.22	0.35	
GaAs	<0.5	25	3.24	1.21	
Si	0.1	20	2.0	0.78	54
Si	0.1	60	13.1	0.70	
Si	0.1	120	43.0	0.76	
InP	0.11	50	3.7	1.4	55

be limited by the secondary electron range in the resist (which is on the order of 3 nm) [21]. This limitation is avoided by the use of inorganic materials such as sodium chloride (NaCl), from which small holes can be evaporated by means of an intense beam exposure. These materials, however, have been found to be too slow for practical use. Practical EBL machines, of course, do not have zero spot size. Typical minimum spot sizes for a high-resolution Gaussian beam are about 5 and 10 nm, respectively, for field emission LaB6 sources. Specially built machines with smaller spot sizes have long been available for experimental work [22,23].

PMMA is the positive resist that gives the smallest features and 10-nm lines were obtained in 1983 [24]. Since then, smaller features have been achieved, such as 7-nm PMMA lines subsequently etched into Si [25] and 3-nm NiCr wires made in Glasgow [26]. The latter are formed as a constriction between two larger wires and arbitrary 3-nm features are not possible. Evidence is available that 10-nm features reported are smaller than the PMMA pattern, and are caused by the metal, usually gold, "balling up" on the substrate because of poor adhesion. Edge roughness at linewidths below 30 nm is a further issue. The preferred edge roughness should be less than 10% of the designated CD, but typical values are on the order of 2 to 5 nm and are considered to be caused by aggregates of polymer molecules [27]. Hydrogen silsquioxane (HSQ) is an example of a negative resist with a small polymer size that has been developed to reduce edge roughness fluctuations [28]. Recently other resists have been reported that are capable of resolution below 20 nm.

Pattern positional accuracy is achieved by laser interferometer control of the stage. Typical stage interferometer wavelength (λ) accuracies vary from values of 1/1024 to 1/120 (0.6 to 5 nm), a benchmark against which the deflection system is calibrated. Positional accuracy for electron beam lithography is usually subdivided into stitching and registration accuracy. Absolute positional accuracy is also significant in applications, such as in mix-and-match lithography, in which some levels are written with an electron beam tool and others by optical lithography. Positional errors arise from sources including scanfield distortion, temperature- and charging-induced drift, and height errors. Scanfield distortion is caused by the nonsquareness of the field. Corrections are usually made by adding small additional deflections to the beam as it is scanned across the field. These corrections are calculated by calibrating the deflection field against the stage interferometer and are applied dynamically as the beam is scanned. Temperature-induced drift can come from many places in the tool, and is minimized by applying rigorous thermal management. A quartz substrate, for instance, has a coefficient of thermal expansion of 10^{-6} °C^{-1}; this means that a 1°C temperature change results in errors of 150 nm across a 150-mm reticle. Charge-induced drift can arise from any insulating surfaces in the beam path. When surfaces near the beam become charged, a voltage-induced beam shift arises, which can be on the order of 100 nm. Electrons striking conducting surfaces in a vacuum eventually produce insulating layers of contamination caused by the breakdown of residual pump oil. Charge-induced drift is reduced by careful column design, improved vacuum systems, and by regular cleaning of the column.

Commercial vector scan Gaussian beam tools have overlay and stitching specifications as low as 40 nm (mean + 3σ), a twofold improvement in 10 years to date. Further improvements have been claimed, for example, 20 nm by means of a Leica EBPG tool [29]. Faster, shaped beam machines have less accuracy but an overlay figure of 50 nm and stitching of 20 nm (both mean + 3σ) has been reported for 1 Gbit DRAM manufacture [30]. These actions also quote 20 nm (3σ) CD control across wafers for 150-nm features.

Recent workers have cited even smaller values, for example (overlay), below 10 nm reported by Yamazaki in 1998, and 20 nm for stitching by Koek in 1994. Between 1989 and 1999, available equipment has improved from, respectively, 80 nm (mean + 2σ) to 40 nm (mean + 3σ).

12.4.2 Etch Tolerances

Etch tolerances, in both the vertical and lateral directions, have a key role. Table 1 shows typical values for lateral dimensions that can be achieved for various material systems. Industrial lithographic processing works on a 10% lateral size control tolerance. For mass-produced 0.25-μm silicon devices, this condition implies a 25-nm feature size control across a wafer. Depth control is achieved by the use of etch stop layers, by simple time reckoning, or by the use of laser reflectometry. Etch stop layers are placed under the material to be etched. An etch is then selected (or developed) with a large differential etch rate between the etch stop layer and the layer to be etched. Almost perfect control may then be achieved, although such conditions can be difficult to obtain for etching into bulk material. Accuracies of approximately 10% can be achieved, by measuring the etch rate and then timing an etch. For greater accuracy, laser reflectometry which makes use of interference patterns obtained between the etched material and the mask, can yield accuracies up to two percent.

12.5 DESCRIPTION OF INDUSTRIAL MICROMACHINING EQUIPMENT

12.5.1 Steppers for Photolithography

See Figure 3 for a photolithography illustration. The step-and-scan system is used for DUV steppers for photolithography. The size and complexity of i-line tools may be noted from their footprint of about 6 m^2 and cost (several million U.S. dollars). The lens system is made of quartz, although CaF_2 may be used for DUV tools, and can contain more than 100 kg of glass.

12.5.2 Electron Beam Lithography

A high-resolution Gaussian beam vector scan system is described here. Many of its features are similar to those of other types of EBL tools. Figure 6 shows a simplified diagram of a typical Gaussian beam EBL tool. The main components are the column, plinth, pattern generator, and master computer.

The column contains the following elements. First, there is the electron gun, the source of the electrons. Older tools typically use a single crystal LaB6 source whereas newer systems, since the mid-1990s, employ much brighter field emission sources. In either case a beam of electrons is drawn from the tip by means of strong electric fields. Most of the column is at ground potential but the electron source itself is at a raised potential to provide the beam energy. Beam voltages are between 10 and 100 kV with higher values being used for improved resolution. Extraction currents are on the order of 10 to 100 μA. As most of this current is discarded on the way down the column, typical beam currents at the substrate are between 0.1 and 100 nA. Lens L1 and the blanker are now considered. Electron beam lenses perform the same function as the familiar optical lens, although they have a different working principle [31]. They use a magnetic field and their focal length can be controlled by varying the current passing through a coil. The blanker consists of two parallel plates; when a voltage is applied across the plates, the beam is de-

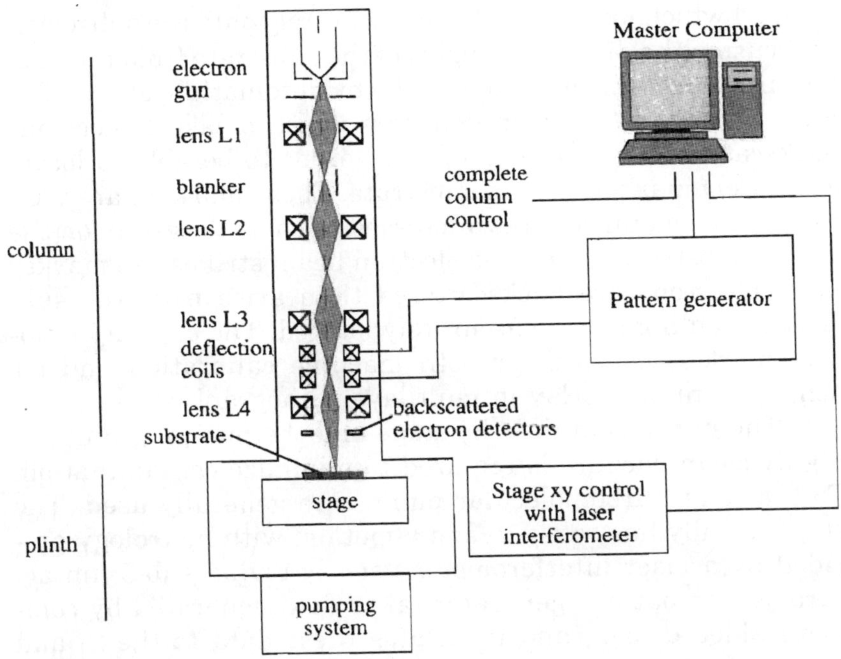

Figure 6 Schematic illustration of commercial electron beam lithography equipment.

flected into a beam dump thus switching off the beam at the substrate. Beam blanking times can be less than 100 ns. Lens L1 focuses the beam to a crossover point in the center of the blanker, to ensure that when it blanks it does not also move on the substrate. The next elements are lenses L2 and L3 which form a zoom lens system but with fractional magnification: the focal plane of L3 is kept constant; by varying the current through L2 and L3 different amounts of demagnification are introduced into the system. This procedure allows the spot size and beam current at the substrate to be varied. Further down the column sit the deflection coils which electronically scan the beam in the x- and y-directions. These coils are controlled by the pattern generator. Below the coils is the final lens, L4 which focuses the beam accurately onto the substrate. In focusing the distance (height) of the substrate from the final lens needs to be measured, usually by automatic systems. The final item is the electron detectors which measure electrons backscattered from the sample, in order to be able to locate accurately markers on the substrate. These markers are typically heavy metal squares or crosses, such as tungsten, on the silicon substrate. When the electron beam strikes the marker more electrons are backscattered than arise from the substrate enabling it to be accurately located. The capacity to locate markers is vital for both machine calibration and for achievement of overlay in multilayer lithography.

The plinth contains the stage and the pumping units. A highly clean vacuum is required to minimize contamination. To that end, turbomolecular pumps are generally used. The stage usually has solely x- and y-motion, with metrology provided by a laser interferometer usually giving sub-5-nm accuracy. The pattern generator takes data generated by computer-aided design, and translates these data to the x- and y-deflections necessary for writing a pattern on the substrate. Typical maximum scan rates are between 10 and 25 MHz with 15 or more bit accuracy in both x- and y-directions.

The master computer controls all aspects of pattern writing. The first stage, calibration of the system, includes focusing the spot, adjusting the field size, and measuring the beam current in order to ensure the correct dose. These procedures can be done automatically. Then the pattern is written on the substrate which can often take several hours. At intervals during the writing process the machine is recalibrated in order to minimize drift in position, focus, dose, and field size.

12.5.3 Dry Etch Kit

Figure 7 shows a basic RIE chamber. The wafer is placed on the bottom electrode, the system is pumped down, and etch gasses are passed through the chamber. The RF frequency applied to the bottom electrode strikes a plasma between the

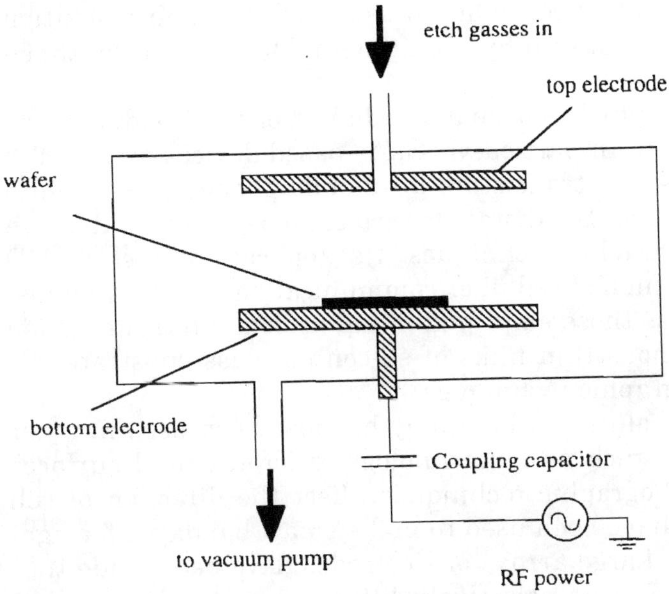

Figure 7 Schematic illustration of a reactive ion etch system.

electrodes and the ions are accelerated across a dark space onto the wafer. Recently developed equipment is more complex, and will be likely to incorporate an airlock for rapid wafer exchange, microprocessor control, and possibly etch depth monitoring apparatus. ECR and ICP etch systems are now more common and incorporate the basic RIE apparatus with additional facilities to increase the plasma density.

12.6 APPLICATIONS

The production of silicon chips will continue to be the main application for the lithography outlined in this chapter. Silicon chips are used in diverse ways. Typical examples are personal computers, high fidelity sound reproduction, car engine management, wristwatches, traffic lights, and medical systems.

Such use has all been made possible by the high resolution machining of material by the lithographic processes outlined here.

Lithography has been applied to other semiconductors as well as silicon. In particular, GaAs-based devices are used to make high-speed transistors for mobile phones and satellite communications. Solid-state laser technology makes extensive use of lithographic techniques. Its applications include CD players and high-speed fiber communications. Flat panel displays, such as those employed in laptop computers, are made from amorphous thin films of silicon on glass substrates by use of lithographic techniques.

High-resolution lithography has also been used in other applications, such as biotechnology. Microtextured surfaces made by lithographic techniques affect the direction of cell growth and have been used to make smart bandages for ligament repair. Large arrays of electrodes have been made that can be used to propel selectively different bioparticles, for fractionating blood into different components [32]. Another application is bespoke microscopy. An atomic force microscope essentially consists of a sharp tip which is scanned across a surface, as described more fully in Chapter 2. By carrying out lithography on the tip, functional probes can be made, such as thermometers to measure temperature or small holes for scanning nearfield optical microscopy (SNOM). Figure 8 shows a 70-nm thermocouple fabricated on the top of a 20-μm high pyramid on a cantilever designed for use as a scanning thermal probe [33].

12.7 DEVELOPMENT

The SIA has produced a lithography "roadmap" to highlight the achievements needed in order to maintain the momentum of Moore's law until the year 2012. If circuit sizes continue to decrease at the current rate, by the year 2012, minimum dimensions in mass-produced transistors will have reached 35

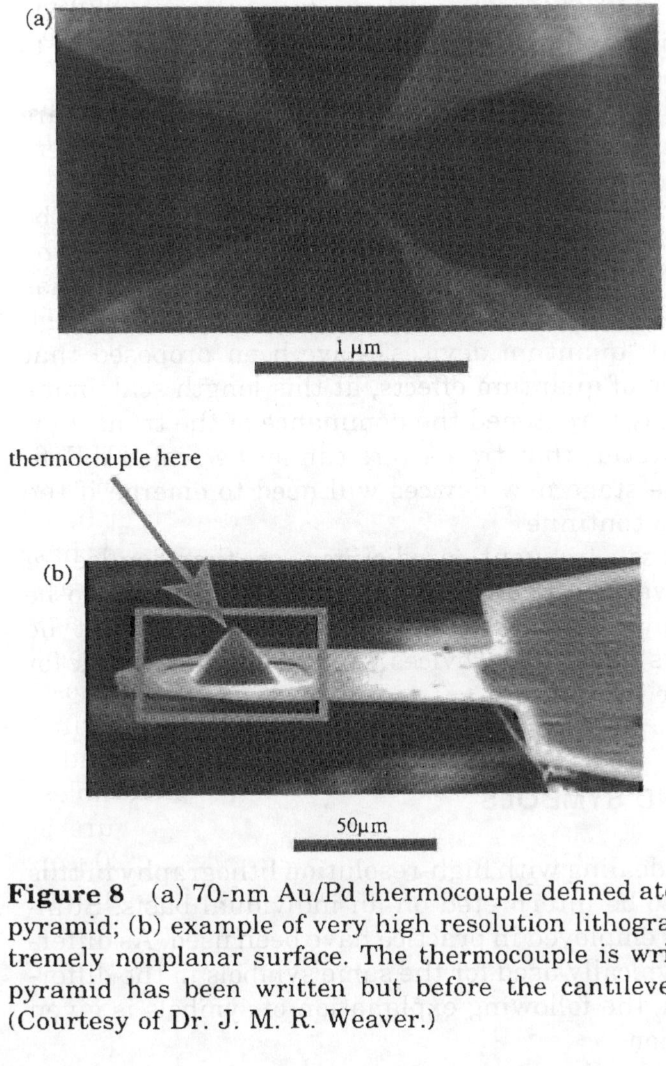

Figure 8 (a) 70-nm Au/Pd thermocouple defined atop 20-μm high pyramid; (b) example of very high resolution lithography on an extremely nonplanar surface. The thermocouple is written after the pyramid has been written but before the cantilever is released. (Courtesy of Dr. J. M. R. Weaver.)

nm in circuits containing more than 10^8 transistors. Such conditions pose major difficulties for the semiconductor industry. It is widely believed that optical lithography will be superseded before this stage is reached, perhaps ending with the 70-nm generation in the year 2006. Contenders for replacing optical lithography include X-ray, projection electron beam, and extreme UV lithographies. Technological obstacles still need to be overcome before any of these techniques can be selected. It is noted that on a logarithmic scale, production devices are now more than one-half way towards the use of atomic dimensions since the formulation of Moore's law in 1965.

Physical as well as technological drawbacks have to be faced as devices continue to shrink. Atomic structure on the 35-nm length scale no longer appears continuous, which has severe implications for transistor performance. Although many so-called "quantum devices" have been proposed that take advantage of quantum effects, at this length scale none has yet seriously threatened the dominance of the transistor. It is clear, however, that transistors cannot be scaled indefinitely. At some stage new devices will need to emerge if the shrinkage is to continue.

Nonetheless, frequent predictions of the demise of Moore's law over the past 20 years have been proven to be unfounded. Thus despite the formidable obstacles that lie ahead, it seems likely that devices will continue to shrink for the foreseeable future.

NOTATION AND SYMBOLS

The equations dealing with high-resolution lithography in this chapter need to be interpreted on an individual basis. Standard equations employed in practice have been used. As different units are typically used for the same symbols in the difference equations, the following explanation of symbols is given for each equation.

(1) K is a constant of proportionality. Because of this condition, the units are arbitrary. However, for a given resist when K is determined, the units should not be changed.

(2) T is time (seconds)

F is fill factor (unitless number from 0 to 1)
A is area (cm^2)
S is sensitivity ($\mu C/cm^2$)
i is beam current (μA). (Here i has to be in μA to give the correct value for T. This equation simply states that the time is given by dividing the total number of Coulombs by the current.)

(3) i is current (A)

B is $A/cm^2/sr$
D is in cm
α is in radians.

(4) R is resolution (μm). Thus wavelength λ needs to be in μm

NA is dimensionless
K is dimensionless.

(5) α, β are quoted in μm

r is in μm
η is dimensionless.

AFM	Atomic force microscopy
ArF	Argon fluoride
B	Electron source brightness
CD	Critical dimension
CVD	Chemical vapor deposition
DQN	Diazonapthoquinone
DRAM	Dynamic random access memory
DUV	Deep ultraviolet
EBL	Electron beam lithography
EBPG	Electron beam pattern generator

ECR	Electron cyclotron resonance
EUV	Extreme ultraviolet
eV	Electron volts
FF	Fill factor
Hg	Mercury
HMDS	Hexamethyldisilizane
KrF	Krypton fluoride
kV	Kilo volts
ICP	Inductively coupled plasma
λ	Wavelength (nm)
LaB6	Lanthanum boride
MHz	Mega Hertz (10^6 Hertz)
$\mu C/cm^2$	Units of electron beam exposure dose (10^{-6} Coulombs per square centimeter)
μm	Micrometers (10^{-6} meters)
NA	Numerical aperture
nm	Nanometers (10^{-9} meters)
P(MMA-MAA)	Copolymer of methyl methacrylate and methacrylic acid
PBS	Poly(butene-1-sulfone)
PEB	Postexposure bake
PMMA	Polymethyl methacrylate
PVD	Physical vapor deposition
RH	Relative humidity (%)
RF	Radio frequency (Hz)
RIE	Reactive ion etching
SCALPEL	Scattering with angular limitation in projection electron beam lithography
SNOM	Scanning nearfield optical microscopy
SS	Spin speed (revs per min)
VLSI	Very large scale integration

REFERENCES

[1] S. D. Berger and J. M. Gibson, New approach to projection-electron lithography with demonstrated 0.1 μm linewidth. *Applied Physics Letters* (57), 153–155 (1990).

[2] T. Dresel and J. Schwider, Fabrication of optical components by laser lithography. *Applied Surface Science* (106), 379–382 (1996).

[3] M. A. McCord and R. F. W. Pease, Lithography with the scanning tunneling microscope. *J. Vacuum Science and Technology* (B4), 86–88 (1986).

[4] D. L. Spears, *Electronic Letters* (8), 102 (1972).

[5] S. Y. Chou and P. R. Krauss, Imprint lithography with sub-10nm feature size and high throughput. *Microelectronic Engineering* (35), 237–240 (1997).

[6] B. G. Casey, W. Monaghan, and C. D. W. Wilkinson, Embossing of nanoscale features and environments. *Microelectronic Engineering* (35), 393–396 (1997).

[7] D. S. Macintyre and S. Thoms, The fabrication of high resolution features by mould injection. *Microelectronic Engineering* (41)42, 211–214 (1998).

[8] S. Gray and A. L. Bogdanov, High speed scanner for deep X-ray lithography with the third generation 1.5 GRV synchotron source. *Microcircuit Engineering* (1999).

[9] M. D. R. Thomas, D. G. Hasko, H. Ahmed, D. B. Brown, and B. F. G. Johnson, Electron-beam exposure characteristics of a novel Ru-PMMA composite resist. *Microelectronic Engineering* (41)42, 327–330 (1998).

[10] T. Ishii, H. Nozawa, and T. Tamamura, A nano-composite resist system: A new approach to nanometer pattern fabrication. *Microelectronic Engineering* (35), 113–116 (1997).

[11] D. Macintyre and S. Thoms, High resolution electron beam lithography studies on Shipley chemically amplified DUV resists. *Microelectronic Engineering* (35), 213–216 (1997).

[12] W. M. Moreau, *Semiconductor Lithography: Principles, Practices and Materials*. Plenum, New York (1988).

[13] J. Pcansky and J. R. Lyeria, Photochemical decomposition mechanisms for AZ-type photoresists. *IBM J. Research Development* (23), 42–54 (1979).

[14] C. Y. Chang and S. M. Sze, *ULSI Technology*. McGraw-Hill, New York (1996).

[15] C. P. Ausschnitt, Microlithography in midlife crisis. *Microelectronic Engineering* (41)2, 41–46 (1998).

[16] Y. Trouiller, N. Buffet, T. Mourier, P. Schiavone, and Y. Quere, 0.12 μm optical lithography performance using an alternating DUV phase shift mask. *Microcircuit Engineering* (41)42, 61–64 (1998).

[17] S. J. Holmes, P. H. Mitchell, and M. C. Hakey, Manufacturing with DUV lithography. *IBM J. Research and Development* (41), 7–19 (1997).

[18] D. F. Kyser and N. S. Viswanathan, Monte Carlo simulation of spatially distributed beams in electron-beam lithography. *J. Vacuum Science and Technology* (12), 1305–1309 (1975).

[19] S. A. Rishton, S. P. Beaumont, and C. D. W. Wilkinson, Exposure range of low energy electrons in PMMA. *J. Electrochemical Society* (129)6, C234 (1982).

[20] H. P. Chang, Proximity effects in electron-beam lithography. *J. Vacuum Science and Technology* (12)6, 1271–1275 (1975).

[21] S. A. Rishton, S. P. Beaumont, and C. D. W. Wilkinson, In *Proc. Microcircuit Engineering 82*, Grenoble, Sitecmo Dieppe, Paris, 341–345 (1982).

[22] Z. W. Chen, G. A. C. Jones, and H. Ahmed, Nanowriter—A new high voltage electron beam lithography system for nanometre-scale fabrication. *J. Vacuum Science and Technology* (B6)6, 2009–2013 (1988).

[23] S. Thoms, S. P. Beaumont, and C. D. W. Wilkinson, A JEOL 100 CXII converted for use as an electron beam lithography system. *J. Vacuum Science and Technology* (B7), 1823–1826 (1989).

[24] H. G. Craighead, R. E. Howard, L. D. Jackel, and P. M. Mankiewich, 10-nm Linewidth electron beam lithography on GaAs. *Applied Physics Letters* (42)1, 38–40 (1983).

[25] W. Chen and H. Ahmed, Fabrication of 5–7 nm wide etched

lines in silicon using 100 keV electron beam lithography and polymethylmethacrylate resist. *Applied Physics Letters* (62)13, 1499–1501 (1993).

[26] D. R. S. Cumming, S. Thoms, J. M. R. Weaver, and S. P. Beaumont, 3 nm NiCr wires made using electron beam lithography and PMMA resist. *Microelectronic Engineering* (30), 423–425 (1996).

[27] M. Nagase, H. Namatsu, K. Kurihara, K. Iwadate, K. Murase, and T. Makino, Nano-scale fluctuations in electron beam resist pattern evaluated by atomic force microscopy. *Microelectronic Engineering* (30), 419–422 (1996).

[28] H. Namatsu, Y. Takahashi, K. Yamazaki, T. Yamaguchi, M. Nagase, and K. Kurihara, Three-dimensional siloxane resist for the formation of nanopatterns with minimum linewidth fluctuations. *J. Vacuum Science and Technology* (B16), 69–76 (1988).

[29] B. H. Koek, T. Chisholm, J. Romijn, and A. J. van Run, Sub 20 nm stitching and overlay for nano lithography applications. *Japanese J. Applied Physics* (33), 6971–6975 (1994).

[30] K. Nakajima, H. Yamashita, Y. Kojima, S. Hirasawa, T. Tamura, Y. Yamada, K. Tokunaga, T. Ema, K. Kondoh, N. Onoda, and H. Nozue, 0.15 μm Electron beam direct writing for Gbit dynamic random access memory fabrication. *Japanese J. Applied Physics*, Part 1 (36), 7535–7540 (1997).

[31] P. Grivet, *Electron Optics*. Pergamon, Oxford (1972).

[32] N. G. Green, M. P. Hughes, W. Monaghan, and H. Morgan, Large area multilayered electrode arrays for dielectrophoretic fractionation. *Microelectronic Engineering* (35), 421–424 (1997).

[33] H. Zhou, A. Midha, G. Mills, S. Thoms, S. K. Murad, and J. M. R. Weaver, Generic scanned-probe microscope sensors by combined micromachining and electron-beam lithography. *J. Vacuum Science and Technology* (B16), 54–58 (1998).

[34] Tserepi, E. Gogolides, C. Cardinaud, L. Rolland, and G. Turban, Highly anisotropic silicon and polysilicon room-tempera-

ture etching using fluorine-base high density plasmas. *Microelectronic Engineeering* (41)2, 411–414 (1998).

[35] K. Paul and I. W. Rangelow, Fabrication of high aspect ratio structures using chlorine gas chopping technique. *Microcircuit Engineering* (35), 79–82 (1997).

[36] P. A. Lewis, H. Ahmed, and T. Sato, Silicon nanopillars formed with gold colloidal particle masking. *J. Vacuum Science and Technology* (B16), 2938–2941 (1998).

[37] J. C. T. Lee, N. Layadi, K. V. Guinn, H. L. Maynard, F. P. Klemens, D. E. Ibbotson, I. Tepermeister, P. O. Egan, and R. A. Richardson, Comparison of advanced plasma sources for etching applications v. polysilicon etching rate, uniformity, profile control, and bulk plasma properties in a helical resonator plasma source. *J. Vacuum Science and Technology* (B14), 2510–2518 (1996).

[38] S. Penner, M. Fallahi, and O. Nordman, Electron cyclotron resonance reactive ion etching of GaAs in chlorine-methane. *Microelectronic Engineeering* (41)2, 383–386 (1998).

[39] S. Murad, S. P. Beaumont, M. Holland, and C. D. W. Wilkinson, Selective reactive ion etching of InGaAs and InP over InAlAs in SiCl4/SiF4/HBr plasmas. *J. Vacuum Science and Technology* (B3), 2344–2349 (1995).

[40] E. W. Berg and S. W. Pang, Electrical and optical characteristics of etching InGaAs. *J. Vacuum Science and Technology* (B16), 3359–3363 (1998).

[41] H. Ohtake, S. Samukawa, H. Oikawa, and Y. Nashimoto, Enhancement of reactivity in Au etching by pulse-time-modulated C12 plasma. *Japanese J. Applied Physics* (37), 2311–2313 (1998).

[42] J. Jacobs, K. Saito, and J. Yamamoto, High density Al etcher process window characterization. *Japanese J. Applied Physics* (37), 2359–2368 (1998).

[43] R. J. Shul, M. E. Sherwin, A. G. Baca, and D. J. Riege, Etching of sub 0.5 μm W/WSix bilayer gates. *Electronic Letters* (32), 70–71 (1996).

[44] E. Gat, F. Bounasri, M. Chaker, M. E. Ravet, M. Moisan, and J. Margot, Temperature effects on tungsten etching. *Microcircuit Engineering* (30), 337–340 (1996).

[45] Y. Chen, Private communication.

[46] J. Hong, J. A. Caballero, E. S. Lambers, J. R. Childress, and S. J. Pearton, Patterning of thin film NiMnSb using inductively coupled plasma etching. *J. Vacuum Science and Technology* (B16), 3349–3353 (1998).

[47] H. Elsner, R. Ijsselsteijn, W. Morgenroth, H. Roth, and H. G. Meyer, Submicrometer patterning of $YBa_2Cu_3O_7$-x. *Microelectronic Engineering* (41)2, 407–410 (1998).

[48] N. Ikegami, A. Yabata, G. L. Liu, H. Uchida, N. Hirashita, and J. Kanamori, Vertical profile control in ultrahigh aspect ratio contact hole etching with 0.05 μm-diameter range. *Japanese J. Applied Physics* (37), 2337–2342 (1998).

[49] Midha, S. K. Murad, and J. M. R. Weaver, Anisotropic pattern transfer of fine resist features to silicon ritride via an intermediate titanium layer. *Microcircuit Engineering* (35), 99–102 (1997).

[50] H. Shiomi, Reactive ion etching of diamond in O_2 and CF_4 plasma, and fabrication of porous diamond for field emitter cathodes. *Japanese J. Applied Physics* (36), 7745–7748 (1997).

[51] S. Thoms, Private communication.

[52] S. M. Sze, *Semiconductor Devices Physics and Technology*. John Wiley, New York (1985).

[53] S. A. Rishton and D. P. Kern, Point exposure distribution measurements. *J. Vacuum Science and Technology* (B5)1, 135–141 (1987).

[54] L. D. Jackel, R. E. Howard, P. M. Mankiewich, H. G. Craighead, and R. W. Epworth, Beam energy effects in electron beam lithography: The range and intensity of backscattered exposure. *Applied Physics Letters* (45), 698–700 (1984).

[55] D. M. Tennant, G. E. Doran, R. E. Howard, and J. S. Denker,

Electron scatterning distribution in InP at 50 kV. *J. Vacuum Science and Technology* (B6), 426–431 (1988).

[56] T. Nishihano et al., *Solid State Technology*, June, 90 (1994).

[57] J. T. Clemens, Silicon microelectronics technology. *Bell Labs Technical J.* Autumn, 76–102 (1997).

Appendix

Micromachining by Finishing Techniques

Joseph McGeough
The University of Edinburgh, Edinburgh, Scotland

A.1 INTRODUCTION

The main text of this book deals with the principal methods of micromachining that are currently attracting industrial and research interest.

Other material removal processes used primarily for finishing and polishing of surfaces are also occasionally considered to act in the manner of micromachining. The purpose of this appendix therefore is to draw attention to and describe some techniques that may be regarded in this category. Some of these methods are already well established and have had to be adapted to meet the requirements of microengineered components.

A useful guide to the place of these techniques may be derived from the review by Komanduri et al. [1]. They classify wide groups of material removal processes. Well-established techniques such as lapping, honing, finishing, and polishing are now being given new emphasis, especially for applications in which precision is paramount. (See, for example, [2,3].) Shaw has given a useful account of the principles of precision finishing [4]. Thus, conventional lapping is used for form accuracy such as flatness or sphericity in the cases of, respectively, planar objects or balls. Honing, like lapping, is used for form and shape accuracy and for generating topography, for example, to trap lubricant. The term "polishing" usually implies best finish without regard to shape or form accuracy; "finishing" is a general term describing any or all of these processes.

A.2 MICROLAPPING

The principles of some of these techniques are now described. We begin with the well-established lapping process, which is now being developed for special micromachining applications.

A.2.1 Principles

In lapping an abrasive such as fine grit diamond is applied in a slurry or paste in water, grease, or oil. A common type is ball-lapping in which balls are placed between cast iron laps into which are machined circular grooves to control the movement of the balls. Rotation of one of the laps results in a circular movement of the balls around the laps onto which is superimposed a spinning motion of the balls to ensure their complete coverage. The hard abrasive becomes embedded in the soft lap and acts as a cutting point.

Material removal by lapping is very slow. The technique is mainly used to improve surface finish. The surfaces produced depend on the material removal mechanism but are generally isotropic. Lapping is used to produce smooth flat parts such as gauge blocks or sealing surfaces and in bearing technology for finishing of steel and ceramic balls.

A.2.2 Applications of Microlapping

Magnetic Materials

A magnetic material extensively used in the electrical industry is Mn–Zn ferrite. Its finishing by conventional abrasive grinding including lapping is found to give rise to microcracking, chipping, and damage to the surface layers. In order to understand the effects that arise, Touge and Matsuo [5] report high-precision lapping of Mn–Zn polycrystalline ferrite with 0.5 to 2.0 μm diamond abrasive grains under loads of 42.7 and 90.5 kPa. They propose that the motion of abrasive grains has a direct effect on process efficiency and the surface characteristics of lapped surfaces. Material is removed by two types of grains: fixed grains trapped on the lapping plate and loose grains contained in the lapping fluid. They conclude that work material is mainly removed by loose grains.

Air Bearing Surfaces

Another application for lapping is in the machining of air bearing surfaces (ABS) of the read-write head for magnetic disk data storage devices; surface roughnesses of 5 mm or better is required. Small abrasive sizes are used to ensure slow material removal rate and satisfactory surface roughness. Chang et al. [6] have studied the mechanisms involved and have classified the mechanisms into four states. They reported transitions for three-body brittle to two-body ductile machining in the first 90 s of lapping with 3-μm diameter grains of $A\ell_2O_3$ abrasive. Acoustic emission (AE) was proposed as a basis for in-process monitoring for microlapping.

Silicon Wafers

Advances in lapping for silicon wafer applications by a rapidly renewable lap technique that separates functions of figure and finish control for a broad range of lapping and polishing processes have been described by Evans and coworkers [7]. Advanced theoretical analyses have been presented by Cai et al. [8].

A.3 MICROHONING

Honing is an abrasive machining process used primarily to improve the surface characteristics of a working surface. Although honing is considered to be a minimal material removal technique it is being increasingly employed as a final machining step to obtain or to correct a part shape and dimensional accuracy. When the technique is used to obtain a very smooth surface, it is often termed "superfinishing"; c.f. conventional honing which usually yields higher material removal and rougher finish.

The process may be categorized into short- and long-stroke honing [9,10].

A.3.1 Short-Stroke Honing (SSH)

In short-stroke honing a superfinishing stone is shaped to conform to the component. An oscillatory motion is applied to the stone so that it interacts at moderate pressure with the workpiece which rotates at relatively low speed. Figure 1 illustrates the relatively large contact area between superfinishing

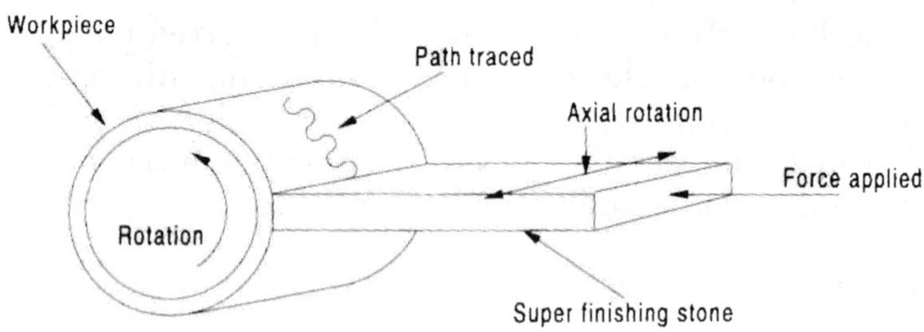

Figure 1 Main features of stone-superfinishing. (From Ref. 10.)

stone and workpiece. "Neat" honing oil is supplied into the working area to (i) provide lubrication over the large contact length in order to reduce stone wear and (ii) flush away any adherent chips that may cause stone loading. Neat honing oils are petroleum-based such as paraffin oils and kerosene and are undiluted with water. The neat oil often has included extreme pressure additives (EP) to prevent welding of chips to the stone surface. In high-precision applications, the lubricant can also be used to control the workpiece temperature in order to achieve finer dimensional accuracy. The superfinishing stone is often impregnated with a lubricant additive such as wax or sulphur directly into the stone structure so that it is directly available at the interface between stone and workpiece.

Oscillation of the superfinishing stone can reduce loading and increase the rate of material removal. The oscillation speed can be controlled to produce a certain surface characteristic, that is, a cross-hatched pattern. The motion of the honing stone resulting from oscillation of the stone and rotation of the workpiece gives rise to a sinusoidal path around the workpiece, the angle of the cross-hatch being dependent on the relative speed of the workpiece and its oscillation.

Typical applications of short-stroke honing are found in the bearing industry, in which "plunge" and "through-feed" technique honing are used. In the former method, an oscillating honing stone of correct form is plunged onto the surface of a rotating ring. Plunge honing can be used for rough and fine finishing with a single stone, by change in process variables. For example, rough honing can be achieved by slow rotation of the workpiece with fast-stone oscillation and finishing, then accomplished by high workpiece speed and low oscillation rates.

However, surface finish and material removal are mainly dependent on grit size, and double-stage honing is used for rough and fine surface finishing. An example of through-feed honing is shown in Figure 2. The workpieces are rotated and fed to different honing stations by drive rolls which are set at

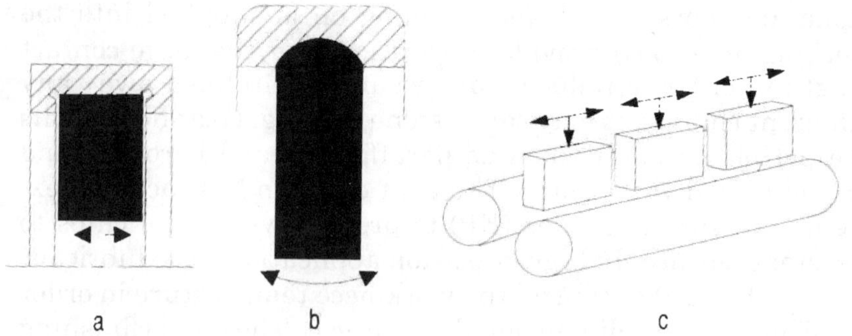

Figure 2 (a,b) Plunge and (c) through-feed honing. (From Ref. 10.)

a slight angle to produce the feeding motion. A coarse grit can be used initially to achieve a required material removal, and fine grits are employed at the end of the operation for surface finishing.

A.3.2 Long-Stroke Honing (LSH)

This technique is used for finishing of bores of components such as bushings, hydraulic cylinders, connecting rods, and cylinder liners. Usually, the honing head is applied to the bore of the component. The honing head, which is expendable, rotates and is reciprocated or stroked through the bore; this motion provides for a cross-hatch pattern as already discussed for short-stroke honing.

The honing head may comprise one or several (for instance, two, four, or six) hones, the characteristics of which (e.g., abrasive type, grit size, or bond) influence the roundness, cylindricity, roughness, and material removal from the bore, as do operating variables. Long-stroke honing can improve bore form errors, such as “out-of-roundness,” waviness, taper, and barrelling resulting from earlier processing such as heat treatment and machining.

In the single-stroke honing of bores diamond and CBN-plated tools are used. The tool has an electroplated sleeve, which is expandable to compensate for tool wear although no expansion occurs during the operation. A desired amount of material is removed in a single pass in order for the hole to be correctly sized and surface finished; the grit size has to be judiciously chosen.

A.3.3 Industrial Applications

As noted above, typical applications for microhoning are found in the bearing and automotive industry and for finishing of components such as hydraulic cylinders. An innovative hybrid microhoning technique has recently been described which has proved useful in machining of internal combustion parts. A laser is used to produce deep valleys, 5-μm deep, 30 μm-wide with 600-μm separation and cross-hatch angle of 30°, followed by honing to yield a low surface roughness of 1 μm Rz. A pattern of slots or pits has also been produced by this technique. Reduction in oil consumption, emission, wear, friction, and fuel is claimed [10].

A.4 SUPERFINISHING

Honing stones are utilized widely in conventional superfinishing practice, although they suffer from the disadvantage of lack of consistency and controllability. They tend to glaze or dull or become loaded such that material removal and finish vary. Satisfactory finishing becomes highly dependent on the skill of the operator [10].

These difficulties may be overcome by use of coated abrasives, which are designed to present a controlled open and "aggressive" cutting surface to the workpiece. This configuration combined with continuous or step feed and oscillation gives a consistent and predictable high-quality surface finish.

The basic structure of a coated abrasive consists of a thin, strong backing material onto which an adhesive "make-coat" is applied. A single layer of abrasive is then electrostatically coated, so that the sharp cutting edges are exposed in a direction perpendicular to that of the surface. A second layer of adhesive, the "size-coat" is then applied to secure the abrasive to the backing. Often fillers such as alumina bubbles (of aluminum oxide), cryolite, or pyrite are added to the make-coat to improve performance. For very fine grit abrasives, a slurry coating is used; the grits become covered and some wear is needed for effective working.

The type of abrasive and its grain size greatly influence surface finish and removal rate. Higher material removal and poorer surface finishing are generally obtained with coarse grains and vice versa for finer grains.

Grains are more precisely graded with a narrower distribution of grain sizes so that when coated onto a backing they give a more uniform abrasive height above the tape thereby reducing deep scratches. The fine abrasive grain sizes used for finishing are so small that they are not classified by the usual mesh or grade number but by the actual grain size (μm). The abrasives used include fused and sintered alumina, sol gel, and superabrasives such as CBN and diamond.

Conventional monolayer-coated abrasives can wear leading to variations in surface finish, especially during the initial break-in period, as much as 30% of the belt life. Clustered-coated abrasives, consisting of fine abrasive grains bonded together are now available which are coated onto the backing in order to extend the life of the abrasive tape. A three-dimensional structure of any shape (e.g., pyramids, cones, or cubes) is produced on the tape. Within these shapes is an abrasive composite of bond and fine abrasive grains. These shapes wear uniformly exposing fresh unused abrasive. A more consistent surface finish and higher rate of cut is maintained for longer times.

Onchi and coworkers describe several typical characteristics of superfinishing shown in Figure 3 for CBN and aluminum oxide superfinishing stones [11].

1. Cutting (CC): high stock removal and rough surface finish

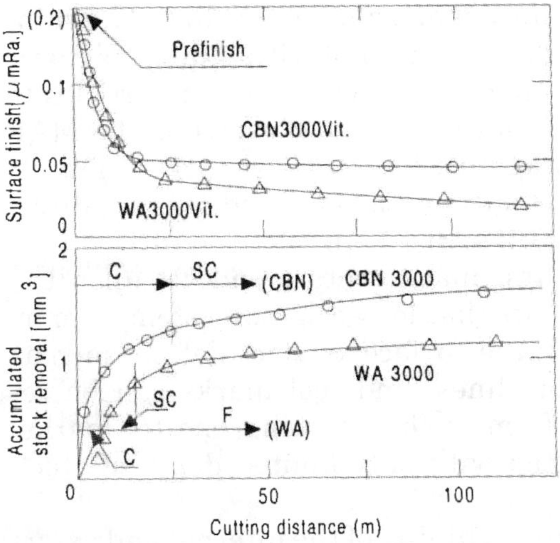

Figure 3 Characteristics of superfinishing. (From Ref. 11.)

2. Mirror finishing (MF): low stock removal and fine mirror finish
3. Semicutting (SC): intermediate behavior.

Further work by Bordeianu and Marinescu on conventional stones of 1000 grit alumina and SiC concludes that alumina should be used for the rough superfinish stage due to its high material removal, whereas SiC is preferable for fine superfinishing [12,13]. Their work was performed on DGBB inner rings of outer diameter 14.7 mm with initial surface finish of 0.1 μm. With SiC a surface finish of 0.025 μm Ra was achieved in 6 s.

A.4.2 Magneto-Abrasive Micromachining and Finishing (MAF)

A related mechanical abrasive technique is MAF. Magneto-abrasive brushes are energized electromagnetically across a small machining gap formed between the work surface and magnetic poles. The cutting agents, abrasive grains of aluminium oxide or boron nitride, are sintered with a ferromagnetic iron base. They arrange themselves with the iron particles to conform with the contour of the work surface while the MAF particles are held firmly against the rotating work surface. A rapid short-stroke oscillatory motion is carried out in a direction parallel to the work axis. See Figure 4.

The MAF brushes first make contact and act upon the surface protruding elements that may originally stem from irregularities in the form of the surface. Surface defects such as scratches, hard spots, lay lines, and tool marks can be removed. Irregularities in form, such as lobbing, chatter marks, and taper can be corrected with only limited depth of stock removal, for example, 20 μm.

MAF is used in micromachining of the internal and external surfaces of materials ranging from carbon and stainless steel to ceramics and thermosetting plastics. In one application to bearing surfaces, surface roughness was lowered from 0.5 to 0.6 μm to 0.05 to 0.06 μm in 30 to 60 s [14,15]. Varghese and Malkin discuss enhanced stock removal for superfinishing, mainly by ultrasonic vibration to reduce loading [16].

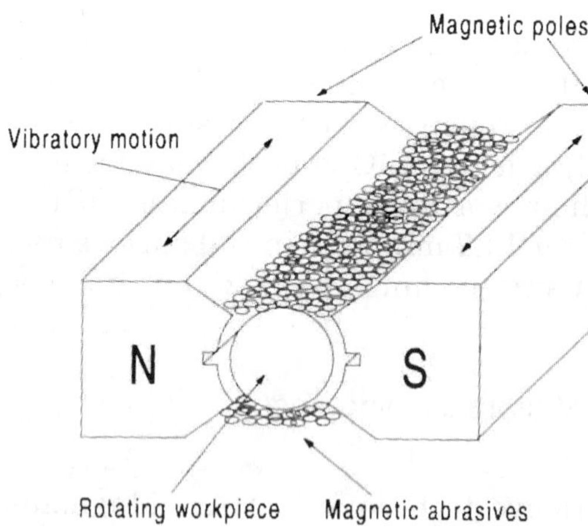

Figure 4 Apparatus for magneto-abrasive finishing. (From Ref. 15.)

A.5 SUPERPOLISHING

The close relationship between superfinishing and polishing of surfaces gives rise to discussion of developments in superpolishing technology, again much of which has been driven by the needs of the electronics industry.

A.5.1 CHEMICALLY ASSISTED MECHANICAL POLISHING (CMP)

CMP combines simultaneous chemical and mechanical action for polishing and producing a plane surface. A rotating polishing pad is pressed against a workpiece. The pad is soaked with a slurry composed of (i) a chemically active liquid such as hydrogen peroxide or ammonium hydroxide and (ii) an abrasive, for example, fine-grained alumina or diamond. Surface finishing is achieved by a cyclic process in which a metal passivation layer is produced and then subsequently removed by the action of the abrasive. A balance has to be obtained between the high oxidation rate needed for rapid metal removal and the etching rate which can lead to selective etching of different phases or components within the workpiece. For example, planarization of metalization structures on silicon wafers may consist of insulating SiO_2 and conductive Cu tracks which can be etched at different rates.

CMP is used extensively in the electronics industry in the manufacture of wafers, flat panel displays, and thin film magnetic heads, as is now illustrated.

As noted above Komanduri et al. have described technological advances in fine abrasive processes [1]. They give detailed consideration to CMP for finishing of silicon wafers in the semiconductor industry. They draw attention to the scaling down of device dimensions to the range 0.1 to 0.5 μm, and the growing importance of local and global planarization.

Other workers have also drawn attention to planarization of submicron dielectrics and metallic layers used in semiconductor chip technology. Jeong and coworkers [17] discuss integrated planarization by use of ultrafine abrasive machining for silicon substrates, blanket films of oxide, and wafers with pattern topography. Loose abrasive superpolishing methods such as CMP and also bonded abrasive methods such as ductile mode grinding with ultrafine abrasives are discussed. Chemical–mechanical grinding combined with the recently developed electrolytic grinding process (ELID) yielded a 0.3-μm flatness and surface roughness of 4 nmRa for a 150-mm substrate wafer (see also Chapter 5 and Section A.6.1). ELID grinding plays a significant role in achieving high flatness. Subsequent chemical mechanical grinding can realize a mirror surface almost the same as CMP, with only 2-μm surface removal. (See the discussion on ELID in Section A.6.)

Further work on planarizing silicon wafers is reported by Byrne et al. [18] who discuss wear of the polishing pad.

A.5.2 Mechanochemical Polishing

An alternative to the above technique is mechanochemical polishing. Abrasive powders are used. These are mechanically softer than the workpiece but can chemically react with it. Examples are $BaCO_3$ powder which is used for polishing of silicon wafers or Cr_2O_3 oxide for polishing SiC and Si_3N_4.

Attractions of the process are low mechanical damage and no scratch marks. The process can be dry so that cleanliness is improved. Surface finishes better than 1 nmRa are claimed [19].

A.6 ELID-GRINDING (ELECTROLYTIC IN-PROCESS DRESSING)

A.6.1 Introduction

Advanced precision grinding processes are playing a major part in competing with, and replacing, conventional loose abrasive processes such as lapping. For example, Zhong and Venkatesh [20] provide a thorough account of semiductile grinding of silicon, germanium, and glass. The need for new advanced manufacturing processes for semiconductors, magnetic leads, and optical components has prompted Ohmori and Nakagawa [21] to devise new mirror surface grinding techniques that use metallic bond grinding wheels with electrolytic in-process dressing. It was devised through application of cast-iron bonded diamond (CIB-D) wheels for grinding of silicon wafers (see Figure 5). In further developments a metal bonded or conductive resin bonded grinding wheel uses diamond or CBN for ferrous materials.

The principles of ELID are as follows. The wheel is the anode and is spaced about 0.1 mm from a secondary electrode acting as the cathode (graphite, stainless steel or copper). A chemical solution type of grinding coolant diluted by water is used both as grinding fluid and electrolyte. The metallic bond wheel is predressed electrolytically before grinding during which the working current is found to decrease. This decrease arises owing to an insulating oxide/hydroxide layer which is generated on the wheel surface. The friable oxide formed prevents grinding chips from adhering to the wheel. As the cutting grains become worn, the insulating oxide layer also becomes worn. Fresh oxidation of the wheel occurs in a self-

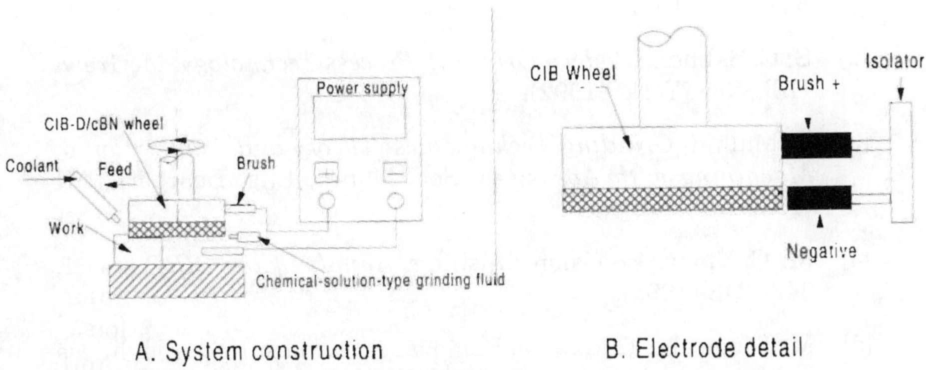

Figure 5 Principles of ELID grinding. (From Ref. 22.)

regulatory fashion, providing continued exposure of fresh cutting points, so that the grinding force is kept low. Knight [10] reports surface finishes of R_{max}22 nm and R_a of 4 nm with 3.0-μm or smaller grits, for a cast-iron fiber-bonded diamond grinding wheel.

Most ELID applications lie with ceramics, optical materials, and, recently, bearing steels [22–24]. The surface finishes achieved are usually 0.011 to 0.036 μm *Ra*, and 0.08 to 0.18 μm Rz. ELID cycle times are about 20 s.

A.7 CONCLUSIONS

As micromachining moves steadily toward nanotechnological applications, increasing effort will continue to be devoted to the place of finishing by the above techniques, their development, and their successors [25,26]. Many of these advances will be driven by the needs of the electronics, medical, and optical industries.

REFERENCES

[1] R. Komanduri, D. A. Lucca, and Y. Tan, Technological advances in fine abrasive processes. *Annals of the CIRP* (46) 2, 545–596 (1997).

[2] S. C. Salmon, *Modern Grinding Process Technology*. McGraw-Hill, New York, (1992).

[3] S. Malkin, *Grinding Technologies: Theory and Application of Machining with Abrasives*. Soc. Manuf. Eng. Dearborn, MI (1989).

[4] M. C. Shaw, Precision finishing. *Annals of the CIRP* (44) 1, 343–348 (1995).

[5] M. Touge and T. Matsuo, Removal rate and surface roughness in high precision lapping of Mn-Zn ferrite. *Annals of the CIRP* (45) 1, 307–310 (1996).

[6] Y. P. Chang, M. Hashimura, and D. A. Dornfield, An investigation of the AE signals in the lapping process. *Annals of the CIRP* (45) 1, 331–334 (1996).

[7] C. J. Evans, R. E. Parks, D. J. Roderick, and M. L. McGlaufin, Rapid renewable lap: Theory and practice. *Annals of the CIRP* (47) 1, 239–244 (1998).

[8] C. Q. Cai, Y. S. Lu, R. Cai, and H. W. Zheng, Analysis on lapping and polishing pressure distribution. *Annals of the CIRP* (47) 1, 235–238 (1998).

[9] H. Gerhard and K. Barton, Characteristics of honing. SME Report MR 93–147 (1993).

[10] P. Knight, Ph.D. Thesis, Cranfield University, Bedford, UK (2000).

[11] Y. Onchi, N. Ikawa, S. Shimada, and N. Matsumori, Porous fine CBN stones for high removal rate superfinishing. *Annals of the CIRP* (44) 1, 291–294 (1995).

[12] I. D. Marinescu and E. Bordeianu, Optimisation of the parameters in the superfinishing of the bearing ball track. Soc. Manuf. Eng. (SME) Report MR94–166 (1994).

[13] E. Bordieanu and I. D. Marinescu, Superfinishing of bearing rings. SME Report MR 95–206 (1995).

[14] M. D. Krymsky, Magnetic abrasive finishing. *Metal Finishing* July, 21–25 (1993).

[15] M. Fox, K. Agrawal, T. Shinmura, and R. Komanduri. Magnetic abrasive finishing of rollers. *Annals of the CIRP* (43) 1, 181–183 (1994).

[16] B. Varghese and S. Malkin, Experimental investigation of methods to enhance stock removal for superfinishing. *Annals of the CIRP* (47) 1, 231–234 (1998).

[17] H. Jeong, H. Ohmori, T. K. Doy, and T. Nakagawa. Integrated planarization technique with consistency in abrasive machining for advanced semiconductor chip fabrication. *Annals of the CIRP* (45) 1, 311–314 (1996).

[18] G. Byrne, B. Mullany, and P. Young. The effect of pad wear

on the chemical mechanical polishing of silicon wafers. *Annals of the CIRP* (48) 1, 143–146 (1999).

[19] N. Yasunaga. Recent advances in ultraprecision surface finishing technologies in Japan. In *Proc. Int. Symp. On Advances in Abrasives Technology*, July 18–27 (1997).

[20] Z. Zhong and V. C. Venkatesh, Semi-ductile grinding and polishing of opthalmic aspherics and spherics. *Annals of the CIRP* (44) 1, 339–348 (1995).

[21] H. Ohmori and T. Nakagawa, Analysis of mirror surface generation of hard and brittle materials by ELID (electrolytic in-process dressing) grinding with superfine grain metallic bond wheels. *Annals of the CIRP* (44) 1, 287–294 (1995).

[22] H. Ohmori, Ultraprecision grinding of optical materials and components applying ELID (electrolytic in-process dressing). *SPIE* (2576), 26–45 (1995).

[23] I. Nobuhide et al. Grinding characteristic of hard and brittle materials by ELID lap-grinding using fine grain wheels. *Materials and Manufacturing Processes* (12) 6, 1037–1048 (1997).

[24] J. Qian et al. Cylindrical grinding of bearing steel with electrolytic in-process dressing. *Precision Engineering* (24), 153–159 (2000).

[25] J Corbett, P. A. McKeown, G. N. Peggs, and R. Whatmore, Nanotechnology: International developments and emerging products. *Annals of the CIRP* (49) 2, 523–546 (2000).

[26] T. Masuzawa, State of the art on micromachining. *Annals of the CIRP* (49) 2, 473–488 (2000).

Index